SALESFORCE APEX WITH CHATGPT

A GUIDE TO SMARTER DESIGN, FASTER CODING, AND AI-DRIVEN DEVELOPMENT WORKFLOWS

MATTHEW MORRIS

ANDY FORBES

Roadmap Press
Hillsboro, Virginia, United States

ISBN: 978-1-969489-15-0 (paperback)
ISBN: 978-1-969489-07-5 (ebook)

Cover design by: Roadmap Press

To the people at Salesforce - thank you for your vision, your leadership, and your relentless drive to push the platform, and the industry, forward.

To our coworkers - your willingness to tackle difficult problems, embrace new ideas, and take on cutting-edge work for clients inspires us every day. This book exists because of the energy, curiosity, and commitment you bring to the craft.

To the wider Salesforce ecosystem - architects, admins, developers, product owners, partners, evangelists, and community leaders - thank you for welcoming us into a community that never stops learning, sharing, and building. We're grateful to be part of this journey with you.

And to Stephen Stanley – thank you for taking the time to do a detailed review and making this a better book.

Matthew Morris and Andy Forbes

The art of life lies in a constant readjustment to our surroundings.

- Kakuzō Okakura

CONTENTS

PREFACE

This book was written across the fall of 2025. As a result, it reflects the state of Salesforce, Apex, and the surrounding AI tooling as they existed during that period. Salesforce evolves continuously, and new platform features and capabilities introduced after that timeframe are not covered here. Where that matters, the intent is not to be exhaustive, but to focus on durable patterns, approaches, and ways of thinking that remain useful even as the platform changes.

While ChatGPT is a central tool throughout this book, this is not simply a guide to "how to use ChatGPT." The deeper focus is on **best practices**: how to think about Apex design, testing, performance, maintainability, and governance, and how to use ChatGPT *in service of those practices*. Used well, AI can amplify good engineering judgment. Used poorly, it can just as easily reinforce bad habits. The goal here is to help you stay on the right side of that line.

This book assumes a technically experienced audience. It is written for readers who already know their way around Apex, are comfortable reading Salesforce documentation, and are familiar with the broader tooling ecosystem. References to APIs, limits, patterns, and official documentation are made without step-by-step guidance. If you are new to

Apex, this book will likely feel opinionated and incomplete. That is by design.

The book itself was written with the help of ChatGPT. In some places, that collaboration is explicit; in others, it is simply part of the workflow. One important caveat is worth stating clearly: what ChatGPT "knows" about Apex is based on public documentation, examples, blog posts, and common usage patterns reflected in its training data and interface behavior. In practice, this means that popular or widely discussed approaches sometimes surface more readily than less common, but potentially better, solutions. Where that distinction matters, the text tries to call it out and explain the trade-offs.

While ChatGPT is a powerful assistant, it is not an authority. Its responses reflect patterns in data, documentation, and common practice, and those responses can occasionally be incomplete, outdated, or subtly incorrect. Readers should apply the same common sense and professional skepticism they would use with any secondary source. One simple but surprisingly effective technique is to challenge an answer directly by asking follow-up questions such as "Are you sure?" or "What are the edge cases or exceptions?" Doing so often prompts deeper reasoning, clarifications, or alternative approaches. Treat ChatGPT as a collaborative partner, not an infallible source, and always validate important guidance against official documentation, platform behavior, and your own experience.

When working with ChatGPT, the quality of the output is heavily influenced by the context you provide. Vague or underspecified questions tend to produce generic answers, while well-framed prompts lead to responses that are more precise, practical, and aligned with real-world constraints. One simple but effective technique is to explicitly set the role or perspective you want ChatGPT to assume. For example, asking it to "answer as a Salesforce Technical Architect" or "respond from the perspective of a production support engineer" helps steer the model toward architectural considerations, trade-offs, and platform limits rather than surface-level explanations. This kind of role framing mirrors how experienced practitioners naturally think about problems and can significantly improve the usefulness of the guidance you receive.

Equally important is providing situational context along with the question itself. Details such as the size of the data set, whether the code runs synchronously or asynchronously, which Salesforce features are in play, and whether the solution is intended for a one-off script or a long-lived production system all shape what a "good" answer looks like. Including this information up front reduces the risk of receiving advice that is technically correct but inappropriate for your use case. Over time, you will find that investing a few extra seconds in framing the question, setting expectations, and clarifying constraints pays dividends in the quality and reliability of ChatGPT's responses.

You may also notice that certain ideas and explanations appear more than once. This repetition is intentional. The book is structured to be used as a reference, not necessarily read straight through from cover to cover. Many readers will arrive with a specific problem to solve and dip into individual sections accordingly. Repetition helps make each section usable on its own.

It is worth noting that, at least at the time this book was written, different AI tools excelled at different tasks. Some were better at reasoning through complex logic, others at summarizing documentation, generating examples, or exploring alternatives. No single model was universally best. Part of becoming effective with AI as a development companion is learning which tool to use for which kind of work, and when to trust its output versus when to challenge it.

Please note: Due to the tools used to publish this book, the code samples are indented with **Nonbreaking spaces**. This may impact you if you have an electronic version of this book and cut and paste the code sample.

Taken together, this book is meant to be practical, opinionated, and grounded in real-world Apex development. If it helps you write better code, make better decisions, and use AI more deliberately in your daily work, it has done its job. While someday AI may replace what you and the authors currently do, that day is not here yet.

1

FOUNDATIONS OF EFFECTIVE APEX DEVELOPMENT

In any successful Salesforce project, well-crafted Apex code is a cornerstone. This chapter lays out why Apex development standards are essential for project success. When teams adopt consistent practices, they not only write cleaner code but also ensure smoother deployments and fewer surprises in production. A strong foundation in Apex best practices prevents the chaos of unmaintainable code and sets the stage for scalable, high-quality Salesforce implementations. By investing in good habits early, projects avoid accumulating crippling technical debt that can slow down or even derail progress later on.

At the heart of effective Apex development are four key principles: *clarity*, *simplicity*, *maintainability*, and *consistency*. Clarity means code that any team member can easily understand – variables have descriptive names, and logic is transparent. Simplicity means avoiding unnecessary complexity; the best code accomplishes requirements in the cleanest way without clever tricks that confuse future readers. Maintainability ensures that when requirements change (as they always do), the code can be modified with minimal risk and effort. Consistency ties everything together – using standard patterns and style across the codebase so that different modules feel familiar and predictable. These cornerstones make Apex code robust and ready for growth.

Following best practices isn't just academic – it directly benefits every stakeholder on a Salesforce team. **Developers** experience faster debugging and easier feature addition when code is clean. **Testers** can more readily understand how the system is supposed to work, which helps in creating effective test cases and spotting issues early. **Architects** gain confidence that the system's design will hold up over time, as consistent code reduces the chance of one module breaking another. Even **product owners and project managers** feel the impact: consistent, high-quality Apex means more reliable delivery of features, predictable timelines, and less budget spent on firefighting bugs. In essence, good Apex practices help align technical implementation with business value by reducing friction and surprises.

A unique aspect of this book is the integration of ChatGPT as a development assistant. Throughout this chapter (and the rest of the book), we will highlight how ChatGPT can help you deepen your understanding of Apex best practices, refine your approaches, and keep your coding standards consistent. Think of ChatGPT as an ever-available pair programmer or code reviewer – you can ask it to explain a concept, review a snippet for issues, or even generate examples of better approaches. For instance, if you're unsure about the optimal way to bulkify a piece of code or enforce security in a trigger, you can **ask ChatGPT in natural language** and get quick, tailored insights. This might be through text, or even through non-text interactions like voice queries or by uploading a code file for analysis (just remember to enable file privacy settings when you do). By leveraging an AI assistant, every role on the team – from a seasoned architect to a new developer – can accelerate learning and enforce best practices with greater consistency.

Over the next several sections, we'll cover the fundamental topics that form the bedrock of effective Apex development. We'll start with making sure everyone is on the same page with coding standards and why they matter. Then, we'll review code readability and why writing *clear* Apex code pays off. Building on that, we'll explore techniques to keep Apex logic simple and manageable rather than overly complex. A critical part of any enterprise-grade code is scalability, so we'll discuss how to design Apex code that can handle growth in data and users without breaking a sweat, alongside performance considerations and working within Salesforce's limits. We'll emphasize weaving security into your code from the beginning – protecting data by design. We'll also demystify Salesforce's governor limits (the platform's built-in limits on resource usage) and strategies to respect them. Finally, we'll tie it all together with practices for continuous improvement, ensuring that code quality remains high throughout the project lifecycle. These topics – readability, maintainability, scalability, performance, security, governance, and code quality – constitute the "foundations" that every Apex developer and architect should master.

While this chapter provides a broad foundation, each concept we introduce here can be explored in greater depth. Indeed, subsequent chapters of this book will dive deeper into many of these areas, offering advanced techniques and patterns once these basics are in place. Think of the guidance here as laying the groundwork: by internalizing these core best practices, you prepare yourself to tackle more complex Apex challenges with confidence.

With these fundamental ideas in mind, let's begin our journey into Apex best practices. One of the first and most important steps to writing quality Apex is establishing clear coding standards that everyone follows. Ensuring consistency in how code is written sets the tone for everything else – after all, it's hard to maintain clarity or security if every developer writes code in a completely different style. With that in mind, we'll start by examining the importance of Apex coding standards and how they help keep a Salesforce codebase clean and effective.

UNDERSTANDING THE IMPORTANCE OF APEX CODING STANDARDS

Before writing a single line of code, teams should agree on **Apex coding standards** – a set of guidelines that define how code should be written and organized. Coding standards encompass things like naming conventions (for classes, methods, variables), how to format code (indentation, brackets on new lines or same line, etc.), and best practices for structuring code (such as one trigger per object or avoiding hard-coded values). In the context of Salesforce, Apex coding standards also often include patterns specific to the platform, like how to handle trigger logic (using handler classes or frameworks) and how to write test methods. Establishing these standards is critical in Salesforce development because it creates a common language for the team. When every developer follows the same playbook, the codebase feels cohesive rather than a patchwork of individual styles.

The **role of coding standards in Salesforce development** cannot be overstated. Apex is used to implement business logic that many people will touch over the lifetime of an org – developers will enhance it, testers will validate it, and admins or architects will review it. If each segment of Apex code follows different rules, maintenance becomes a nightmare. By clearly defining and enforcing standards, teams achieve multiple practical benefits:

Ease of maintenance: When code is consistent, a developer can jump into any Apex class and understand it faster. Standard patterns mean less time deciphering what the previous developer intended, so bug fixes and enhancements can be done with confidence.

Faster debugging: Debugging is easier when you can predict where to look. For example, if your standards say that all database operations in triggers go through a helper class, you immediately know to inspect that class when something with DML goes wrong. A consistent structure narrows down the search for issues.

Efficient onboarding: New team members ramp up faster if the codebase is orderly. A junior developer or a new hire can follow the established style and patterns by looking at existing examples. They spend less time guessing how to write code that fits in, and more time delivering features.

Reduced technical debt: Technical debt often accumulates from quick-and-dirty code written without guidelines. Over time, such debt can slow development to a crawl. Using standards from the start prevents many common pitfalls (like thousands of lines in one class, or inconsistent error handling) that would otherwise require painful refactoring later. It's an upfront investment that pays off by avoiding future rework.

To illustrate, imagine a Salesforce project that lacks coding standards. Developer Alice writes her Apex triggers with logic directly inside them and names variables like x and y. Developer Bob, meanwhile, uses a trigger handler framework and verbose variable names, but leaves huge comment blocks of commented-out code in his classes. When Alice goes on leave and Bob has to fix an issue in her trigger, he spends hours understanding her style (or lack thereof). The inconsistency leads to delays and bugs. This scenario is all too common when standards aren't in place. By contrast, a team with a **defined Apex style guide** – for example, specifying that all triggers delegate logic to handler classes, and that variables should have meaningful names in camelCase – avoids that confusion. Bob could open Alice's code and immediately grok it because it follows the agreed structure.

Establishing coding standards should be a collaborative effort. A great way to kickstart this process is by using ChatGPT as a facilitation tool. For instance, the team can have a brainstorming session on what standards to include (naming conventions, trigger patterns, error handling practices, etc.) and then prompt ChatGPT to help flesh this out. **Example Prompt:**

User: "We are creating coding standards for our Salesforce Apex development team. What are important elements to include in an Apex coding standard?"

ChatGPT: "Key elements of an Apex coding standard include naming conventions for classes, methods, variables, and test classes; trigger handling patterns (e.g., one trigger per object with logic in handler classes); formatting rules like indent size and bracket placement; guidelines for comments and ApexDoc usage; best practices for SOQL/SOSL (to avoid governor limits); error handling conventions (using exceptions vs. error messages); and testing requirements (such as code coverage and test data handling). You might also include specific patterns to avoid – for example, 'do not perform DML inside loops' – which overlap with best practices. Each guideline should have examples of correct and incorrect usage for clarity."

This kind of response gives the team a solid starting checklist. The team could then refine and say, "ChatGPT, generate a draft document of Apex coding standards based on those elements," and it would produce a nicely formatted list that the team can review and tweak to fit their context. Teams can even **upload a sample Apex class file** (ensuring they have file privacy enabled in ChatGPT settings) and ask ChatGPT to evaluate it against the proposed standards, to see if the guidelines are comprehensive.

For example, suppose you have a snippet of Apex code that you want to check for adherence to standards:

```
public with sharing class AccountTriggerHandler {

  public static void handleAccounts(List<Account> accList) {

      for(Account a : accList) { if(a.Name != null && a.Name != '') { a.Status__c = 'Active'; update a; } }

  }

}
```

At first glance, several issues stand out: there's an update DML inside a loop, the if statement isn't using braces or line breaks, and maybe the naming of handleAccounts could be more specific. You could ask ChatGPT to review this snippet for coding standard violations and improvements. **Example Prompt:**

User: "Review the following Apex method for compliance with best practices and coding standards, and suggest improvements:\napex\npublic static void handleAccounts(List<Account> accList) {\n   for(Account a : accList) { if(a.Name != null && a.Name != '') { a.Status__c = 'Active'; update a; } }\n}\n"

ChatGPT: "First, for readability and maintenance, the code should be formatted consistently. The if logic should be expanded with proper indentation and braces. For example:

```
for (Account a : accList) {

  if (a.Name != null && a.Name != '') {

    a.Status__c = 'Active';

    recordsToUpdate.add(a);

  }
```

}

update recordsToUpdate;

Also, performing a DML operation (update) inside a loop is against best practices – it's better to accumulate records and update them in bulk after the loop to avoid governor limits. In terms of naming, handleAccounts is okay, but if this method is meant for activating accounts with a Name, perhaps a name like activateNamedAccounts would be clearer. Overall, apply consistent formatting (spacing, line breaks) as per your style guide, and avoid inline DML in loops."

Notice how ChatGPT's feedback covers both stylistic improvements (formatting, clearer naming) and best practice violations (DML in loop) – coding standards often blend style and technique. This interactive review not only flags issues but also educates the team on *why* they are issues, reinforcing the standards you want to uphold.

Architect Tip: It's wise for a Salesforce **architect** to spearhead the creation of an **enterprise-wide Apex style guide**. By standardizing naming conventions, patterns (like using service layers or framework X for triggers), and even things like how to structure test classes, an architect ensures consistency across multiple projects or orgs. This governance early on prevents divergence when teams grow. Tools like static code analyzers (PMD, SonarQube) can be configured to enforce many of these standards automatically. An architect can even use ChatGPT to **consolidate best practices** from Salesforce documentation and community blogs into a draft guide, then refine it for the company's needs. Once the standards are in place, periodic reviews (possibly with ChatGPT's help) can keep the team on track.

By clearly defining and following Apex coding standards, a team creates a uniform development environment. Everyone knows the expectations, and the codebase doesn't depend on individual quirks. This consistency is the fertile soil out of which maintainable and readable code grows. In fact, with standards in place, developers often find their code naturally becomes more **readable** – the next topic we will explore. Just as standards set the baseline for consistency, focusing on readability takes code quality to the next level, ensuring that anyone can quickly grasp what the code is doing and why.

PRINCIPLES OF READABLE APEX CODE

Have you ever opened a piece of code and struggled to understand what it was doing, even if it was syntactically correct? Code like that might work, but if it's hard to read, it becomes a long-term liability. **Readable Apex code** is code that conveys its purpose clearly to any reader (not just the original author). Readability is arguably one of the most important qualities of a codebase, because code is read far more often than it's written. Every time someone new joins the team, every time a bug needs fixing, or when adding a new feature, developers and testers will read existing code. If that code is clear and straightforward, these activities will be faster and less error-prone. Thus, readability directly impacts productivity (developers can implement faster when they don't have to decode cryptic logic), maintainability (the easier it is to understand code, the easier it is to modify safely), and collaboration (team members can review each other's code effectively and work in the same area without confusion).

What practices make Apex code more readable? Let's break down a few key ones:

Meaningful naming conventions: Choose clear, descriptive names for classes, methods, variables, and triggers. A class named OpportunityService immediately tells you it contains logic pertaining to Opportunity records (perhaps business processes or calculations), whereas a class named OppHelper2 is vague. Variable and method names should reflect their intent (e.g., calculateRenewalAmount is better than calcAmt). Salesforce has some recommended naming conventions (for example, class names in PascalCase, variable names in camelCase), and following a consistent convention is part of coding standards. But beyond the format, the *content* of the name matters. Avoid abbreviations that aren't universally known, and include context in the name if it helps ("tempList" is less informative than "expiredContractsList"). A good rule of thumb is that anyone reading the code should get a hint of what a variable or method is for just from its name.

Thoughtful commenting (and self-documenting code): Comments in Apex should clarify the *why* behind code, not restate the *what*. For example, a comment like // increment i above a line i++ is useless;

anyone can see that. But a comment like // assuming all accounts passed are Active; otherwise, they will be skipped provides context or business reasoning. Use comments to explain non-obvious decisions, complex algorithms, or important ramifications. At the same time, strive for *self-documenting* code – code that is so clear that it doesn't need much commentary. This can be achieved by using well-named methods to break down complex processes (so the method name serves as documentation), and by avoiding overly clever hacks. Salesforce also supports **ApexDoc** (similar to Javadoc) for documenting classes and methods; using it can generate useful documentation for other developers. A balanced approach is to write code clearly and add comments only where they add value, such as describing the intent or citing a reference for a formula.

Proper formatting and structure: Consistent indentation, spacing, and bracing style make a world of difference in readability. Indent your code blocks (e.g., code inside a loop or if-statement) by the same amount each time (typically 4 spaces in Apex, though 2 is also common – the key is consistency). Use blank lines to separate logical sections of a method. For instance, if a method does three main things in sequence (validate input, perform calculation, update records), you might separate those with a blank line and perhaps a comment for each section. Ensure every if/else, loop, or method has matching braces and that they're aligned – most IDEs like Visual Studio Code can auto-format Apex code. Adhering to one statement per line and avoiding very long lines (Salesforce's style guide suggests a max of 120 characters per line) helps prevent horizontal scrolling, which no one enjoys. By structuring code neatly, you reduce the cognitive load on the reader. They can follow the logical flow visually.

Clarity in logic structure: Write logic in a way that's easy to follow. For example, deeply nested if-else statements can often be restructured for clarity – sometimes using a switch or by returning early from a method to avoid nesting. Consider this: which is easier to read, a chain of conditions nested four levels deep, or a flat sequence of checks that each handle a case and return? Often, refactoring complex conditionals into helper methods or breaking a large method into smaller ones improves clarity (we'll talk more about simplifying complexity in the next section).

Another aspect is to group related operations. If your code queries some data, then processes it, then updates something, keep those stages distinct and perhaps even in separate well-named methods. That way, a reader can skim method names and understand the high-level flow, drilling into details only as needed.

To see readability in action, let's take a poorly written snippet and improve it. Imagine we have an Apex method that updates some account statuses based on their last activity date. Here's an *unreadable* version:

```
public static void upd(List<Account> a) {

  Datetime t = Datetime.now().addDays(-30);

   for(Integer i=0;i<a.size();i++){Account rec=a[i]; if(rec.LastActivityDate
!= null && rec.LastActivityDate < t){rec.Status__c='Inactive'; update rec;}}

}
```

At a glance, this code is doing something with accounts and last activity, but it's hard to parse. The method name upd is cryptic. The variable t is not clearly named. The loop uses an index and manually retrieves a[i] which isn't necessary in Apex (we can directly loop over accounts). There's an update inside the loop, which is not only a performance/governor limit issue but also makes the intent harder to see. Let's improve the readability step by step:

```
public static void markInactiveAccounts(List<Account> accounts) {

  // Define the cutoff date as 30 days ago

  Datetime cutoffDate = Datetime.now().addDays(-30);

  List<Account> accountsToUpdate = new List<Account>();

  for (Account acct : accounts) {

            if (acct.LastActivityDate != null && acct.LastActivityDate <
cutoffDate) {

        acct.Status__c = 'Inactive';

        accountsToUpdate.add(acct);

    }
```

```
    }
    if (!accountsToUpdate.isEmpty()) {
        update accountsToUpdate;
    }
}
```

Now the code is much more approachable. We renamed the method to markInactiveAccounts, which clearly indicates its purpose. We use cutoffDate instead of t to convey meaning. The loop uses a readable for-each style. We accumulate records to update in a list, and then do one bulk update outside the loop – this not only is better for limits (as we will discuss later) but also actually improves readability by separating the "determination of which accounts to update" from the "performing the update". Comments are added to explain the critical part (the cutoff date definition). A fellow developer or tester can now glance at this method and understand what it does and why each step is there.

Readable code also fosters collaboration. Code reviews, for instance, are far more productive when the code under review is clean. Reviewers can focus on the logic and potential bugs, rather than getting distracted by deciphering messy code. Testers often read code to understand edge cases or to figure out why a test might be failing; if the code is straightforward, they can pinpoint issues faster or identify missing test scenarios. Even product owners or business analysts, though not coding, might occasionally read Apex (or pseudocode generated from it) to verify business rules – clarity helps them confirm that the implementation matches the requirement.

One of the best ways to achieve and maintain readability is through continuous feedback, and ChatGPT can assist here as well. Developers can use ChatGPT as a quick sounding board for code clarity. For example, if you're unsure whether a piece of logic is easy to follow, you can ask ChatGPT to explain the code back to you. If the explanation comes back incorrect or confused, that's a sign the code might be written in a confusing way! This acts like an AI rubber duck debugging session. In fact, you can use voice input to literally *ask* ChatGPT (using the voice feature) something like, "Explain what this Apex method does," and hear how it interprets your code. If ChatGPT's explanation misses the mark, perhaps the code's intent isn't as self-evident as you thought.

Example Prompt (Readability Analysis):

User: "Here's an Apex method. Can you identify anything that makes it hard to read or suggest how to make it cleaner?\napex\npublic static Decimal calcDisc(Decimal p, Decimal d){Decimal r=p; if(d!=null){r = p - (p*d);} return r;}\n"

ChatGPT: "The method calcDisc is calculating a discount, but there are a few readability issues. The name calcDisc could be clearer (perhaps calculateDiscountedPrice). The parameters p and d are not descriptive – maybe use price and discountRate. Inside, the logic is crammed into one line. It would be easier to read as:

```
Decimal result = price;

if (discountRate != null) {

   result = price - (price * discountRate);

}

return result;
```

This expansion makes it clear what the calculation is. Also, consider clarifying what d represents (is it 0.1 for 10%? If so, maybe name accordingly or add a comment). Overall, using descriptive names and breaking the calculation into a few lines greatly improves readability."

This example shows how ChatGPT not only points out issues but also demonstrates a more readable form of the code. You as the developer could then implement those suggestions. You can even follow up with ChatGPT, perhaps by asking it to rewrite the function entirely with best practices applied. Additionally, ChatGPT can help enforce formatting if you paste in poorly indented code – it will often return the code properly indented (acting like a formatter).

Developer/Testers Tip: Emphasize readability during code **reviews and testing.** For developers, when writing a piece of Apex, consider if *future you* or a colleague could quickly grasp it. If not, take a moment to refactor for clarity – your future self will thank you when debugging.

For testers, don't hesitate to read through the Apex code related to features you're validating; if something is unclear, raising a question can sometimes lead to a realization of a bug or an opportunity to improve the code's clarity. In fact, testers can use ChatGPT to generate high-level summaries of Apex code logic to ensure it matches expected behavior. A side benefit: focusing on readability often flushes out hidden assumptions or errors. When code is written clearly, it's harder for a bug to hide in it. Always remember, **clear code = fewer bugs = faster projects.**

Readable code and coding standards go hand in hand. If you have a standard that enforces naming conventions and formatting, you're already halfway to achieving readability. The next step after making code *readable* is to ensure that the code's structure itself isn't needlessly complicated. In the following section, we'll look at managing and reducing complexity in Apex logic. After all, even well-formatted and well-named code can be overwhelming if it tries to do too much at once. Keeping Apex logic simple and focused will further enhance maintainability and set the stage for scalable, high-quality applications.

SIMPLIFYING APEX LOGIC AND COMPLEXITY

Software has a tendency to become more complex over time as new features and fixes are added. In Salesforce Apex development, it's common to encounter classes or methods that attempt to handle every imaginable scenario – they become large, unwieldy, and difficult to understand. **Complex Apex logic** can manifest as a single method that spans hundreds of lines, deeply nested loops and conditionals, or classes that do too many things. This complexity is the enemy of maintainability. When code is overly complex, a developer might be afraid to change it ("if I touch this, will I break something?"), bugs find hiding spots in edge-case logic, and onboarding someone new onto that part of the system becomes a lengthy process. Therefore, a critical best practice is **simplifying Apex logic** wherever possible, breaking down complexity into manageable pieces.

There are several strategies to keep Apex code simple and clear in its logic:

Use smaller methods and functions: Adhere to the principle of *single responsibility* at the method level. A method should ideally do one clear thing (or a logical unit of work). If you find yourself writing a method that handles multiple distinct tasks – for example, first it validates inputs, then calculates a value, then updates the database and sends an email – that's a sign it should be broken into smaller methods. By splitting logic, each method becomes easier to understand, test, and reuse. In Apex, where unit tests are required, having smaller methods can also make it easier to write targeted tests for specific functionality. If a method exceeds, say, 50-60 lines, scrutinize it to see if it can be divided. Even within a method, you can use helper methods: e.g., a complex calculation might be moved to a helper calculateXYZ() method, so the main method reads almost like a summary of steps. This approach also naturally creates points of abstraction, where you can later modify one part (like the calculation) without touching the surrounding logic.

Introduce helper classes or services: Beyond individual methods, consider the organization of your classes. If one Apex class is trying to handle everything for a given object, it might be time to introduce a helper class or a service layer. For instance, you might have an AccountTriggerHandler class that processes account triggers. If that class starts ballooning with logic not directly related to triggers (say, complex business rules about account scoring), you could refactor those rules into an AccountService class. The trigger handler then just calls into the service. This separation of concerns keeps each class focused. Salesforce developers often use design patterns like the **Trigger Handler pattern** or **Domain layer pattern**, where triggers delegate work to classes designed for specific tasks (validation, calculation, database operations, etc.). Utilizing these patterns prevents any single part of the code from becoming too complex. It's much easier to manage a collection of specialized classes than one monolithic do-it-all class.

Leverage clear abstraction layers: Abstraction is about hiding complexity behind a simpler interface. In Apex, this could mean having well-defined layers in your application architecture. For example, you might have a controller (for a Visualforce page or LWC, if doing UI), which calls a service class (business logic), which in turn maybe calls a repository class or uses SOQL to get data. Each layer has a defined role.

This way, when reading code, you can understand the high-level flow (controller -> service -> database) without getting lost in the weeds of each query or each calculation at once. If you need to dive deeper, you go into the specific class. Abstraction also applies to using Apex features smartly – like using custom metadata or custom settings to hold configuration instead of peppering logic with literal checks. That moves complexity out of code and into data, which is easier to adjust. Similarly, if you have repetitive code, consider abstracting it into a reusable method or even a base class. For instance, if multiple classes do similar logging, create a utility or a superclass that handles logging, so the subclasses remain focused on core logic.

Let's demonstrate simplification with a concrete example. Suppose you have a single Apex method that processes an opportunity close operation. In one method, it calculates the sales rep's commission, updates the opportunity status, creates a task for follow-up, and sends a notification. This is a lot for one method. It might look like:

```
public static void closeOpportunity(Opportunity opp) {

  // calculate commission

  Decimal commission = 0;

  if (opp.Amount != null) commission = opp.Amount * 0.1;

  // update opp status

  opp.StageName = 'Closed Won';

  opp.ClosedDate = Date.today();

  update opp;

  // create follow-up task

  Task t = new Task();

  t.OwnerId = opp.OwnerId;

  t.Subject = 'Follow up with customer';

  t.WhatId = opp.Id;

  insert t;
```

```
  // send notification
   Messaging.SingleEmailMessage email = new Messaging.SingleEmail-
Message();
  //...set email properties...
    Messaging.sendEmail(new Messaging.SingleEmailMessage[] { email
});
}
```

While this might work, it's doing several distinct things. A more simplified, modular approach would be:

```
public static void closeOpportunity(Opportunity opp) {
  applyBusinessUpdates(opp);
  updateOpportunityRecord(opp);
  createFollowUpTask(opp.Id, opp.OwnerId);
  sendCloseNotification(opp);
}
```

Then each of those helper methods (applyBusinessUpdates, updateOpportunityRecord, createFollowUpTask, sendCloseNotification) would be defined elsewhere, handling one aspect of the process. The closeOpportunity method now reads like a clear sequence of steps, and if any step needs changes, you go to that specific method. Also, if one of those steps grows more complex (say commission calculation becomes multi-step), it's isolated in its own function. This makes both reading and maintaining the code easier.

ChatGPT can assist significantly in the refactoring process to simplify code. If you have a complex method and are unsure how to break it apart, you can describe the code to ChatGPT or even paste it (if it's not too large) and ask for a refactoring suggestion. **Example Prompt:**

User: "I have an Apex method that is 200 lines long and hard to follow. It does validation, some calculations, calls a couple of SOQL queries, updates records, and sends an email. How can I refactor this to be simpler?"

ChatGPT: "To refactor a large Apex method, identify logical sections of the code and turn each section into its own method (or even class, if appropriate). For example, separate the validation logic into a method validateInput(data). Isolate calculations into methods like calculateMetrics(data) that return results. Encapsulate the SOQL queries perhaps into a data layer method like fetchRelatedRecords(ids) so the main method isn't cluttered with query details. The update of records can be its own method (especially if it might be reused or needs special handling) say updateRecords(List<sObject> records). Email sending definitely can be a separate method sendNotification(emailInfo) or even moved to a utility class if it's generic. After refactoring, your main method will essentially become a sequence of high-level calls: validate, calculate, fetch data, update, notify. Each sub-method should have a clear name that documents its purpose. This makes the code more readable and each piece easier to test. Additionally, if some of these operations aren't always needed, you can incorporate conditions at a high level rather than burying them deep in the logic. If the logic naturally groups into a few classes (e.g., a class dedicated to validation or to notifications), consider that as well. Using a service class to coordinate these methods might simplify the structure if the logic spans multiple domains (like it touches Opportunities, Contacts, Tasks, etc.)."

With such advice, you could then proceed to break your code apart. You might even ask ChatGPT to actually perform a refactor on a given snippet. Many developers have found success in copying a long method and saying "refactor this into smaller methods with explanations", and ChatGPT will do a decent job of it, providing new code with helper methods.

Another technique is to use ChatGPT for *explaining* complex code – sometimes understanding it fully is the first step to refactoring. If you feed a complicated method into ChatGPT (again, mindful of token limits, perhaps piece by piece or summarizing parts), and ask "explain what this code does," the explanation can reveal the distinct steps that naturally become candidates for separate methods or classes.

Product Owner & Architect Insight: From a **product owner's** perspective, it might seem advantageous to cram as much logic as possible to quickly deliver a feature. However, this often results in brittle implementations. It's important for product owners to understand that taking a bit of extra time to simplify and structure the code is an investment that leads to more stable features and faster changes down the road. Encourage a development atmosphere where **refactoring for simplicity** is considered part of the task, not a separate "nice to have." For **architects**, always be on the lookout for areas of the codebase that are becoming too complex. Regular design reviews or architecture checkpoints can catch complexity creep. If one feature's code is getting overly intricate, an architect might suggest a design pattern (like Strategy, Command, or Factory patterns in Apex) to manage it. ChatGPT can be a brainstorming partner here too – for example, an architect could ask, "What design pattern might simplify managing multiple discount calculation strategies in Apex?" and use the answer as a springboard. Simpler code is also easier to **delegate** or split among team members, which is an architectural concern for parallel development. In sum, product owners should value maintainability in their acceptance criteria, and architects should champion simplicity as a fundamental design goal.

By simplifying Apex logic, we make the codebase not only easier to work with today but also resilient for the future. Simpler code tends to have fewer bugs and is easier to adapt when requirements change. This sets a strong foundation – think of it as reducing the cognitive load for everyone dealing with the code. With readability and simplicity in place, our Apex code is much more maintainable. Next, we need to ensure that this clean and simple code will also **scale** effectively as data volumes grow and usage increases. In the following section, we'll shift focus to planning for scalability from the very start of development, ensuring our Apex solutions remain efficient and effective even as our Salesforce org expands.

ENSURING SCALABILITY FROM THE START

One of the unique challenges (and opportunities) of Salesforce development is designing Apex code that can **scale**. Scalability means your code continues to perform well as the number of users, the amount of data, and the complexity of operations increase. In the context of Apex, scalability often boils down to handling larger data volumes and higher transaction counts without hitting walls – whether those are performance bottlenecks or Salesforce governor limits. It's much easier to bake scalability into your code from the beginning than to retrofit it later when your org has grown. Planning for scalability early in the Apex development cycle can save enormous time and cost down the road. If you consider things like bulk processing and efficient queries upfront, you'll avoid frantic last-minute refactors because a trigger started timing out in production or a batch job couldn't handle end-of-month data volumes.

So, what does designing for scalability entail? Here are some core concepts and practices:

Bulkification: This is a Salesforce-specific term for writing code that can handle multiple records at once. Salesforce often processes records in batches – for example, when a user uploads data via Data Loader or when several records are updated at once, a single trigger invocation might receive 200 records. Bulkification means your code is prepared for that: using collections (Lists or Maps) to process records together, and not assuming a single-record context. Concretely, if you have a trigger on Contact, don't write it to expect only one Contact; write it to loop over all trigger.new contacts. If you need to query related data, query it once for all relevant IDs, instead of inside a loop for each record. If you need to update records, add them to a list and do one update at the end, rather than multiple update calls. Bulkification is probably the #1 Apex best practice for scalability because it aligns your code with how the platform dispatches work. Well-bulkified code can handle 1 record or 200 with equal grace. We saw a glimpse of this in previous sections: e.g., accumulating accounts to update in a list and doing one DML operation is bulkification in action.

Governor limit awareness: Salesforce enforces various limits (which we'll cover in detail in the next section) on resource usage per transaction, such as how many SOQL queries you can run or how much memory you can use. Scalable Apex code is written with these limits in mind so that it doesn't come close to them under normal operation. For example, if you know there is a limit of 100 SOQL queries per transaction, you design your code to use perhaps a handful of queries at most, even in bulk scenarios. That might involve writing more complex WHERE clauses to retrieve all needed data in one go, or restructuring logic to avoid querying in inner loops. Another example: limit of CPU time (e.g., roughly 10 seconds for sync transactions). A scalable design avoids extremely large loops or inefficient algorithms that could consume that CPU time on large datasets. Being limit-aware doesn't mean constantly counting every single query in your head, but it does mean choosing patterns that are inherently efficient. For instance, using Map<Id, SObject> to avoid repeated lookups is both a code clarity improvement and a governor-savvy move.

Performance-oriented coding: Beyond just avoiding limits, think about general performance. One aspect is query selectivity – writing queries that efficiently target records. If a query returns 50,000 records when only 100 were relevant due to a missing filter, that's both a potential limit breaker and a performance drag. Use filters that leverage indexed fields (like primary keys, foreign keys, or marked indexed custom fields) when possible, so Salesforce can execute the query faster. Another aspect is algorithmic complexity: if you have logic that compares lists of records, using nested loops results in O(n^2) operations which might be fine for 10 records but not for 10,000. Instead, use maps or sets to achieve near O(n) performance (e.g., put all records from one list in a Set and then loop over the other list checking contains in the set, which is constant time lookup). Additionally, consider the user experience: sometimes doing everything synchronously is not necessary. If an operation on 1000 records is slow, a performance-oriented approach might use **asynchronous Apex** (like running a Batch or Queueable) to process in the background, freeing up the user and ensuring the main thread isn't overloaded. Scalability isn't just back-end; it often translates to front-end speed and user satisfaction.

A common mantra in Salesforce is "build for the **worst case** scenario." Even if today you only process opportunities one at a time, imagine an integration tomorrow dumps 500 opportunities in one go. Will your trigger or Apex logic handle it? If you designed with bulkification and limits in mind, likely yes. If not, that integration might start failing or slow everything down.

Let's consider a scenario and how to ensure scalability: Suppose you have an Apex trigger that, whenever an Account is updated, goes through all its related Contacts and updates some fields. A naive approach might be:

```
trigger AccountAfterUpdate on Account (after update) {
  for (Account acc : Trigger.new) {
      for (Contact con : [SELECT Id, Status__c FROM Contact WHERE AccountId = :acc.Id]) {
        con.Status__c = 'Active';
        update con;
    }
  }
}
```

This code will work for a single account update. But if 200 accounts are updated at once (bulk update), it will run a SOQL query for each account (potentially 200 queries, hitting the governor limit of 100 easily) and do an update DML for each contact (maybe hundreds or thousands of DML calls). It will not scale. A scalable rewrite would be:

```
trigger AccountAfterUpdate on Account (after update) {
  Set<Id> accountIds = new Set<Id>();
  for (Account acc : Trigger.new) {
    accountIds.add(acc.Id);
  }
  // Bulk query all contacts related to these accounts
    List<Contact> contactsToUpdate = [SELECT Id, Status__c FROM Contact WHERE AccountId IN :accountIds];
  for (Contact con : contactsToUpdate) {
    con.Status__c = 'Active';
```

```
    }
    if (!contactsToUpdate.isEmpty()) {
        update contactsToUpdate;
    }
}
```

Now we have 1 SOQL query total, and 1 DML operation total, regardless of whether 1 account or 200 accounts triggered it. This is bulkified and limit-aware. It's also likely faster because the database can retrieve all contacts in one go and one large update is generally more efficient than many small ones.

ChatGPT can help identify such scalability issues and suggest changes. You could show it the first version of the trigger and ask for improvements. In fact, ChatGPT is very good at recognizing the "SOQL in a loop" anti-pattern and will almost always advise to bulkify it as shown. **Example Prompt:**

User: "Review my Apex trigger for any scalability or governor limit problems:\napex\ntrigger AccountAfterUpdate on Account (after update) {\n   for (Account acc : Trigger.new) {\n   for (Contact con : [SELECT Id FROM Contact WHERE AccountId = :acc.Id]) {\n   con.Status__c = 'Active';\n   update con;\n   }\n   }\n}\n"

ChatGPT: "Your trigger as written has significant scalability issues. Notably, it performs a SOQL query inside a loop (for (Contact con : [SELECT...]) inside the Account loop) which will execute one query per Account in Trigger.new. This will quickly run into the governor limit of 100 SOQL queries if more than about 100 Accounts are updated in one go. Additionally, it performs an update DML inside the inner loop, potentially doing one DML per Contact record. This could hit the DML limit (150 DML operations per transaction) and is inefficient. The better approach is to bulkify:

- Collect all Account Ids from Trigger.new in a Set.
- Query all Contacts related to those Account Ids in a single SOQL query, retrieving all necessary fields.

- Iterate over the contacts and set the Status__c field as needed.
- Perform a single update on the list of contacts.

This way, no matter how many accounts, you only ever do one query and one update. This will handle bulk updates gracefully and avoid governor limits. Also consider filtering Contacts in the query if not all need updating (for example, only query contacts that aren't already Active, if that's relevant) to reduce the processing load. By bulkifying the trigger, it becomes both more efficient and more in line with Salesforce best practices for large scale data operations."

Such feedback not only gives the solution but explains the reasoning (which is great for learning). You can follow these suggestions and adjust your code accordingly. ChatGPT can also highlight more subtle performance issues; for instance, if you wrote an inefficient algorithm (maybe sorting a list with a custom bubble sort?), it might suggest using a more efficient approach or leveraging built-in methods.

Sometimes ensuring scalability involves considering an entirely different approach. For example, if you need to perform a complex calculation on every record nightly for millions of records, doing it in a single Apex transaction might not be feasible – you might need a Batch Apex that chunks the work, or even move some logic to a scheduled job. ChatGPT can help here as well; you could ask "I need to process 1 million records in Apex, what's the best approach?" and it would likely discuss batch processing and maybe even using skinny tables or other platform features. It might mention using the **Queueable** interface for chaining jobs if needed.

Non-text interactions can also come into play for scalability considerations. Imagine using **ChatGPT's Advanced Data Analysis (formerly code interpreter)**: you could feed in a log file or CSV of performance timings or limit usage from your Apex runs (after anonymizing data) and ask it to plot or analyze where the bottlenecks are. For example, if you export aggregate information about how many queries or what CPU time each execution is using (Salesforce debug logs can provide this), ChatGPT could potentially crunch those numbers and identify patterns. With the new **Operator** mode, one could envision ChatGPT automatically observing code commits and flagging if a change might introduce a scalability problem (almost like an AI code reviewer integrated into your development pipeline). These are emerging possibilities as the tools evolve.

Developer Tip: Always **test your code with bulk data** early in the development process. Don't wait until QA or production to discover that your code fails with 201 records. You can simulate bulk operations by writing an Apex test that inserts, updates, or deletes 200 sample records to see if your triggers and classes handle them. Additionally, use Salesforce's developer console or VS Code debuggers to keep an eye on governor limits during test runs (the debug log will show how many SOQL queries were used, etc.). By proactively testing for scalability, you build confidence that your code will work on real-world volumes. Also, be mindful of *query/selectivity* issues: test on data that mimics production sizes. A query that is fast on 100 records might slow down on 100,000 records if not indexed properly. If you suspect such cases, consider creating test indexes or ask Salesforce support about large data handling. In summary, think big: design and test as if your data is 100x larger than today.

With scalability built into your Apex from the get-go, you minimize unpleasant surprises as your Salesforce usage grows. Fast and bulk-safe code keeps users happy and admins free from limit exception emails. However, in our quest for performant, scalable code, we must not overlook another absolutely critical foundation: **security**. No amount of speed or elegance matters if your code inadvertently exposes sensitive data or allows unintended access. In the next section, we'll discuss how to embed security into Apex development practices from day one, ensuring that our robust and scalable code is also safe and compliant.

EMBEDDING SECURITY INTO APEX FOUNDATIONS

Salesforce applications often handle sensitive business and customer data, so **security in Apex development** is paramount. It's far better to build security checks and balances into your code from the beginning than to patch holes later (or worse, suffer a data breach). When we talk about security in Apex, we are referring to both controlling access to data (ensuring users can't see or modify records they shouldn't) and writing code that is robust against common vulnerabilities. Embedding security into the foundation of your Apex code means making it a habit to consider "Who should be able to do this?" and "Could someone misuse this code to get data they shouldn't?" for every feature you implement.

Here are core security concepts and practices you should incorporate into Apex development:

Enforce CRUD and FLS: CRUD stands for Create, Read, Update, Delete permissions on objects, and FLS stands for Field-Level Security. By default, Apex running in system context can bypass these security restrictions (especially if a class is declared "without sharing"). This means if you're not careful, your code might perform operations the user isn't actually allowed to do via the normal Salesforce UI or APIs. For example, a user without edit permission on Account should not be able to edit an account via an Apex-triggered process. To embed security, you have a few options:

- Use **With Sharing** on your classes so that record-level access (sharing rules, etc.) is enforced automatically. This ensures that, for instance, queries in that class only return records the user has access to. Be mindful though: some system processes or utilities might need without sharing, but then you must manually handle security.
- Manually enforce object and field permissions in code. Salesforce provides Describe calls and the Schema.sObjectType descriptions to check, e.g., Schema.Account.fields.Status__c.isUpdateable() before performing an update, or Schema.Account.isAccessible() before reading records. This can be verbose, but it's the way to ensure the running user has the rights.
- Use newer built-in security enforcement methods. As of recent Salesforce API versions, methods like Security.stripInaccessible are available. For example, after a query, you can call Security.stripInaccessible(AccessType.READABLE, results) to remove fields from the result that the user shouldn't see.

Similarly for data about to be updated, you can strip out fields they can't edit.

A good practice is to create a utility that does these checks, or to consistently apply them around your SOQL and DML operations. When writing triggers or batch processes that run as admin, you might need to think carefully: should this ignore user permissions (perhaps for an admin-level data cleanup) or should it respect them (for business logic running in user context)? Most business logic should respect user permissions to avoid privilege escalation.

Secure coding practices (preventing injection and leaks): While Apex doesn't face all the same risks as, say, a web app (because Salesforce does a lot under the hood), there are still vulnerabilities to consider. One is **SOQL injection** – if you ever construct dynamic SOQL queries or SOSL searches using user input, you must sanitize that input. The best way is to use bind variables in dynamic SOQL. For example, don't do: String soql = 'SELECT Id FROM Account WHERE Name = \'' + userInput + '\''; Database.-query(soql); because if userInput contains something like a' OR Name != 'a, you have an injection issue. Instead use Database.query('SELECT Id FROM Account WHERE Name = :userInput'), which safely binds the variable. Another aspect is not exposing sensitive information in exceptions or debug logs that could be visible. For instance, if a catch block emails an admin with an error, avoid including someone's personal data in that error message.

In fact, when working with dynamic SOQL, consider using Database.-queryWithBinds instead of Database.query to enhance security and code clarity. Database.queryWithBinds allows you to separate your SOQL query string from the bind variables, making it easier to avoid SOQL injection vulnerabilities and ensuring user input is safely parameterized. This method also improves maintainability, as you can manage the query and its parameters more cleanly within your Apex code.

Also, be careful with **sharing data between users** in Apex. If you have a web service or a schedulable job, ensure it's not inadvertently running under an admin and touching data it shouldn't for other users. Use the principle of least privilege: run code in user mode when possible.

Additionally, avoid hardcoding secrets (like API keys, passwords) in Apex. Use Protected Custom Metadata, Encrypted Custom Settings, or Named Credentials for any integration secrets, so that even if someone reads your code, they won't see plaintext credentials.

Protect against common vulnerabilities: Beyond injection, consider things like **cross-site scripting (XSS)** if your Apex outputs data to a Visualforce page. Always HTML-encode output if it includes user-entered data. If you're writing Apex REST endpoints, validate the input payloads – don't assume the JSON will only contain what you expect (someone could call your endpoint with extra fields or malicious data). Also ensure you enforce object/field permissions in those APIs if they operate in user context.

For triggers that run on records, be cautious with recursion or triggers that do "dangerous" things like delete or modify records extensively – ensure appropriate checks so that one user's action doesn't cascade into unintended consequences for data they shouldn't touch.

Logging and monitoring is another part of security: use the platform's logging to catch unexpected behavior. For example, if an Apex method is critical, you might log an error if something suspicious occurs (like an unexpected null where there shouldn't be, which could indicate someone tampered with data).

By instilling these security practices early, you create Apex code that upholds Salesforce's robust security model rather than accidentally undermining it. Let's consider an example of where not thinking about security could cause a problem:

Imagine a piece of Apex code in a controller for a Visualforce page that allows managers to update an employee's salary custom field on a User record. Perhaps the developer wrote:

```
public with sharing class SalaryController {

  public User employee { get; set; }

  public void save() {

    update employee;
```

```
    }
}
```

At first glance, "with sharing" is there, which is good for record access. But what about field-level security? If a manager using this page doesn't have edit permission on the Salary__c field (which they probably shouldn't normally), this code will still execute the update because Apex in system mode doesn't enforce field security by default. The result: a user without proper permission can change a sensitive field. The fix would be to explicitly check, before update, that Schema.User.field-s.Salary__c.isUpdateable() for the running user. Or use Security.strip-Inaccessible on the User record to strip out Salary__c if not editable – then if it's stripped, you know not to proceed or throw an error. Alternatively, design the page such that it runs as a specific user with proper permission and controls the logic; but even then, performing security checks is a good practice.

Let's show how ChatGPT could help ensure security in code. **Example Prompt:**

User: "Here's an Apex snippet. Do you see any security issues or missing checks?\napex\npublic without sharing class CaseHelper {\n   public static void escalateCase(Id caseId) {\n   Case c = [SELECT Id, Status, Account.Name, Account.Secret_Field__c FROM Case WHERE Id = :caseId];\n   c.Status = 'Escalated';\n   update c;\n   }\n}\n"

ChatGPT: "There are a couple of potential security concerns in this code. First, the class is declared without sharing, meaning it runs in system context and does not enforce record-level sharing rules. If this method can be invoked by a user (say through AuraEnabled or as part of a global class in an API), that user could escalate a Case they don't have access to view or edit. It might be safer to use with sharing here, unless there's a deliberate reason to bypass sharing. Second, the SOQL query is selecting Account.Secret_Field__c. If the running user doesn't have field-level access to Secret_Field__c on Account, normally they wouldn't be able to retrieve it. But because this is without sharing (and FLS isn't automatically enforced), the code will fetch it anyway. This could expose sensitive data. To fix this, you'd want to check field accessibility. For example, use Schema.Account.fields.Secret_Field__c.isAccessible() to guard the query, or better, don't select fields the user shouldn't see. Also, ensure the user has permission to update the Case's Status (CRUD on Case and edit on the field). In summary, likely this method should run with sharing and explicitly handle field security for that Account secret field. If truly needed to run as admin, then ensure it's not callable by arbitrary users or that it manually filters based on user permissions."

This answer highlights both sharing and FLS issues. As a developer, you could take that advice: switch the class to with sharing (or perhaps keep without sharing but add your own sharing check logic if needed), remove or protect the sensitive field selection, etc.

Another area ChatGPT can help is **security review**. Salesforce has a Security Review process (for AppExchange apps) and general guidelines. You could ask ChatGPT "What security considerations should I check for in my Apex code?" and it might give you a nice checklist: enforce CRUD/FLS, use sharing appropriately, avoid SOQL injection, etc. Using such a checklist, you can do a self-audit of your code or even feed segments into ChatGPT to see if it catches anything you missed.

Architect/Testers Tip: Make security a routine part of development, not an afterthought. **Architects** should define a clear policy: for example, "All Apex classes that deal with record data must run with sharing unless there is a documented reason not to, and all service methods must check for user permissions on objects and fields before DML." Automate what you can: use tools like Salesforce's **ESAPI** library or the built-in security scanner (Salesforce has a tool called Checkmarx or similar that scans code for vulnerabilities). You can even set up a CI step to run static analysis for security. **Testers (QA)** can incorporate security testing by creating users with restricted profiles and ensuring that, when those users execute functionality (either via UI or directly calling Apex if possible), they are prevented from actions they shouldn't do. Testers can also use ChatGPT by describing a test scenario: "User A with no edit rights tries to do X, what should happen?" and ensuring the application indeed behaves that way. Regularly reviewing code for security – perhaps scheduling a monthly security review session – is a good practice. ChatGPT's operator mode could possibly help by scanning all Apex classes for certain patterns (like queries that don't use with sharing or direct field references without permission checks). But even with AI assistance, human due diligence is key. Everyone on the team should feel responsible for data protection. Ultimately, a secure foundation in your Apex code protects your company's data and maintains trust with users and customers.

With security measures woven into your code, you have a solid safety net: your Apex respects Salesforce's security settings and guards against common exploits. We've covered a lot of ground on writing Apex that is clean, simple, scalable, and secure. Another piece of the puzzle – which we touched on under scalability – is working within Salesforce's unique system constraints known as **governor limits.** Even well-designed code must live under these limits. In the next section, we'll take a closer look at Salesforce governor limits and how to manage them effectively as part of foundational best practices.

MANAGING SALESFORCE GOVERNOR LIMITS

If you've been working with Salesforce for a bit, you've likely encountered or heard of **governor limits.** Governor limits are Salesforce's way of enforcing the fair use of resources in its multi-tenant environment – they're like strict rules that prevent any one piece of code from hogging the shared resources (CPU, memory, database, etc.) on the Salesforce servers. For Apex developers, understanding and respecting these limits is part of daily life. It might sound restrictive, but in practice governor limits guide us to write efficient code. They are not arbitrary hurdles; they often align with good coding practices (like bulkification, as we saw). A foundational skill in Apex is **managing governor limits** so that your code stays within these limits no matter what dataset or scenario it encounters.

Let's provide an overview of some key Salesforce governor limits (note: these are as of current platform guidelines, always check the latest in Salesforce documentation as they can evolve):

SOQL Queries: You can execute a maximum of 100 SOQL queries per synchronous transaction (that is, say, during a single trigger execution or an HTTP request), and up to 200 in an asynchronous transaction (like in a Batch or Queueable). This means if your code tries to run the 101st query, it will hit a runtime error. The intent is to prevent overly chatty database interactions. We manage this by combining queries, using IN clauses, and so on.

DML Operations: You can perform a maximum of 150 DML statements (like insert, update, delete) in a transaction. Again, this encourages bulk operations: instead of 150 single-record updates, do maybe 1 update of 150 records.

Records Retrieved or Processed: A single SOQL query can fetch up to 50,000 records. There's also a limit of total records retrieved via SOQL in a transaction (which is quite high, in the millions, so typically if you hit that you should reconsider your approach). For DML, you can insert/update up to 10,000 records at once. For sending emails, 10 emails per transaction via code, etc. There are limits for almost everything – heap size (6MB for synchronous Apex, 12MB async), CPU time (generally 10 seconds sync, 60 seconds async), callouts (100 callouts to external services per transaction), etc.

Concurrent and Misc Limits: For example, you can only have so many Batch Apex jobs running at the same time (like 5 active), or only 50 Queueable jobs chained from one to another, etc. Those are more advanced scenarios.

The significance of these limits is that if you don't heed them, your code might work fine in small tests but then **fail in production when conditions are heavier**. It's quite literally Salesforce saying, "We will *stop* your code if it uses too much." When a governor limit is hit, an exception is thrown and the whole transaction is rolled back. That's obviously something we want to avoid, as it can disrupt business operations.

So, how to proactively manage and design around these limits:

Efficient loops and logic: Ensure you aren't doing redundant work. For instance, if you need to process 100 records, do it in a single loop, not nested loops that multiply the operations unnecessarily. Avoid nested loops over large collections whenever possible, or break out of loops when a condition is met to not waste iterations. Use collections (Set/Map) to streamline lookups as mentioned earlier, which also saves CPU time. If you find yourself needing a nested loop (like comparing every record of one list to every record of another), think if there's a better way (maybe group by key in a map).

Minimize database calls: This echoes bulkification – try to consolidate your database interactions. Instead of querying inside a loop (which can blow through the SOQL limit), gather the keys you need and do one query. Instead of updating inside loops, accumulate and do one update. Sometimes you can even avoid queries by using relationships in SOQL (subqueries or parent field access) to get data in one shot. Also, consider caching within a transaction: if you query something and will need it again for another record in the same transaction, store it in a map to reuse, rather than querying again.

Effective SOQL: Write selective queries. For example, if you have a large org with millions of records, a query like [SELECT Id FROM Order__c] (no WHERE clause) will try to pull all orders – likely hitting the 50k limit or at least being extremely slow. Always try to query with a WHERE clause that limits scope to what you need. Salesforce can use indexed fields to speed up queries; common indexed fields include Id, Name, OwnerId, lookup fields, some formula fields (if marked index), etc. If you filter on non-indexed fields on a huge dataset, you could hit "non-selective query" issues which, while not exactly a governor limit, can cause your query to timeout or be blocked. Using tools like Query Planner or asking Salesforce for large data strategies might be needed in extreme cases. But for most, just avoid the obvious pitfalls (don't query everything just because you can).

Leverage Asynchronous processing: If something might hit limits in a single transaction, consider moving it to a batch. For instance, a mass update of tens of thousands of records – doing that in a loop in a trigger is not possible without hitting limits, but a Batch Apex can handle it by processing in chunks of 200 records per batch execution. Each chunk is its own transaction with fresh limits. So, asynchronous Apex (Batch, Queueable, Future, Scheduled) can help distribute work to stay within per-transaction limits. (Keep in mind, async has its own set of limits like number of jobs, but those are easier to manage by design.)

Use Limits Apex methods for awareness: Apex provides the Limits class. For example, Limits.getQueries() tells how many SOQL queries have been used so far in this transaction, and Limits.getLimitQueries() gives the max (like 100). Similar methods exist for CPU, DML, etc. You can use these to add assertions in test code or even runtime checks if you want to guard against approaching a limit (though usually better to design not to get close). It's sometimes useful in debugging to log how close you came to limits.

Let's illustrate with a concrete potential governor issue and resolution. Imagine you have a piece of Apex that needs to create a series of records and then send an email for each record created. Perhaps something like:

```
for (Order__c order : ordersToProcess) {
    OrderSummary__c summary = new OrderSummary__c(Order__c = order.Id, Status__c = 'New');
  insert summary;
   Messaging.SingleEmailMessage mail = new Messaging.SingleEmailMessage();
  //...set up mail to notify someone about this order summary...
  Messaging.sendEmail(new Messaging.SingleEmailMessage[] { mail });
}
```

This code will break limits quickly if ordersToProcess is large. It's doing one insert per iteration (DML limit risk) and one email send per iteration (email limit risk, which is 10 per transaction for sendEmail if not using a bulk email approach). A better approach:

Collect all OrderSummary__c records in a list and insert them in one go (or a few goes if the list exceeds 10k).

Use the result of that insert or the list itself to then prepare a list of emails. Actually, Messaging.sendEmail allows an array of messages, but still max 10. If we need more, we might have to send fewer, or use a single email with all notifications combined, or use a different mechanism (maybe a future call for each chunk of 10).

Alternatively, mark those summaries and have a scheduled job send emails in batches later.

By understanding the limits (150 DML, 10 email sends), we adapt our design: bulk insert, and throttle email notifications.

ChatGPT can definitely help spot potential governor issues. It often knows the common limits and patterns that lead to them. **Example Prompt:**

User: "This Apex method is sometimes failing in production with a governor limit error. Can you spot why?\napex\npublic static void processTasks(List<Id> taskIds) {\n for (Id tid : taskIds) {\n Task t = [SELECT Id, Status, WhatId FROM Task WHERE Id = :tid];\n if (t.Status != 'Closed') {\n t.Status = 'Closed'; update t;\n }\n if (t.WhatId != null) {\n // do something with related record\n }\n }\n}\n"

ChatGPT: "The method as written has a couple of issues that could lead to governor limit exceptions. The for loop is querying a single Task in each iteration using [SELECT... WHERE Id = :tid]. If taskIds contains more than 100 Ids, this will exceed the SOQL query limit of 100 per transaction (since it's doing one query per Id). Additionally, it's performing an update inside the loop for each Task that is not Closed, which could hit the DML limit if the list is large (though 150 DML is a higher threshold, it's still inefficient). The repeated query is the primary culprit. To fix this, the code should bulk query all Tasks at once, e.g., List<Task> tasks = [SELECT Id, Status, WhatId FROM Task WHERE Id IN :taskIds]; outside the loop. Then loop through that list of Tasks in memory, update statuses as needed, and do a single update tasksToUpdate at the end. That way you use 1 query regardless of list size and 1 DML operation. This will avoid the query governor limit and also be more efficient. Also consider if you expect a very large list, if so maybe doing it in chunks or using Batch Apex might be safer, but for moderate sizes bulkifying as described is sufficient."

This identifies the classic query-in-loop error and the remedy, showing awareness of the "100 queries" limit. As a developer, you implement the suggestion: one bulk query, one update.

Another scenario: if a developer is trying to process way too many records in a single transaction (like iterating 50k records in one go), ChatGPT might mention breaking into batches.

Non-text tools can assist here too. For example, if you have debug logs of a transaction, you can copy the log (with the limits usage summary) to ChatGPT and ask it to interpret which limits were nearly hit. Or use voice: e.g., quickly ask "What's the limit for CPU time in a transaction?" and get the answer by voice without searching docs.

Developer/Testers Tip: Don't treat governor limits as an enemy – treat them as design guidelines. **Developers** should proactively use tools like the Developer Console's Execution Overview, which visualizes how many queries, DML, etc., a run used. After writing a complex trigger or class, run a test scenario and check the log's "Limits" section. If you see, for instance, "SOQL queries: 5 out of 100" and you only processed a small batch, consider how that scales – maybe okay. But if you see "SOQL queries: 98 out of 100" for just moderate input, that's a red flag to optimize. Also leverage test methods to simulate limit conditions – e.g., create 101 dummy records and verify that your trigger doesn't hit the 101 query situation. **Testers** in a Salesforce context (especially if doing specialized testing like performance or load testing) can help by running scenarios with large data volumes and monitoring for limit errors. In user acceptance testing, sometimes users might do unexpected bulk operations; having testers mimic those can catch limit issues. Salesforce admins can also set up alerts for when certain limits are hit (there are some AppExchange tools that monitor org for frequent limit errors). Early detection allows the team to refactor before it becomes a serious problem. Ultimately, well-written Apex rarely comes close to governor limits in normal operation, because it's been built with a bulk & efficient mindset from the start.

By conscientiously managing governor limits, you ensure your Apex code is robust under the hood of Salesforce's multi-tenant architecture. Together with scalability, this means your code runs smoothly even as workload increases, without running into the hard walls Salesforce sets.

Now we've addressed code quality (standards, readability), structure (simplicity and layering), performance (scalability and limits), and security. There's one more foundational aspect to cover: making sure that as time goes on, these best practices remain in place and the codebase keeps improving rather than degrading. In the final section of this chapter, we will discuss setting up a culture and process for **continuous Apex**

improvement – how to keep code quality high through ongoing practices and how ChatGPT can assist in that journey as well.

SETTING UP FOR CONTINUOUS APEX IMPROVEMENT

Writing good code is not a one-and-done activity – it's an ongoing process. Salesforce orgs are living systems; they evolve with new business requirements, integrations, and updates from Salesforce itself. Without a mindset of continuous improvement, even a well-built Apex codebase can slowly degrade as changes accumulate (often called software entropy or "code rot"). This is why embracing **continuous Apex improvement** is a best practice in its own right. It's about having practices and culture in place that constantly polish and reinforce code quality, ensuring the foundations we laid (standards, readability, etc.) are maintained over time and enhanced as needed.

What are some of the iterative improvement practices that a Salesforce team should adopt?

Regular code reviews: Set up a process where Apex code changes (whether via Pull Requests in source control or just informal reviews if not using a VCS) are reviewed by peers or a tech lead. Code reviews are opportunities to catch issues (logic errors, missed best practices, security holes) before they hit production. They also spread knowledge among the team. When reviewing, look not just for "does it work" but "is this code following our standards? Is it readable? Is it as simple as it could be? Does it account for bulk and limits? Security?" Basically, all the topics we've covered. Over time, team members learn from each other's suggestions, and the overall code quality bar rises. It's important to keep reviews constructive and educational, not just nitpicky. Many teams create a checklist of things to verify in a review (one could derive such a checklist from this chapter's headings, for example).

Pair programming and mentoring: Pair programming (two developers coding together, even remotely using screenshare) can be great for quality. It naturally enforces a review (since two sets of eyes are on the code as it's written) and often leads to better solutions through discussion. A senior developer pairing with a junior one can mentor best practices in real-time, which is often more effective than post-factum code review comments. Even if pairing is not constant, doing it for tricky tasks or initial frameworks can ensure a solid baseline. Similarly, have architecture/design discussions as a group for significant features, so everyone's on the same page about how to approach it cleanly.

Leverage feedback loops: Agile methodologies encourage retrospectives and continuous feedback. In a Salesforce context, this could mean after each sprint or major release, the team discusses what could be improved in terms of code quality or process. Maybe in the last release, a last-minute bug was due to a governor limit – in retro, note that and plan to add an automated check or update the coding guideline so it doesn't happen again. If testers or users found something confusing, see if it was due to code complexity and plan a refactor. Encouraging a culture where it's okay to refactor code that "works" for the sake of maintainability is key. Technical debt should be acknowledged and addressed regularly, not just at project end (which may never come).

Automated tools for code quality: In addition to human processes, use tools to keep quality high. Salesforce has static code analysis tools (like PMD with Apex rules, SonarQube with Apex plugins) that can scan code for known issues or deviations from standards. Set these up in your CI pipeline if you have one. For example, you can run a static analysis on each commit to catch things like "SOQL inside loop" or "empty catch block" or style violations, etc. You can enforce code coverage minimums with each deployment (Salesforce already has a 75% org-wide requirement, but you might enforce, say, each class must have at least 90% coverage or meaningful tests). If you have a large team, consider technologies like **Unlocked Packages** to modularize the org and test components in isolation, which can improve overall code management.

Continuous learning and improvement culture: Encourage team members to stay updated with Salesforce releases (maybe a new feature can simplify some code, e.g., if Salesforce introduces a new Apex method or a setting that reduces the need for custom logic). Maybe someone attends a Dreamforce session or reads a blog about a better way to handle something – have them share with the team and consider adopting it. Keep your best practices document alive; update the coding standards as new patterns emerge or if certain rules were found to be impractical. Essentially, treat the way you write code as something that can always be refined.

Now, how can ChatGPT support ongoing code quality efforts? We've already seen it can act as a code reviewer or a brainstorming partner for improvements. This can be integrated into continuous improvement in several ways:

On-demand code reviews: A developer can paste a snippet or describe a change and ask ChatGPT for a review. This is especially useful for lone developers or when seeking a second opinion at 2 AM when no team members are available. While ChatGPT might not understand the full business context, it's excellent at spotting generic issues (like "you forgot to bulkify this" or "this name is confusing" or "this could be simplified"). Using ChatGPT doesn't replace human reviews, but it can augment them. It's like having an automated reviewer that never gets tired. Some teams even script this: e.g., using the ChatGPT API to comment on GitHub pull requests with suggestions.

Refinement prompts: If a piece of code works but is messy, a developer can ask ChatGPT, "How can I improve this code?" and get immediate refactoring suggestions. This encourages a habit of not settling for the first solution that works, but iterating to make it better. For example, after passing all tests, a dev might still spend an extra hour polishing the code with ChatGPT's help. Over time, this yields a cleaner codebase.

Voice and interactive assistance: Imagine during a code review meeting, someone uses a voice interface to quickly query, "ChatGPT, what are common pitfalls in implementing a custom sharing rule in Apex?" The team might get a quick rundown from ChatGPT, sparking a discussion if they handled those pitfalls. Or a tester might use voice to ask, "Explain the flow of this new feature's Apex logic" to verify understanding. These non-text interactions can make engaging with the knowledge base (ChatGPT's learned info or documentation) more fluid in a live setting.

File uploads for analysis: With ChatGPT's ability to handle file inputs (especially using the Advanced Data Analysis tool or similar), you could feed in an entire Apex class or even a zip of the codebase and ask for an analysis. For example, "Analyze this Apex class for potential issues and suggest improvements." ChatGPT might output a report highlighting sections of code and recommendations. Always ensure you're comfortable with the privacy of doing so – if not, you can sanitize or use dummy code structures to get generic advice. If you do upload real code, as we noted, turn on any data privacy features (like not saving chat history, etc.) or use the API with data retention off.

Operator mode automation: As a forward-looking idea, ChatGPT's Operator mode (which allows it to perform multi-step actions autonomously) could potentially be set up to monitor your code repository. For example, it might scan for any commit that reduces code coverage or adds a new method without a corresponding test, and then automatically flag it or even write a first draft test. Or it might detect when multiple classes have similar code and suggest a refactor to eliminate redundancy. This is still speculative, but the technology is heading in that direction – AI assistants that proactively maintain code quality.

Crucially, continuous improvement also involves roles beyond developers:

Architect/Product Owner/Manager Tips: **Architects** should champion processes like regular refactoring sprints or allocating time for improving "areas of concern" in the code. Maybe designate a part of each sprint for technical debt cleanup. Track metrics like code complexity (some static analysis tools can measure cyclomatic complexity), code coverage, bug rates, etc., to identify trends – e.g., if bug rates are creeping up, maybe quality is slipping and time to reinforce best practices is needed. **Product owners** and project managers play a role by acknowledging that time spent on quality is valuable. They should avoid scheduling every moment for new features with no allowance for improvement of the existing system. A product owner can frame refactoring or adding tests as risk reduction and enabler for future features (which it is). If needed, ChatGPT can help translate the tech debt into business terms: "What is the impact of not addressing technical debt in software projects?" – the answer can help POs justify that work in planning meetings. **Middle managers (e.g., development managers)** should ensure the team has the tools and training to enforce quality. Investing in things like a CI/CD system with automated tests and static analysis, or sending team members to training (or bringing in an expert) to learn new best practices, all pay off. Managers can use ChatGPT to keep themselves informed too: if they're not hands-on in code, they might ask it for summaries or explanations of the team's technical discussions to better support decisions. For instance, if devs are debating using a Trigger framework, a manager could quietly ask ChatGPT, "What are the benefits of using a trigger framework in Salesforce?" and then better understand the context.

By setting up these cultural and procedural supports, a Salesforce team ensures that code quality isn't a one-time initiative but a continuous journey. Your Apex code can actually *improve* over time – imagine that! – as opposed to degrading. Clean sections stay clean, and any messy sections are gradually cleaned up. This results in a codebase that is easier to work with year after year, yielding long-term efficiency.

As an example of continuous improvement in action: consider a team that inherited an older Salesforce org with some messy code. Instead of rewriting everything at once (often impractical), they might establish a rule: whenever you touch an old piece of code to add a feature or fix a bug, leave that code better than you found it. One bug fix might be the opportunity to split a giant method into two, add a few comments, and maybe add a missing test. Over a year, with dozens of changes, that ugly duckling of a codebase transforms significantly. ChatGPT can help by quickly suggesting improvements each time a piece is touched, making it feasible to do this within the scope of regular work.

To wrap up this section, continuous improvement ensures that the best practices covered in this chapter remain living principles. It closes the loop: we start with a solid foundation and we keep reinforcing that foundation at every step. It's like preventative maintenance on a machine – you avoid the big failures by regularly tuning and oiling the parts. With people, process, and tools (including AI assistants) all contributing, continuous Apex improvement becomes second nature.

Having journeyed through coding standards, readability, simplicity, scalability, security, governor limits, and continuous improvement, we've established a comprehensive set of foundations for effective Apex development. These aren't just theoretical ideals; they are practical habits and techniques that yield better Salesforce applications. In the next chapter, we will build upon these foundations and dive into structuring and organizing Apex code for maximum effectiveness, exploring patterns and strategies to further enhance the clarity and robustness of your Salesforce codebase. But before we move on, let's recap the key learnings from this foundational chapter.

SUMMARY

In this opening chapter, we explored the **foundational best practices of Apex development** and saw how they benefit everyone involved in a Salesforce project. We began by emphasizing the importance of establishing Apex coding standards to ensure consistency across the codebase. By defining clear guidelines for naming, formatting, and structuring code, teams can significantly ease maintenance, speed up debugging, onboard new developers faster, and prevent technical debt from accumulating. We saw that a consistent style is the bedrock upon which readable and maintainable code is built.

We then reviewed the **principles of readable Apex code**. We highlighted how meaningful naming, thoughtful commenting, and proper formatting make code understandable at a glance. Readability is not just about aesthetics – it directly correlates with productivity and fewer bugs. We demonstrated how even a working piece of code can be refactored into a much cleaner version and how ChatGPT can assist by pointing out readability issues and suggesting improvements, effectively serving as an AI code editor.

Next, we tackled **simplifying Apex logic and managing complexity**. We recognized that keeping methods and classes focused and bite-sized, using helper classes and clear abstraction layers, leads to a codebase that is easier to navigate and modify. Complex, monolithic routines were broken down into smaller components, illustrating the power of the single-responsibility principle in Apex. Both architects and product owners were reminded that simplicity in design is not a luxury but a necessity for sustainable development. ChatGPT's refactoring suggestions proved useful in turning convoluted code into structured, modular pieces.

With a firm grasp on code clarity, we shifted to **scalability and performance** considerations. Planning for scalability from the start means writing bulkified code that can handle large volumes of data without issue. We reviewed Salesforce's approach to bulk operations and saw how to avoid pitfalls like SOQL-in-loops. The concept of being governor limit-aware was introduced, reinforcing that efficient queries and minimal round-trips to the database keep our code within Salesforce's operational guardrails. Through examples, we transformed non-scalable code into bulk-safe implementations. We also noted that ChatGPT (and its advanced modes like Operator) can help identify performance bottlenecks and even analyze code in aggregate to foresee scalability issues.

Security was our next focus, underlining that **embedding security in Apex** from the beginning is non-negotiable. We went over enforcing user permissions (CRUD/FLS) and using sharing rules correctly, ensuring our Apex doesn't inadvertently grant data access beyond what's intended. Secure coding practices, such as guarding against SOQL injection and protecting sensitive information, were discussed. We learned how tools like with sharing and Security.stripInaccessible act as safeguards. ChatGPT helped simulate a security review, catching missing permission checks and suggesting remedies. The takeaway: robust Apex code not only needs to be fast and clean, but it must also be safe and compliant with Salesforce's security model.

Understanding and managing **Salesforce governor limits** came next. We summarized key limits (queries, DML, CPU time, etc.) and why they exist. More importantly, we covered strategies to design within these limits – from writing efficient loops and combining SOQL calls to leveraging asynchronous Apex for large jobs. By treating governor limits as design constraints, we naturally wrote better, more optimized code. Examples showed how to rework code that would hit limits into code that operates comfortably below them. We saw that by using best practices like bulkification and efficient querying, many limit issues can be avoided entirely. Developers and testers were encouraged to proactively test for and monitor limit usage, catching issues early. ChatGPT served as an advisor in pinpointing where code might be flirting with a limit and how to address it.

Finally, we looked at establishing a practice of **continuous improvement** for Apex code. We acknowledged that maintaining high code quality is an ongoing effort, supported by regular code reviews, pair programming, and iterative refactoring. The value of tools (static analysis, CI pipelines, automated tests) and processes (agile feedback loops, learning culture) was highlighted as means to keep the codebase healthy as it grows. Crucially, we discussed how ChatGPT can be woven into these processes – as a constant code review buddy, as a source of quick advice on best practices, or even as an automated quality gate. Tips were provided for architects, product owners, and managers to foster an environment where technical excellence is valued and resourced. By making continuous improvement a habit, teams can ensure that their Apex code doesn't just stagnate or decay after initial development, but actually gets better and more robust with time.

In summary, this chapter equipped us with a holistic set of practices for writing Apex that is clean, efficient, secure, and maintainable. We also saw that modern tools like ChatGPT can amplify our efforts, providing guidance and checks at every step – from brainstorming coding standards to reviewing a final pull request. By internalizing these foundations, you as a reader (whether an architect, developer, tester, or product stakeholder) can approach Salesforce Apex development with a mindset of quality and foresight.

As we move forward in this book, these foundational principles will recur and be reinforced in more specialized contexts. In the next chapter, we will build upon what we've learned here and dive deeper into **Apex code structure and organization.** We'll explore how to organize classes and triggers in a Salesforce org, effective naming conventions (taking the brief mention here to a more advanced level), and design patterns that can further improve the clarity and modularity of your Apex code. With the solid groundwork from Chapter 1 in place, you'll be well-prepared to tackle the structural aspects of Apex development that Chapter 2 brings into focus. Keep these best practices in mind – they will continue to be our compass as we navigate the broader landscape of Apex development excellence.

2

APEX CODE STRUCTURE AND ORGANIZATION

Effective code structure and organization form the backbone of sustainable Salesforce applications. In Apex development, how you arrange and compose your code can be just as important as the features it delivers. A poorly organized codebase can lead to confusion, bugs, and technical debt that slow down development. In contrast, a well-structured Apex codebase brings clarity. It allows developers to quickly navigate and understand the system, makes it easier to extend functionality, and reduces the risk of regressions when changes are introduced. By the end of this chapter, you will appreciate why investing effort in Apex code organization upfront pays dividends throughout the life of a Salesforce project.

Thoughtful Apex code structure directly translates to reduced technical debt and easier maintenance. When classes and methods are logically arranged, new team members can onboard faster, and seasoned developers can modify or debug code without wading through chaos. Consistent organization also fosters smoother collaboration: architects can define clear patterns that developers follow, testers can trace logic flows without guesswork, and product owners gain confidence that the implementation aligns with design. A structured approach means fewer surprises in later stages of a project – performance issues, limits errors, or security gaps become less likely because good organization often accompanies good practices in those areas.

In this chapter, we dive into techniques for mastering clear organization of Apex code. You will learn proven conventions for naming Apex classes, methods, and variables to maximize readability and consistency. We will explore strategies to structure Apex classes logically, ensuring that each piece of code has its rightful place and purpose. The chapter covers how to leverage helper classes and utility methods to avoid duplication, and how to use object-oriented principles like inheritance and interfaces appropriately for maintainable code. We'll also address controlling dependencies between classes to keep the system flexible (a critical aspect when your Salesforce org grows large) and discuss how packaging and namespacing can be used to modularize a codebase in an enterprise environment. Finally, we underscore the importance of documentation and annotations in Apex – because even the best-structured code benefits from clear explanation for current and future stakeholders.

Throughout the discussion, we will highlight how these practices benefit different roles on a Salesforce team. **Architects** will find that a well-structured Apex architecture is easier to govern and evolve at the enterprise level. **Developers** will notice improvements in daily productivity when code is organized and straightforward to navigate. **Testers** benefit from predictable patterns that make it clear where to write and find test cases or how to mock certain components. **Product owners** and **project managers** may not write Apex code, but they reap the rewards of faster delivery and lower maintenance costs when best practices are followed. We will include special tips for each role, calling out how an optimized code structure supports their goals – from big-picture system design to nitty-gritty testing efficiency.

Crucially, you'll see how *Salesforce Apex Best Practices with ChatGPT* isn't just a theoretical theme of this book – ChatGPT can be a practical assistant every step of the way. We'll incorporate extensive ChatGPT prompt examples illustrating how to get AI help in organizing your Apex code. For instance, you can ask ChatGPT to review an Apex class and suggest a better structure, or even use voice input to brainstorm a naming scheme during a meeting. ChatGPT can analyze uploaded code files (ensure you enable file privacy in settings before sharing sensitive code) and provide feedback on consistency and design. It can generate diagrams or images on demand to visualize class relationships, using advanced tools or the new ChatGPT "Operator" mode to interact with files and documentation beyond plain text. By leveraging these capabilities, teams can enhance their understanding and enforcement of best practices collaboratively.

Consistency is a recurring theme when structuring Apex in enterprise projects. Large Salesforce implementations often involve multiple teams and long lifecycles; a consistent organizational approach keeps everyone aligned. As you read on, consider how each principle ties back to consistency – naming conventions, class design, utilities, use of inheritance, dependency management, packaging, and documentation all interlock to form a coherent style. Embracing these practices uniformly across the codebase means any developer can step into any part of the system and quickly grasp what's going on. This chapter will prepare you with detailed guidance and examples to build that level of consistency in your Apex code.

With these foundations laid, we can now move from theory to practice. The first step in structuring Apex code effectively is deciding how to name everything in a clear and consistent way. Let's begin by examining **naming conventions** – the simple, but powerful rules that make Apex code readable and maintainable from the ground up.

EFFECTIVE NAMING CONVENTIONS

Apex code organization starts with the names we choose for classes, methods, variables, and other identifiers. Adopting and consistently applying clear naming conventions significantly impacts code readability and long-term maintainability. When names accurately describe purpose, anyone reading the code can understand intent without delving into implementation details. In contrast, inconsistent or cryptic naming forces readers (and even the original author, months later) to play detective to figure out what something represents. By establishing effective naming conventions, a Salesforce team creates a common language within the codebase that makes collaboration and hand-offs much smoother.

Why Naming Conventions Matter: Consider an Apex class named Calc with a method m1(). These names are so vague that a reader must open the method to decipher its function. Now imagine instead a class OpportunityRevenueCalculator with a method calculateAnnualRevenue(). Immediately, it's clearer what the code does. Good naming reduces cognitive load on developers – they can spend more time focusing on *what* the code should do rather than *what the code is*. In a large enterprise Salesforce org, dozens of developers work on intersecting features; consistent naming ensures that a variable or class named similarly means the same thing everywhere. This consistency prevents misunderstandings that could introduce defects or duplicate logic. Moreover, naming conventions help tools and scripts (like static code analyzers or search utilities) to easily find relevant pieces of code. For example, if all test classes end with Test, a simple search or filtering by name quickly isolates all tests.

Salesforce generally recommends following Java-style naming conventions for Apex identifiers. Apex is case-insensitive in its execution, but developers **should** use consistent capitalization styles to convey meaning. The standard conventions include:

Classes: Use PascalCase (each word capitalized) and typically use nouns or noun phrases. For example, AccountController or InvoiceService. This clearly identifies the entity or service the class represents. Class names should ideally be unique and not conflict with Salesforce standard classes.

Methods: Use camelCase (first word lowercase, subsequent words capitalized) and use verb phrases that describe the action or calculation. For example, calculateDiscount() or sendEmailNotification. A method should sound like a command or query it performs. Using a verb (like "calculate", "get", "send") at the start sets an expectation that the method *does* something.

Variables: Use camelCase as well, and choose meaningful, descriptive words that indicate what data the variable holds. Avoid single-letter or overly abbreviated names except in trivial contexts (like i in a short loop). Instead of cnt or x, write invoiceCount or annualRevenue – the extra clarity far outweighs a few extra characters. A good guideline is that someone should guess a variable's role from its name without needing a comment.

Constants: For constants or final static variables, many teams adopt ALL_UPPERCASE with underscores. For example, MAX_RETRY_ATTEMPTS might be a constant in a utility class. This differentiates constants from other variables at a glance.

Interfaces: By Java convention, interface names are often adjectives or descriptive nouns (like Sortable, NotificationSender). Some teams prefix interface names with I (e.g., ISortable), though this is less common in Apex. The key is to be consistent within your codebase if you choose a prefix.

In practice, naming conventions often extend to include **suffixes** or **prefixes** that convey a class's role or layer in the application. For example, by convention many Apex developers name controller classes with a Controller suffix and extension classes with ControllerExt. A Visualforce controller for an Account page might be AccountController, possibly with an accompanying extension AccountControllerExt for extended functionality, and their test classes would be AccountControllerTest and AccountControllerExtTest. Similarly, classes that act as services or domain logic might include Service or Manager in the name (e.g., OrderService for business logic around orders). Trigger handler classes often carry the TriggerHandler suffix to distinguish them from the trigger itself. For instance, if you have an Apex trigger on the Account object, a common pattern is to name the trigger **AccountTrigger** and implement all logic in a separate class **AccountTriggerHandler.** This way, the trigger's name is simple and tied to the object, while the handler class' name clearly indicates its purpose. Test classes for these would mirror the names (e.g., AccountTriggerTest and AccountTriggerHandlerTest). Adopting such suffix conventions makes it immediately apparent what each class is for. A developer scanning the project can guess that anything ending in "Controller" is part of the UI layer or business logic invoked by a UI, "TriggerHandler" classes contain trigger logic, and classes ending in "Test" are test classes (which is also enforced by Salesforce—test classes must have "Test" in their name or be in a*.test* file).

Consistent Naming in Practice: When the team agrees on naming patterns, it's important everyone sticks to them. For example, if the convention is to use singular for class names (e.g., InvoiceService not InvoicesService), then all follow that. Even seemingly small choices like whether acronyms are uppercased (do you name a class XMLParser or XmlParser?) should be standardized. A good practice is to document the naming conventions in a wiki or a short guide for your project, and even use automated code scanning tools to catch deviations. ChatGPT can help in both documenting and enforcing these rules. A team could prompt ChatGPT to **generate a naming conventions document** by describing their preferences, and use that as a starting draft for their official guidelines. For instance, you might say:

Prompt: "ChatGPT, help me draft Apex naming conventions for our team. We want classes in PascalCase with a noun, services ending in 'Service', triggers ending in 'Trigger', test classes ending in 'Test', and variables in camelCase. Include examples."

ChatGPT can then produce a nicely formatted reference that the whole team reviews and agrees upon. Once established, developers can even use ChatGPT as a spot-check tool. For example, you could paste a piece of code and ask: *"Do the identifiers in this snippet follow our naming conventions?"* ChatGPT will point out if a variable looks out of place (perhaps tempVar where a more specific name was expected) and suggest improvements.

In modern AI-enhanced development, remember that non-text interactions are available too. If you're brainstorming names for a new class or method, you could use voice input to **ask ChatGPT aloud** while you're away from the keyboard. For example, a product owner in a meeting might use a voice query: *"What would be a clear name for a class that handles quarterly financial calculations in Apex?"* ChatGPT's voice response can give suggestions (like "QuarterlyFinancialCalculator or Finance-QuarterProcessor") without even writing anything down, making your planning sessions more interactive. If you already have some Apex code written with placeholder names, you could **upload the file** to ChatGPT and request it to review naming consistency – just ensure you have your file upload privacy settings enabled to protect any sensitive code. ChatGPT will scan the file and highlight any names that deviate from the guidelines you've provided.

Architect Tip: Define and enforce naming conventions at the **start** of a project. It's much harder to rename classes and variables later on without ripple effects. Many architects include naming standards as part of the project's development handbook. This upfront discipline creates a uniform style that new developers will adopt naturally.

Developer Tip: Embrace the team's naming conventions even if they feel unusual at first. Consistency is more important than personal preference. If you're unsure about a name while coding, use ChatGPT as a quick thesaurus – describe what the variable or method does and ask for a name suggestion. Often an outside perspective (even from an AI) can propose a name that's both descriptive and aligns with standards.

Clear, descriptive naming sets the stage for every other best practice in Apex. Once everything in the code **sounds** like what it represents, the next step is to ensure each piece of code is in the right place. In the following section, we turn to organizing Apex classes themselves: how to structure your classes and logic in a way that further enhances clarity and maintainability without creating monoliths or spaghetti code.

ORGANIZING APEX CLASSES

Even with consistent naming, code can become hard to manage if it isn't organized well into classes and modules. In Apex, organizing classes means deciding how to break up functionality across multiple classes and how to arrange those classes in a logical structure. Good class organization follows the principle of *separation of concerns*, meaning each class is focused on a specific aspect of the application, and *single responsibility*, meaning each class has one primary reason to change. By organizing Apex classes thoughtfully, you enhance readability, simplify testing, and ease troubleshooting when issues arise. In this section, we explore strategies for class structure, class sizing, and logical layering of code in a Salesforce org.

Separation of Concerns: A robust Apex architecture will divide responsibilities among different classes. For example, one class might handle database operations (queries and DML) for a certain object, another class handles business logic or calculations, and yet another deals with integration calls. This way, when you need to make a change or fix a bug, you know *which* class to look in. A common pitfall is the "God class" – a class that tries to do too much. Imagine a single AccountProcessor class that queries accounts, updates fields, sends emails, and logs changes all in hundreds of lines of code. Not only is that class hard to understand, but any modification risks breaking multiple pieces of functionality. Instead, it would be better to have, say, AccountQueryService (for database interactions), AccountBusinessLogic or AccountService (for core logic like calculating fields), and AccountNotification (for sending emails or notifications related to accounts). Each of these classes addresses one aspect of handling Account-related operations. This is the essence of separation of concerns and adheres to the idea that each class should have cohesive methods related to one purpose.

Keeping methods focused and classes cohesive aligns with known best practices. In fact, two key principles often cited by experienced developers are DRY (*Don't Repeat Yourself*) and separation of concerns. DRY means you should not duplicate logic in multiple places – if two classes need to do the same thing, that's a sign that functionality should be pulled out into a single helper (which we'll cover in the next section). Separation of concerns, on the other hand, guides you on how to partition different kinds of logic. In Apex, you might separate code by layer: one layer for database operations (sometimes called *selector* or *repository* classes), one for domain logic (business rules), one for user interface or endpoint interactions (like controllers or services), etc. By doing so, each class is more modular, and you can modify one layer without heavily impacting others.

Balancing Class Size and Complexity: How big should an Apex class be? There's no hard rule measured in lines of code, but generally if a class grows too large or has too many responsibilities, it's a candidate for splitting. Look for logical groupings of methods. For instance, if within a single class you find half the methods deal with validation rules and the other half deal with formatting output, that class might be doing two jobs – perhaps create a DataValidator class and a separate DataFormatter. On the other hand, one should avoid splitting classes without reason – if two pieces of logic always change together and are small, combining them might be perfectly fine. The goal is to make each class logical and not overly complex. An easy way to gauge this is to see if you or a teammate can describe what a class does in one clear sentence. If you stumble and use words like "and then it also does...", that class might be tackling more than one responsibility.

Salesforce's project structure (especially with modern Salesforce DX) allows grouping classes into folders by functionality or domain for organizational purposes. While all classes reside in the same namespace in a Salesforce org (unless using packages with namespaces), creating a folder structure in your source code repository can help maintain order. For example, you might have folders like classes/Opportunity/ containing all Opportunity-related classes (OpportunityService, OpportunityValidator, OpportunityTriggerHandler, etc.), and another classes/Account/ for Account-related classes. Or you might organize by layers: classes/Domain/, classes/Service/, classes/Util/, etc. This client-side organization does not affect runtime but helps developers navigate the code. As one Salesforce expert noted, you can divide your classes into any structure that makes it easier to navigate – some group by application area (e.g. Marketing vs. Sales), others by type of class (service classes, batch jobs, etc.). The only caution with such structuring is to avoid duplicate class names in different folders, since Salesforce won't allow two classes with the same name even if in different folders in source control. Also, if you eventually deploy packaged modules, remember that class names must remain unique per namespace to avoid conflicts.

Layered Architectures in Apex: In larger enterprise projects, architects often define layers or tiers in the Apex code to enforce separation. A prevalent approach is inspired by *Apex Enterprise Patterns* (such as the fflib library by FinancialForce). In these patterns, you might see:

Domain Classes (sometimes with a naming convention like ObjectNameDomain or a prefix like DOM_): containing behavior and logic closely related to a specific sObject, often including trigger handling logic for that object.

Service Classes (e.g. ObjectNameService or prefix SRV_): containing higher-level operations that might orchestrate multiple domain classes or implement business processes.

Selector/Repository Classes (e.g. ObjectNameSelector or ObjectNameRepository): dedicated to retrieving records (all SOQL queries for that object in one place).

Utility Classes (we'll detail soon): for shared, general-purpose logic not specific to one object.

Not every org uses this exact structure, but the idea is to avoid mixing SQL queries, business logic, and presentation logic all in one class. For example, if a method is performing a SOQL query and then doing calculations and then formatting a string for output, that might indicate it's breaching multiple layers of concern. Instead, you could have the selector class fetch data, pass it to a service class that computes results, and perhaps a separate formatter utility to produce a string or JSON. This kind of layered approach makes the code more maintainable and testable – you can unit test the business logic separately from the database retrieval, for instance.

ChatGPT as a Design Assistant: Organizing classes can sometimes feel like an art – how to break apart logic is not always obvious. This is where a tool like ChatGPT can provide a second opinion. You could present ChatGPT with a summary of what your codebase does and ask for a proposed class structure. For example: *"We have an application that manages Orders: it validates orders, calculates totals, sends confirmation emails, and integrates with an inventory system. How should I organize my Apex classes?"* ChatGPT might respond with a suggestion to create OrderValidator, OrderCalculator, OrderEmailService, InventoryIntegrationService, etc., each handling a portion of the work, and explain the reasoning. If you already have a large class, you could paste key method signatures or describe its contents, then prompt: *"This Apex class seems to be doing a lot. How could I refactor it into smaller classes?"* The AI can identify themes in the methods and propose splits.

Consider also using ChatGPT's ability to handle files: you might **upload an Apex class file** (with privacy settings on) and ask for a critique of its structure. ChatGPT can highlight sections that are unrelated or too large and suggest where to make cuts. Another approach is to **have ChatGPT draft a UML-style outline** of your current classes – just by reading the code, it can enumerate classes, their relationships (if, say, one instantiates another), which can help visualize the current structure. Modern tools or the ChatGPT Operator mode could even navigate a repository for you to gather this info if integrated properly.

Architect Tip: Before coding, try to sketch the system's class design. Some architects use whiteboards or diagrams to map out classes and their relationships. You can leverage ChatGPT to refine this plan: describe your proposed classes and see if the AI identifies any overlapping responsibilities or missing pieces. It's easier to adjust a plan than refactor thousands of lines later. Additionally, consider defining an "architectural blueprint" that all developers can reference – ChatGPT can help generate diagrams or even boilerplate class templates for each layer.

Tester Tip: Well-organized classes mean you can write more targeted tests. If logic is split into small classes, you can instantiate just that class and test its behavior in isolation. Testers (or QA-focused developers) should encourage separation of concerns because it often correlates with more accessible logic to validate. If you find it hard to write a unit test for a class because it does too many things (e.g., needs lots of setup data or triggers multiple actions), that's a flag that the class might need refactoring. Communicate this to the development team – often, testing difficulties shine a light on structural problems.

As you refine your Apex class organization, you'll naturally start noticing opportunities for reuse. Code that is used in multiple places shouldn't be copied in two classes; that violates DRY and makes maintenance harder. This leads us to the next vital practice in Apex code organization: creating and leveraging helper classes and utility methods to centralize common logic and avoid repetition in the codebase.

LEVERAGING HELPER CLASSES AND UTILITIES

Software development often involves writing similar code more than once – perhaps two different triggers need to format a string the same way, or multiple classes perform the same validation check. In Apex, as in any programming language, duplicating code is a red flag. It bloats the codebase, makes fixes or changes laborious (you have to modify every copy of the code), and increases the chance of inconsistencies or bugs. The solution is to **factor out** common tasks into helper classes or utility methods. By leveraging helper classes and utilities, you adhere to the DRY (Don't Repeat Yourself) principle and streamline coding practices. This section introduces the concept of helper classes, discusses when and how to use them effectively, and illustrates how ChatGPT can assist in identifying refactoring opportunities for code reuse.

What Are Helper Classes and Utilities? Helper classes (sometimes called utility classes) are typically classes that provide common, reusable functions which are not tied to a specific business object. They often contain static methods (or sometimes instance methods) that can be called from anywhere to perform generic tasks. For example, you might have a StringUtil class with methods like capitalizeFirstLetter(String s) or maskEmail(String email) that are used in various places in your code. Another example is a DateUtils class that might have methods to calculate the end of the quarter or add business days to a date. In Salesforce Apex context, you might create helpers for common tasks such as running SOQL queries with certain patterns, handling errors in a uniform way, or encapsulating a formula that's used in different classes. The key is that a helper method is written once and used many times, improving consistency and reducing total lines of code.

Benefits of Using Helpers: The immediate benefit is reduced repetition – if you find a bug in a routine that was copied in five places, you only need to fix it in the helper method rather than tracking down all five places. It also means all parts of the application behave consistently; e.g., if you have a standard way to format phone numbers in PhoneNumberUtil.format(phone), you won't have one class trimming to 10 digits while another pads with country code differently. Helpers also make code more readable at the call site. Compare InvoiceService.calculateTax(invoiceAmount, taxRate) to having the tax calculation logic inlined everywhere – the method name itself communicates what is happening, improving readability. Additionally, having a library of well-tested utility methods can speed up development; developers can rely on those building blocks to construct new features rather than reinventing the wheel each time.

Identifying Opportunities for Reuse: How do you spot when a helper class is warranted? One obvious case is when you notice the same block of code in multiple classes. For example, suppose both the Opportunity trigger and the Order trigger have code to split the full name of a Contact into first and last name. That could be a candidate for a ContactUtil.splitName(fullName) static method. Sometimes duplication is not copy-paste, but conceptual repetition: two classes might have different implementations for something that could be unified. For instance, if both LeadConverter and AccountManager classes need to find an available username for a new user (appending numbers until a unique username is found), they might be doing it differently. Extracting that logic to a UserUtil.generateUniqueUsername(baseName) ensures one standard approach. Another clue is if a single class starts accumulating a lot of unrelated utility methods as private – maybe those could live in a separate shared class and even be public for others to use.

However, one should also use judgment: not every piece of code used twice justifies a new utility. If it's a trivial one-liner, sometimes duplication is acceptable or you might just inline it. The overhead of abstraction should not exceed the benefit. A helper should represent a meaningful chunk of functionality or a *unit of logic* that stands on its own. Also, beware of creating "catch-all" utility classes where unrelated functions accumulate – e.g., a single Util class that has string functions, math functions, and trigger handlers all mixed would defeat the purpose. It's better to have multiple focused helper classes (StringUtil, MathUtil, TriggerUtil, etc.) than one monolithic utility class.

ChatGPT as a Refactoring Partner: ChatGPT can be incredibly useful in finding and suggesting where helpers could be used. If you provide pieces of Apex code, it can analyze for duplication. For instance, you could feed two class files (either by copy-pasting or uploading them, making sure to keep them private) and ask: *"Do you notice any duplicate logic between these classes that could be refactored into a utility method?"* ChatGPT might recognize that both classes have a section where they parse an email address or calculate a discount, and it will point that out. It might respond with something like: "Both classes convert a list of records into a map keyed by Id. This conversion could be a utility method, e.g., RecordUtil.listToMap(recordList)." This kind of insight can save you the effort of manually scanning a large codebase for common patterns.

Additionally, you can ask ChatGPT to **create a helper method** for a given task. For example: "Write an Apex utility method to calculate the distance between two coordinates (latitude, longitude) using the Haversine formula." ChatGPT can produce the method code which you can then integrate into a GeoUtil class. You might also get suggestions for a whole helper class if you describe a scenario: "We often need to check user permissions and field accessibility in our code. What could a security utility class look like?" ChatGPT could outline a class SecurityUtil with methods like hasFieldAccess(sObjectType, fieldApiName) or checkCrudPermissions(recordList).

Don't forget non-text means: If your code is too large to conveniently paste, using the **file upload feature** is a safer route. Upload, say, three related Apex classes and then query, *"Analyze these classes for repeated code segments that could be moved to a helper class."* ChatGPT will have the context of all uploaded files to compare and find repetition. Just ensure file privacy – you don't want proprietary code to leak. Also, if you prefer, you can use voice commands to describe your code to ChatGPT abstractly: *"We have two different classes doing similar validation. One is checking for valid email format and another is checking for valid email format in a slightly different way. Can I unify this?"* Even without providing raw code, ChatGPT can discuss the scenario and confirm that yes, a shared EmailValidator.isValid(email) would be beneficial and even provide a quick regex-based implementation.

Developer Tip: When writing new code, pause and think: "Have I solved a similar problem before?" It's tempting to just write it again, but investing a little time to abstract a helper can save a lot more time in the future. If you're not sure whether a utility exists in the codebase, you could ask ChatGPT (if you've previously given it context of your project) or use an internal search. Better yet, maintain a **team utility library** and encourage contributions to it. ChatGPT can assist by generating initial code for utility functions, which you then test and refine.

Tester Tip: From a testing perspective, centralized utilities are a boon. If multiple parts of the application rely on the same helper method, you can write a focused test for that method's logic once. This reduces duplicate test code and ensures consistency. For example, if all date calculations funnel through DateUtil.addBusinessDays(), your tests for date adjustments all target that method's correctness. As a tester, if you see different behaviors in different parts of the app that *should* be the same, it's worth flagging. It might indicate those parts aren't using a common helper when they could. Encouraging the use of a single utility method will also mean you only have to verify that logic in one place.

By utilizing helper classes, your Apex code becomes more modular and clean. We remove clutter from individual classes and promote reuse across the codebase. This lays a strong foundation, but organizing code isn't just about splitting into many classes – it's also about how those classes relate. Next, we'll discuss how to use Apex's object-oriented features like inheritance and interfaces to further enhance code maintainability and extensibility, ensuring that classes can share behavior or adhere to contracts in a structured way without becoming entangled.

APEX INHERITANCE AND INTERFACE USAGE

Apex is an object-oriented language, which means it supports class inheritance and interfaces similar to Java. These features, when used appropriately, can dramatically improve code maintainability by promoting reuse and establishing clear contracts for behavior. Inheritance allows a class to derive from another, inheriting its methods and properties, while interfaces define a set of methods that implementing classes must provide. Proper use of inheritance and interfaces in Apex helps avoid code duplication (by centralizing common logic in a base class) and can make the system more extensible (by letting new classes plug into existing workflows if they adhere to an interface). However, misuse of these features can lead to complexity, so it's important to follow best practices and understand when to apply inheritance or interfaces. In this section, we'll explore practical use-cases for inheritance and interfaces in Apex, guidelines to decide when to use them, and common pitfalls to avoid. We'll also show how ChatGPT can help evaluate your design for these patterns.

When to Use Inheritance in Apex: Inheritance is best applied when you have a clear "is-a" relationship and shared functionality between classes. For example, suppose you're writing an Apex system to handle different types of notifications: Email, SMS, and Push notifications. They all share some common behaviors (like all have a send() method), but also have differences in implementation. You could create a base class Notification with common properties (recipient, message content) and perhaps a default implementation or abstract method for send(). Then EmailNotification, SmsNotification, and PushNotification classes can extend Notification and implement the specifics of sending an email (via Messaging.sendEmail), sending an SMS (maybe via a third-party API), or pushing a mobile notification. In this scenario, inheritance avoids repeating the code for, say, storing the recipient or logging the send action in each subclass – those can be done in Notification once. It also means if you introduce a new notification type in the future (say, FaxNotification), it's straightforward to add; it naturally fits into the class hierarchy.

Apex has some constraints to keep in mind: it does not support multiple inheritance (a class can extend only one other class), and by default classes are not extensible – you must explicitly declare a class as virtual (or use abstract if it's meant to be subclassed but not instantiated itself) to allow others to extend it. Also, methods intended to be overridden should be declared as virtual. One best practice is not to force inheritance where composition could suffice. If classes don't truly share an "is-a" relationship, consider alternative designs. For example, you might be tempted to have a monolithic base class Utility and make various classes extend it to gain a bunch of unrelated helper methods – this is not a good use of inheritance. Those would be better as static methods or a separate helper instance (composition). Inheritance hierarchies should be reasonably shallow in Apex; very deep inheritance chains (multiple levels of subclassing) can become hard to follow and aren't often needed in typical Salesforce scenarios.

Interface Use-Cases: Interfaces in Apex define a contract. If a class implements an interface, it promises to provide the implementations for all methods the interface declares. This is useful for defining roles or capabilities that different classes can fulfill, even if they are in different parts of the class hierarchy. A practical example in Salesforce could be a custom interface Batchable that your classes implement to indicate they can be processed in batches (though Apex already has the Database.-Batchable interface). Or consider a scenario with strategies – e.g., different tax calculation strategies for different regions. You might define an interface TaxCalculator with a method calculateTax(Order o) and then have USTaxCalculator, EuropeTaxCalculator, etc., implement it. The rest of your code can be written to use the TaxCalculator interface, without needing to know which particular implementation is being used. This makes the system extensible: adding a CanadaTaxCalculator class later requires no changes to the code that uses the interface, you just supply an instance of the new class.

Another common use of interfaces in Apex is to facilitate **dependency injection** or mocking for tests. For example, you could have an interface InventoryService with a method checkStock(productId) which is implemented by a RealInventoryService (that calls an API or database) and a MockInventoryService (that returns dummy data for tests). Your business logic classes would depend on InventoryService interface, not the concrete class, so in production you inject a Real one, and in test you inject the Mock. This technique greatly improves testability and decoupling (we will talk more about decoupling in the next section on dependencies). Interfaces can also help with the concept of **interface segregation principle** – designing small, focused interfaces rather than one giant interface. For instance, you might separate a large concept into multiple interfaces if not all implementers need all methods, though in Apex one should be careful not to over-engineer since each interface implementation means more code.

Avoiding Pitfalls: While inheritance and interfaces are powerful, they can be overused. A common pitfall is creating an inheritance hierarchy because it seems conceptually neat, but it doesn't provide real benefit in code reuse or polymorphism. If you find that a subclass overrides almost everything from the base class and shares little actual code, inheritance might not be the right tool – that subclass might as well be independent. Conversely, not using inheritance when you have obvious duplication can also be a missed opportunity; striking the balance is key. With interfaces, a pitfall is creating too many or overly granular interfaces that complicate the design without clear gain. Also, every interface method is essentially abstract – if you add a method to an interface later, all implementing classes must update, which can be disruptive. So design interfaces thoughtfully, anticipating future needs if possible, or create new interfaces for new needs instead of changing old ones.

Apex doesn't allow default implementations in interfaces (unlike Java 8's default methods), so sometimes a hybrid approach is used: an interface for the contract, plus an abstract class that implements the interface with some default common code, and concrete subclasses extend that abstract class. This way, you get both the contract (for flexibility) and some inheritance (for reuse). Use this pattern carefully; it adds complexity but can be useful in frameworks. For example, you might define an interface SchedulableJob with executeJob() method and then an abstract class AbstractSchedulableJob that implements Schedulable (Salesforce's scheduling interface) and inside its execute calls executeJob() – forcing all your custom jobs to just implement executeJob() logic. This is advanced, but demonstrates combining interfaces and abstract classes.

ChatGPT for Design Evaluation: If you're unsure whether to use inheritance or interfaces in a design, you can literally ask ChatGPT. Describe the problem: *"I have three classes that handle file exports: PDFExport, ExcelExport, CSVExport. They have some common code and some differences. Should I use inheritance or interfaces?"* ChatGPT might guide you through the pros and cons. It could suggest an abstract BaseExport class with common methods like exportHeader() and exportFooter(), and abstract exportBody() to be implemented differently by PDF/Excel/CSV subclasses. Or it might note that if they share no code, an interface Exportable that each class implements might be sufficient with a separate helper to orchestrate them. The point is, ChatGPT can function as a design peer review, pointing out issues like "if you need multiple behaviors on one class, interfaces would be better since Apex doesn't support multiple inheritance."

When you have existing code, you could also ask ChatGPT to identify if inheritance is used correctly. For example, you might input a base class and a subclass and ask: *"Is this subclass leveraging the base class appropriately, or is it overriding too much?"* It might respond with observations such as "The subclass overrides five out of six methods, which suggests it doesn't benefit much from the base class; consider refactoring." Alternatively, supply two similar classes and ask: *"Would it make sense to derive one from the other or from a common superclass?"* ChatGPT will weigh in on whether those classes truly share a relationship or should remain separate.

For interfaces, you can ask something like: *"I want to allow different algorithms for assigning leads. Should I use an interface for the strategy pattern?"* ChatGPT would likely confirm that using an interface LeadAssignmentStrategy with implementations like RoundRobinStrategy, TopPerformerStrategy could be a good approach, and might even outline how the code to invoke them would look. If you provide it with a piece of code that uses if/else or switch across types, it might point out that this is an opportunity to use polymorphism (interfaces or inheritance) to remove the need for conditional logic – a hallmark of a good object-oriented design.

Furthermore, ChatGPT can generate small UML-like diagrams or descriptions of class hierarchies. Although it cannot create actual images by itself in this text-only scenario, it could output text diagrams or PlantUML code if asked, which you could then render as an image. For instance, you could ask: *"ChatGPT, given these class definitions, produce a PlantUML class diagram."* This might be beyond what a typical user needs, but it's worth noting as an advanced use: by visualizing the inheritance structure as an image (which you can do by feeding the PlantUML to an image generation tool), you might spot an overly complex relationship or confirm that things are nicely layered.

Architect Tip: Establish guidelines for when to use inheritance vs interfaces in your Apex coding standards. For example, declare that utility functionality should use static methods (composition) rather than inheritance, or that any planned inheritance hierarchy should be reviewed for necessity. An architect should also be vigilant about the ripple effects of changes – e.g., modifying a base class or interface affects many classes. Using ChatGPT to simulate "what-if" scenarios (like *"What if we add a new subtype? Does the current design handle it easily?"*) can be part of an architectural review.

Senior Developer Tip: If you're a lead developer, take the time to mentor others on these concepts with concrete Apex examples. Perhaps create a small demo in a sandbox of a good inheritance example and a bad one. You could even use ChatGPT in a meeting – have it explain the concept of interface vs abstract class in simple terms or an analogy (it's surprisingly good at analogies). This can help the whole team develop an intuition for proper usage. Sometimes the team might lean too heavily on one technique (like writing everything as a separate class with no hierarchy, or conversely making unwieldy base classes). Keeping an open dialogue, aided by ChatGPT's clarifications, ensures patterns are applied consistently and correctly.

By applying inheritance and interfaces judiciously, your Apex code becomes more modular and extensible. Classes can reuse common logic cleanly or adhere to contracts that make the system flexible to change. However, as classes interact and rely on one another, we must be cautious about coupling. Our next focus will be on managing dependencies between Apex classes – ensuring that while classes collaborate, they aren't so tightly bound that a change in one forces changes in others, preserving the flexibility and maintainability we're striving for.

MANAGING APEX DEPENDENCIES AND COUPLING

In a complex Apex codebase, classes will inevitably interact with each other. *Dependencies* refer to these connections – for example, an OpportunityService class might depend on an OpportunityValidator or an AccountService to do its job. *Coupling* describes how tightly bound two classes or modules are. High coupling means changes in one class likely affect the other; low coupling means classes are more independent. The goal in a well-structured Apex codebase is to minimize unnecessary coupling and manage needed dependencies in a way that preserves flexibility. By managing dependencies carefully, you make your code more adaptable to change, easier to test, and less prone to cascading failures. In this section, we will discuss strategies to reduce coupling, such as using interfaces and dependency injection techniques in Apex, and how to structure code so that modules communicate through well-defined boundaries. We'll also see how ChatGPT can help analyze the coupling in your code and suggest decoupling strategies.

Why Minimize Coupling? Imagine you have a class OrderProcessor that directly instantiates and calls a PaymentGateway class to charge a credit card. If tomorrow you need to change the payment provider or add a second provider option, you might have to edit OrderProcessor code. That's a sign of tight coupling — OrderProcessor is coupled to a specific implementation of payment processing. Now imagine instead OrderProcessor calls an interface PaymentService with a method charge(amount) and you supply a StripePaymentService or PayPalPaymentService at runtime. OrderProcessor wouldn't need to change if you swap the implementation; it just knows it has some payment service. This decoupling using an interface makes the system more flexible. In Apex, where direct runtime dependency injection frameworks (like Spring in Java) aren't available out-of-the-box, we often implement a manual form of dependency management, such as passing in collaborators or using a factory pattern to decide which implementation to use.

High coupling also makes testing harder. If ClassA always internally creates an instance of ClassB and calls it, then in a unit test for ClassA you're forced to deal with ClassB's behavior too (unless you use some advanced mocking like the Apex Mocks framework). If instead ClassA depends on an interface BInterface and you can provide a fake implementation during tests, you can isolate ClassA's testing from ClassB entirely. For example, a class that sends emails might depend on an EmailSender interface. In production, an implementation uses the actual email sending mechanism, but in a test, you give it a dummy EmailSender that maybe just records that an email "would have been sent" without actually calling Apex's email methods – thus avoiding hitting governor limits or external services in tests.

Strategies for Low Coupling in Apex:

Use Interfaces or Virtual Classes: As discussed in the previous section, if you program to an interface (or a virtual base class) rather than a concrete class, you decouple the caller from the implementation. This is a form of the dependency inversion principle – high-level code depends on abstractions, not on low-level concrete code. For instance, a Report-

Generator class might use an interface DataSource to fetch data. Whether the data comes from a SOQL query, a CSV file, or an external API can be decided by providing a specific DataSource implementation to ReportGenerator. The ReportGenerator doesn't need to know the details, it just calls dataSource.getData().

Factory Pattern and Inversion of Control: In Apex, a common way to handle choosing implementations is using factory methods or a central registry. For example, you might have a class ServiceLocator or Factory with static methods that return instances of interfaces. If AccountService needs an InventoryService implementation, instead of new Inventory-ServiceImpl(), it asks InventoryService inv = ServiceLocator.getInventoryService(). The ServiceLocator can decide which concrete class to return – perhaps reading from a custom metadata that says "use Mock-InventoryService" or just hardcoding a choice for now. This indirection adds a bit of complexity, but it means AccountService isn't directly tied to InventoryServiceImpl; you could change the ServiceLocator logic to return a different implementation later (for example, switch to a different algorithm or a newer version).

Passing Dependencies via Constructors or Methods: Another straightforward technique is to pass dependencies into a class either through its constructor or via setter methods. For example,

```
public class OrderProcessor {

  private PaymentService paymentService;

  public OrderProcessor(PaymentService svc) {

    this.paymentService = svc;

  }

  public void process(Order o) {

    //...some logic...

    paymentService.charge(o.amount);

  }

}
```

Here, OrderProcessor does not create a PaymentService on its own; it expects one to be given. The calling code can decide which PaymentService to provide. This pattern is a manual form of dependency injection. It works in Apex but requires that whoever instantiates OrderProcessor is aware and passes the dependencies. In a trigger context or utility context, this might mean you have to wire things up explicitly (which can be fine for controllable contexts like tests or single-use utilities).

Loose Coupling via Events or Queues: Sometimes decoupling can be achieved by not calling a dependency directly at all, but by using an intermediate system like a Platform Event or a queue (like a persistent Queueable job). For instance, instead of Class A calling Class B, Class A could publish an event that Class B is subscribed to. This way A doesn't know B at all – it just throws an event into the Salesforce event bus. B, which is listening, will react. This is a much looser coupling (often called *event-driven architecture*). In Apex, Platform Events and even just using the out-of-the-box triggers can achieve something similar (like one object's trigger causing changes in another object without direct code calls, though careful with such designs). However, events might be beyond the scope of code structure and more of an integration pattern. Still, it's a thought for decoupling modules of logic.

Practical Example of Decoupling: Apex common frameworks like fflib (Apex Enterprise Patterns) implement a unit of work and dependency injection mechanism. For example, they often have an Application class that holds singletons of various service classes. If you want to swap a service, you change what Application returns. That library also strongly uses interfaces and classes implementing them, which is a form of dependency inversion. Without going into that specific library, consider something simpler: you have two classes LeadProcessor and ContactCreator. LeadProcessor converts leads to contacts and then should notify someone. Instead of LeadProcessor directly calling an email routine, you could define an interface NotificationService and let LeadProcessor call notificationService.notifyLeadConversion(leadId). Then at runtime, you decide that the implementation will send an email. But in the future, you could change it to send a Slack message or create a task, without touching LeadProcessor. This decoupling is minor but keeps LeadProcessor focused only on lead conversion logic.

Testing Perspective: Managing dependencies goes hand-in-hand with writing testable code. Apex tests cannot directly substitute dependencies unless your code is designed for it (Apex is not dynamic like some languages where you can monkey-patch easily). But one tool available is the **Stub API (ApexMocks)**, which allows creating mock implementations of interfaces on the fly. That only works if your code uses interfaces in the first place. For example, you could mock PaymentService to simulate exceptions or success without actually calling anything. This is extremely valuable for testing error handling and rare scenarios. If code is tightly coupled (like directly calling a static method of another class), you lose that flexibility in tests. So decoupling not only makes production code flexible, it empowers testers to simulate various conditions.

ChatGPT for Coupling Analysis: How can we use ChatGPT to improve here? One approach is to have it review code and identify coupling. You might paste two class definitions and ask: *"Do you see any tight coupling between these classes and how could we reduce it?"* ChatGPT might notice that one class instantiates the other or calls a very specific method. It could suggest introducing an interface or passing the object in. For example, if it sees ClassA doing new ClassB() in a method, it could say: "ClassA is tightly coupled to ClassB. Consider using an interface or a factory to create ClassB outside of ClassA, and pass it in, to reduce coupling."

Another use: if you have a big method that calls many other classes' methods, you can ask ChatGPT, *"Which parts of this method could be refactored to reduce coupling or follow a cleaner separation?"* It might identify that certain calls could be moved behind an abstraction. If you describe an architecture, it can also give tips: *"We have a monolithic class handling both order creation and inventory update by directly calling both Order and Inventory logic. How can we decouple this?"* The AI might propose using events or splitting into two classes coordinated by a third.

You can get creative: perhaps ask ChatGPT to simulate what happens if one component changes. *"If I change the output of ClassB, which classes might be affected?"* If those classes are tightly coupled to ClassB's internal workings, that's a sign of coupling. ChatGPT might list classes that call ClassB's methods. This kind of impact analysis is something architects manually do, but an AI can help if it has knowledge of the codebase structure (which it can gain if you provide summaries or code relationships).

For a high-level approach, you could give ChatGPT an outline of module dependencies (like a list of classes and who calls whom) and ask for a critique. Perhaps feed it something like: "Class A calls Class B and C. Class B calls Class D. Class C calls Class D and E. Class D calls an external API. Is there a coupling issue here?" It might identify that A depends on details of both B and C, which in turn share D, etc., and suggest where an abstraction could simplify this dependency graph.

Developer Tip: One way to check coupling is to see if you can **unit test** a class in isolation. If you find you must load a bunch of other stuff or the class automatically triggers a cascade of operations, that implies tight coupling. Developers should proactively refactor code that is hard to test due to coupling. Use Apex interfaces and stub data to your advantage. If a piece of code queries the database or calls another class, consider wrapping those calls in an interface so you could simulate them in a test. And don't hesitate to ask for help; sometimes another set of eyes, or ChatGPT, can see a simpler way to connect classes.

Tester Tip: If you encounter flakiness in tests or tests that have to do a lot of setup across multiple objects just to exercise one function, give that feedback to the development team. It often indicates that function has tentacles in too many parts of the system (high coupling). For instance, if testing a small validation method requires inserting an Account, a Contact, and setting a bunch of custom settings, it might be doing more than just validation. A decoupled design would allow testing that validation logic by itself with just the necessary inputs. Testers can be strong advocates for decoupling, as it generally leads to more reliable and focused tests.

With mindful management of dependencies and coupling, Apex code becomes easier to maintain and extend. Classes are connected through stable contracts (interfaces or well-defined APIs) rather than brittle hard-coded links. This modularity will serve an enterprise well as the codebase grows. Stepping back even further, beyond individual classes and their relationships, we also need to consider organizing code at a higher level – across multiple projects or packages. Salesforce provides mechanisms like packages and namespaces to structure code in larger units. In the next section, we explore best practices for using packages and namespaces to organize an Apex codebase in an enterprise environment, which is especially relevant for large teams or ISV (independent software vendor) scenarios.

BEST PRACTICES FOR APEX PACKAGES AND NAMESPACES

As Salesforce orgs grow and multiple teams or products share a single environment, maintaining a well-organized codebase goes beyond just classes and interfaces. Salesforce's packaging mechanism allows developers to bundle Apex classes (and other metadata) into packages, which can be managed or unlocked, and to use *namespaces* to avoid naming collisions and clearly separate modules. In an enterprise setting or for AppExchange products, leveraging packages and namespaces is crucial for modular architecture and governance. Even within a single org, thinking in terms of packages can impose structure and boundaries that keep projects manageable. In this section, we discuss the role and benefits of using packages and namespaces in organizing Apex code, guidelines for structuring packages, and the considerations when deploying code in a modular fashion. We'll also mention how tools like ChatGPT can assist in evaluating packaging strategies or naming.

Understanding Packages and Namespaces: A *managed package* in Salesforce is a collection of components (including Apex code) that is tracked as a unit and can be distributed and installed in other orgs. Managed packages have a unique namespace prefix (for example, abc__) that gets prepended to all contained custom components, thus isolating them from other code. An *unlocked package* (part of the second-generation packaging introduced with Salesforce DX) also can have a namespace (though it's optional) and is often used within a single organization to organize code into modules without the full ISV distribution model. Namespaces ensure that if two packages have a class with the same name, they remain distinct (abc__InvoiceService vs xyz__InvoiceService). Namespaces also make it clear which package a component comes from.

For an enterprise that is not distributing apps outside but has a large org, unlocked packages allow you to break the "monolithic" org development into smaller, more modular pieces. Each package can represent a business capability or project (e.g., a Sales package, a Finance package, a Common utilities package). Teams can work somewhat independently on their package, and you gain benefits like versioning and the ability to test/deploy packages separately. Conceptually, it's similar to multi-module projects in other programming environments.

Benefits of Modular Packaging:

Isolation: Packages create boundaries. For example, if you have an HR system and a Sales system in the same org, packaging them separately means changes in one are less likely to impact the other. You can enforce that classes in one package don't directly access classes in another (unless you expose them globally or via explicit package dependencies).

Conflict Avoidance: With namespaces, you won't accidentally have two teams creating a class with the same name or two packages defining the same Apex trigger on an object (each package can have its own trigger on the object without conflict, thanks to namespacing, though their combined logic still runs).

Clear Ownership: It's easier to assign ownership of code when it's grouped in packages. Each package can have designated owners or teams. It also becomes clearer who to consult when a certain part of the code (say, the billing__ namespace package) is causing an issue.

Deployments and Release Management: Especially with unlocked packages, you can version control and release them independently. For example, you might release version 2.3 of the "CommonUtils" package without touching the "Finance" package, as long as the interface between them remains stable. This can reduce risk during deployments – you deploy smaller units rather than one giant metadata deployment.

Org Strategy: Salesforce architecture encourages a package-based development model as a way to handle large orgs (sometimes referred to as *modularization*). In fact, the platform and tooling have matured so that teams can shift towards modular, package-based development if they make the effort. By using packages, you essentially future-proof your org structure to be more scalable in terms of development practices.

Guidelines for Structuring Packages:

Logical Grouping: Decide on a logical way to slice your system into packages. Common approaches are by business capability (e.g., SalesCloudExtensions, ServiceCloudExtensions, CPQCustomizations) or by technical layers (e.g., Base Foundation, Feature A, Feature B if those features are mostly independent). You might start with a base or common package that has utility code and shared data model pieces, then have feature-specific packages that depend on that base.

Minimize Cross-Package Dependencies: While some dependencies are inevitable (and you can declare them so one package can use classes from another), keep them simple. A hierarchical dependency (common base, then feature packages) is easier to manage than a web of interdependencies. Cyclic dependencies (two packages depending on each other) are not allowed by Salesforce anyway.

Namespace Naming: If you use namespaces, choose a prefix that is short, distinct, and meaningful. You cannot change a namespace once set for a package. Many companies use an abbreviation of their name or project. For unlocked packages, you have the option to have no namespace (which means the code just lives in the org's global namespace), but that forfeits the conflict avoidance benefit. Using a namespace even internally can be wise for large orgs to simulate that isolation.

Versioning and Lifecycle: Treat packages as products in a way. They have versions, release notes, and you might even deprecate/replace one package with another over time. Plan for how updates will be handled. For managed packages (ISV scenario), you need to carefully design Apex APIs (global classes and methods) that you expose because once released, you can't easily remove or change them without keeping backward compatibility. For unlocked packages, you have more flexibility but should still avoid unnecessarily moving code between packages once it's in place, to keep version history clean.

If your team is new to packaging, it might start as a cultural shift. Initially, development might feel easier all in one place (the traditional "happy soup" where all code is in one package or no package). But as complexity grows, the benefits of modularization become clear. Many Salesforce developers have reported that after adopting a package-based approach, they manage dependencies more systematically and their deployments are more reliable. Essentially, the Salesforce platform now supports a modular, package-driven development model very well, and it's a best practice for larger projects to utilize that.

ChatGPT and Packaging Decisions: How can ChatGPT help here? If you're designing an org's package structure, you could describe your org's functional areas and ask for suggestions. For instance: *"We have a Salesforce org with marketing, sales, support, and a custom built ERP extension. How might we divide this into packages?"* ChatGPT might suggest a base package for common schemas, then separate packages for Marketing, Sales, Support, and ERP, possibly with some dependency (ERP might depend on Sales if they interact, etc.). It may also remind you of things like "one trigger per object per package" nuance (e.g., you can technically have one trigger in each package for the same object, which isn't advised unless you're careful, but if designing right, you avoid multiple triggers scenario across packages too).

Additionally, ChatGPT can help in generating a plan for migrating to packages. If you have an existing codebase, you could outline the classes and ask: *"Which sets of classes could form a package?"* It could cluster classes by naming (maybe noticing that classes prefixed with "HR" or in certain functional areas go together). It's a high-level advisor here, since actual packaging requires some trial and error.

Another scenario: ensuring you use consistent naming across packages. For example, if not using a namespace, some teams prefix class names with something like HR_ or FIN_ to simulate a namespace. ChatGPT can help enforce or suggest these naming schemes. It might say, "Consider prefixing all classes related to finance with Fin_ to easily identify them." Or if you feed it the class list, it could highlight naming conflicts or overlaps that a namespace would otherwise solve.

If you are using the actual namespace, ChatGPT can help remember to always reference cross-package classes with the prefix. It might not know your prefix unless you tell it, but you can include that in your conversation. This is more of a detail, though.

One neat use could be documentation: have ChatGPT generate a brief documentation or dependency diagram for packages. For instance, after establishing packages, you might ask: *"Summarize the responsibilities of each package and how they depend on each other."* This provides clarity to stakeholders or new team members about the modular architecture.

Architect Tip: When planning an enterprise org, think of packages as the *architecture boundaries*. Decide early what the high-level breakdown looks like (it can be expensive to retrofit after writing a lot of code without packaging). A good pattern is to create at least one "common" or "foundation" package. Use ChatGPT to sanity-check your plan: ask it to critique a proposed module list. Also, ensure you are aware of Salesforce limitations (like a class in one namespace calling a class in another might need that class to be global or an accessible Aura API if doing something like LWC across packages). It might be worth having ChatGPT list considerations for multi-package interplay so you don't overlook something (e.g., "Classes in managed packages must be global to be accessible to other packages or org code.").

Product Owner Tip: While product owners might not get into the weeds of code, understanding the concept of packages can be valuable for release planning. If each feature area is a package, releases might be staggered or independent for those areas. Product owners can coordinate with development on which package (feature set) is ready to push. Think of it like different micro-services or modules delivering value on their own schedule. In conversations, ask the architects to explain the package strategy in business terms (ChatGPT could help create such an explanation). For example, "Our billing enhancements are in a separate module so we can deploy them without affecting sales processes." This can aid in planning training, documentation, and communication around releases.

Using packages and namespaces effectively elevates the organization of Apex code to a higher plane, bringing order to potentially chaotic large orgs. With this modular architecture in place, we've covered structural best practices from the naming of a single variable all the way to partitioning a big project. There's one more piece to rounding out Apex code structure: ensuring the code is well-documented and annotated, so that human readers (and even tools like ChatGPT) can quickly grasp what the code is intended to do. In the next section, we'll look at best practices for documenting and annotating Apex code, which is integral to structured development practices.

DOCUMENTING AND ANNOTATING APEX CODE

Clear documentation and proper annotations are integral parts of a well-structured Apex codebase. Writing code isn't just about instructing the computer what to do; it's also about communicating to future developers (including future you) **why** and **how** the code does what it does. In Salesforce projects, where team members with different roles collaborate, good documentation ensures that architects, developers, testers, and even product owners can follow the logic and intent behind the code. In this section, we emphasize best practices for inline documentation (code comments), class and method annotations, and maintaining consistency in documentation across the team. We also explore how tools like ChatGPT can assist in generating and reviewing documentation, and why a consistently documented codebase is a long-term asset for any Salesforce org.

Code Comments Best Practices: Apex, like Java, allows comments in two forms: single-line // comments and block /*... */ or /**... */ comments. A common framework for thinking about comments is: comments should explain **why**, not **what**. A well-written piece of code with good naming (as we covered earlier) often documents the "what" by itself. For example, a method named calculateAnnualRevenue is self-explanatory in *what* it does. But **why** it does something a certain way, or any non-obvious decisions, should be in comments. If there's a complex SOQL query with particular filters due to business rules, a comment can cite those rules or the reason for filtering. If a workaround is implemented due to a known platform limitation or bug, absolutely comment that with references (like "Workaround for Salesforce issue XYZ – remove when fixed").

At a minimum, every Apex class and trigger should start with a brief description of its purpose. This might include who created it and when (some teams include an author/date, though version control often handles history now), and any other context. Using a standard header comment block at the top of classes is a good practice. Some teams include a modification log in the comment, but with modern source control, that can be redundant. Instead, focus the header on what the class is for, and perhaps which modules or features it relates to.

For methods, especially public methods, a Javadoc-style comment (/**... */) before the method is encouraged. This should outline the method's purpose, description of parameters, and description of return value. For example:

```
/**

* Calculates the annual revenue from a list of monthly revenues.

* @param monthlyRevenues A list of 12 decimal values representing monthly revenue.

* @return The total revenue for the year as a Decimal.

*/

public Decimal calculateAnnualRevenue(List<Decimal> monthlyRevenues) {

  //... implementation...

}
```

Notice the use of @param and @return tags – this follows a JavaDoc/ApexDoc convention that can be used by documentation generation tools (like ApexDoc). By maintaining this style, you could automatically generate documentation pages for your Apex classes, which is useful for large teams or for managed packages delivered to customers.

Inline comments (within method code) should be used sparingly but effectively. They can precede a complex block of logic, summarizing what that block is doing or why it's needed. Avoid commenting every line or stating the obvious ("// increment i" on i++ is not helpful). Over-commenting can clutter the code and sometimes indicate that the code itself is not clear. Ideally, refactor code so that it's self-explanatory, and then comment the tricky parts that remain.

Annotations in Apex: In Apex, the term "annotations" usually refers to the @Something tags you put above classes or methods. Some important ones relevant to code structure:

@AuraEnabled, @InvocableMethod, @InvocableVariable: These expose Apex to external systems (like Lightning Components or Flow). Document why a method is exposed, especially @InvocableMethod which will show up for admin users in Flow Builder.

@TestVisible: Used to expose private methods or variables to test classes. If you use this, add a comment why – perhaps it's to allow testing of a utility method that would otherwise be private. Use it sparingly; it's better than making things public just for tests, but also indicates that maybe the design could be adjusted if too much is needing test visibility.

@IsTest on classes (for test classes) – test classes typically don't need lengthy comments since they ideally mirror the functionality under test. But complex test logic might still get a comment for clarification.

@Deprecated: Mark methods or classes that should no longer be used. If an Apex method is deprecated, put in the comment since when or what to use instead. For example, @Deprecated // Use newMethod() after Summer '25 release as a note. This helps other developers know not to call it. Salesforce will throw a warning if you try to deploy code calling a deprecated method, which is good.

Consistency in comments is key. If one developer writes a beautiful Javadoc for a method and another doesn't comment at all, the codebase will be inconsistent. It can be useful to have a **documentation standard** as part of your team norms. This might define that every class has a header, every public method has a doc comment, non-trivial private methods have either a doc comment or at least a short comment describing them.

Team and Evolving Documentation: Code documentation should evolve with the code. A big danger is outdated comments – a comment that no longer matches what the code does is worse than no comment. It misleads the reader. So whenever code is changed, developers should be diligent to update any relevant comments. In code reviews, reviewers should also check: "Does the method doc still accurately describe the method after this change?"

One strategy is to treat documentation as code – giving it attention and even tasks in user stories ("write Javadoc for new classes"). Some teams incorporate documentation review as an explicit part of code review checklists. It's also helpful to generate the ApexDoc (if using such a tool) and publish it somewhere internally; seeing an official looking document might encourage developers to fill in any gaps they notice.

How ChatGPT Helps with Documentation: ChatGPT can play a surprisingly helpful role in documentation:

Generating Comments: If you have an Apex method with no documentation, you can copy it and ask ChatGPT: *"Write a JavaDoc comment for this Apex method."* It will analyze the code and attempt to describe what it does. It often does a decent job at summarizing the intent, which you can then refine. It might not get everything perfect (especially if the code's intent isn't clear or depends on business context), but it saves time as a starting point. As one developer quipped, feeding functions through ChatGPT to get Javadoc-style comments is a low-effort way to document old code.

Translating Code to Plain Language: ChatGPT can explain Apex code in plain English. This is useful if, say, a product owner or tester wants to understand what a piece of code is doing without reading Apex syntax. They could even use voice input: *"Explain what this Apex class does:"* and read it to ChatGPT, and get a summary that's easier to digest.

Checking for Missing Documentation: You could ask ChatGPT to review a class and see if any public members lack comments. It might not explicitly know your team rules, but generally if a public method lacks a preceding /** block, ChatGPT could flag that.

Consistency and Tone: If you feed it multiple comments, ChatGPT can help standardize wording. For example, you might say: *"Ensure the tone is consistent in these comments and fix any grammatical errors."* It can proofread and homogenize the style a bit (for instance, consistently using third person descriptions like "Calculates the X" vs imperative "Calculate the X").

Embedding Documentation Requests in ChatGPT usage: As an example, imagine a scenario where a tester finds a piece of code confusing. The tester could upload the class file into ChatGPT and ask for an explanation or a summary of logic. This effectively uses ChatGPT as a dynamic documentation reader. We should note to ensure file privacy if doing so, especially if the code contains sensitive logic.

Developer Tip: Make it a habit to write the comment *before* writing the code for a new method (a bit like documentation-driven development). If you can articulate what the method will do in a sentence or two first, it often leads to clearer implementation. Use ChatGPT if you get stuck – even for wording. For example, *"How should I describe a method that calculates late fees based on payment due dates?"* might give you a nicely phrased starting point. And always revisit comments after code changes.

Tester/Product Owner Tip: Don't shy away from reading the code documentation – it's meant for you as well, not just developers. If something is unclear or lacking, raise it. Sometimes QA or product folks ask developers, "Could you explain what's going on here?" That's a sign the code or its comments might need improvement. Also consider that ChatGPT can be your ally: if you have a log or error and you suspect which class is involved, you can query the documentation (if available) via ChatGPT by providing the class name or even content. This can bridge knowledge gaps if you're less comfortable reading raw code.

By maintaining high-quality documentation and comments, you ensure that your Apex codebase is approachable and understandable, which in turn reduces onboarding time and prevents misunderstandings. Annotations used appropriately add clarity and enforce structure (like marking things @Deprecated or @TestVisible where needed). In essence, well-documented code is a sign of a mature and disciplined development practice – it reflects foresight that code is read more often than it is written.

With the exploration of documentation, we have completed our deep dive into Apex code structure and organization best practices. From naming conventions to class design, from utilizing helpers to applying

OOP principles, from managing inter-class relationships to packaging code modules, and finally to documenting the whole lot – these practices all interconnect. In the next part of the book, we will build upon this structured foundation to tackle specific Apex scenarios (like triggers) and advanced patterns. But before that, let's recap what we've learned in this chapter and reinforce how these fundamentals set the stage for successful Apex development.

SUMMARY

In this chapter, we explored the critical role of code structure and organization in Apex development, detailing a spectrum of best practices and strategies. We began by emphasizing how thoughtful structuring of Apex code reduces technical debt and eases maintenance. Through a comprehensive look at naming conventions, we saw that consistent and meaningful naming of classes, methods, and variables greatly improves readability and team communication. With clear naming in place, we moved on to organizing Apex classes: applying principles like separation of concerns and single responsibility to achieve logical class structures. We highlighted how keeping classes focused and layering an application (for example into service, domain, and utility classes) makes the system more navigable and testable. The use of helper classes and utilities emerged as a practical way to enforce the DRY principle, by centralizing common logic and avoiding repetition across the codebase.

Delving into object-oriented design, we discussed inheritance and interfaces in Apex – when to use them to share code or define contracts for extension, and when to be cautious. We recognized that proper use of inheritance can eliminate duplicate code and that interfaces can add flexibility and facilitate loose coupling. This naturally led into managing dependencies and coupling between classes. By designing classes to depend on abstractions (like interfaces) and injecting or controlling dependencies, we can significantly lower coupling, resulting in more modular and adaptable code. Throughout these sections, we saw how an AI assistant like ChatGPT can assist in brainstorming design improvements, spotting duplication, or suggesting patterns to reduce tight coupling, serving as a helpful pair in code reviews or refactoring sessions.

On an enterprise scale, we examined Apex code organization beyond just classes – using Salesforce packages and namespaces. Embracing a package-based development model allows large Salesforce implementations to modularize their code, with clear boundaries and ownership for different functional areas. Namespaces prevent naming collisions and signal which module a piece of code belongs to, enforcing consistency on a grander scale. Tools and practices around second-generation packaging (unlocked packages) have matured to support teams in shifting to this modular approach , and we outlined guidelines for structuring and naming packages for maximum benefit.

Finally, we underscored the importance of documenting and annotating Apex code as part of its structure. Adhering to documentation standards – such as writing Javadoc-style comments for classes and methods – ensures that the intent and usage of code are clear to everyone on the team. We reinforced that comments should focus on the "why" behind code decisions , and we addressed key Apex annotations like @Deprecated and @TestVisible which support structured development and maintenance. With consistent documentation, knowledge transfer becomes easier and the codebase remains accessible and intelligible over time. Even here, ChatGPT can play a role by generating draft documentation or explaining code in simpler terms, bridging gaps between technical and non-technical team members.

Together, these practices form a holistic approach to Apex code organization. By giving attention to everything from a variable's name to the modular packaging of an application, a Salesforce team can dramatically simplify long-term project management. The code becomes self-descriptive, easier to navigate, and safer to change – all of which are invaluable in complex, evolving orgs. Moreover, these structural best practices enable more effective collaboration: architects can set enterprise-wide patterns, developers can code within well-defined guardrails, testers can align tests to clear units of functionality, and product owners or managers can gain confidence in the stability and scalability of the solution they are funding.

Crucially, we've seen that ChatGPT is not just a silent observer but an active assistant throughout these endeavors – from helping brainstorm naming conventions, reviewing class designs, identifying duplicate code for refactoring, suggesting interface patterns, to even aiding in documentation. In practice, ChatGPT serves as a guide and a guardian of best practices, available anytime the team needs advice or a second opinion. Used wisely (with awareness of data privacy and verification of suggestions), it can accelerate the adoption of the structured approaches discussed in this chapter.

Having cemented a strong foundation in code structure and organization, you are well-equipped to tackle more advanced Apex topics. In the next chapter, we will focus on Apex triggers – a powerful feature tightly related to code organization due to their broad impact on application behavior. We will explore how to use triggers strategically and how to control them to prevent common pitfalls, building on the principles of clarity and consistency we've established. The journey continues as we apply our structured thinking to the realm of triggers, ensuring that even as we handle complex event-driven logic, our Apex codebase remains clean, maintainable, and robust.

3

APEX TRIGGERS – STRATEGIC USE AND CONTROL

Apex triggers serve as the nerve center of complex Salesforce automation, injecting powerful, custom logic into the standard flow of data changes. In a Salesforce enterprise architecture, triggers are both **powerful** and **flexible**: they can enforce intricate business rules across objects and ensure data integrity in ways that point-and-click tools sometimes cannot. However, with great power comes great risk. A poorly designed trigger can inadvertently cause recursive loops, slow down save operations, or introduce hard-to-debug errors that ripple across your org. This chapter establishes why mastering triggers is critical – not just writing them, but using them **strategically and with control** to harness their benefits while avoiding pitfalls.

Apex triggers must be managed with discipline. Without careful design, triggers can easily violate governor limits or create unintended side effects. For example, an update in one trigger might fire another trigger repeatedly (trigger recursion), or inefficient queries inside triggers might slow down the entire system. **Disciplined trigger management** means anticipating these challenges and structuring trigger logic to prevent issues like infinite recursion, poor performance from non-bulkified code, and maintenance nightmares due to entangled logic. In large Salesforce deployments, it's common to see triggers accumulate over time – only a strategic approach can keep this under control.

To design and control triggers effectively, you need a **strategic mindset** that emphasizes simplicity, scalability, and maintainability. Triggers should do no more than necessary – ideally containing only straightforward logic or delegating work to helper classes – and they should be built to handle bulk operations from the start. A strategic trigger design looks beyond the immediate fix and considers how the logic will scale with more data, how easily others can understand it, and how reliably it will behave in various scenarios. By focusing on these qualities, you ensure your triggers serve the business needs without becoming a source of technical debt.

Equally important is establishing clear trigger frameworks and structured trigger logic. A consistent pattern for how triggers are written (for example, using one trigger per object with a dedicated handler class) makes it easier for teams of architects, developers, testers, and product owners to collaborate. When everyone follows the same patterns, architects can more easily review implementations against enterprise standards, developers can navigate and update code faster, testers know where to look for logic to validate, and product owners gain confidence that the system's automation is robust and predictable. In essence, a good structure turns triggers from ad-hoc scripts into **first-class citizens of your application architecture**.

Throughout this chapter, we will also highlight how ChatGPT can be leveraged as a helpful assistant in your Apex trigger journey. From explaining trigger behavior to reviewing code for best practices, ChatGPT offers a new way to **enhance understanding and refine trigger implementations**. You can ask ChatGPT to analyze a trigger for potential issues, get suggestions on designing a trigger framework, or even simulate scenarios to see how a trigger would execute. By incorporating ChatGPT's guidance – whether through text, voice queries, or by uploading code files for analysis – development teams can catch pitfalls early and iterate on improvements faster. (For example, a developer might **upload an Apex trigger file** and ask ChatGPT to identify any bulkification or recursion issues, leveraging non-text interactions for efficient code review.)

In this chapter, we'll explore seven key aspects of Apex triggers that embody strategic use and control: **embracing the single-trigger-per-**

object pattern; writing context-aware trigger logic; ensuring triggers are bulkified for performance; managing recursion and cascading effects; implementing trigger handler patterns for cleaner code; selecting and using Apex trigger frameworks; and developing robust testing strategies for triggers. Each section dives deep into one of these areas with best practices, real-world considerations, and plenty of examples – including ChatGPT prompts – to illustrate how these principles can be applied. By the end of the chapter, you'll have a 360-degree view of Apex trigger best practices and how to apply them in an enterprise setting.

With these foundations laid, we begin our journey by examining a core design principle that underpins many trigger best practices: maintaining a single trigger per object and why this simple rule can drastically improve your org's trigger hygiene.

SINGLE-TRIGGER-PER-OBJECT PATTERN

One of the fundamental best practices in Apex development is the **single-trigger-per-object** design principle. In simple terms, this means for any given SObject (standard or custom object), you consolidate all trigger logic into **one trigger file.** This practice leads to cleaner, more maintainable code and is favored by Salesforce experts for good reason. When multiple triggers exist on the same object, Salesforce does not guarantee the order in which those triggers run – effectively, their execution order is non-deterministic. This uncertainty can introduce random behavior: one trigger might set up data that another trigger (running later) expects, but if the order flips, things break. By enforcing a single trigger per object, you eliminate this source of unpredictability. All the logic for that object's data changes is housed in one place, executing in a controlled, logical sequence.

Let's clearly define this principle: a single trigger per object means you never have, say, AccountTrigger1 and AccountTrigger2 in the same org. Instead, you combine their logic into **one** AccountTrigger. Within that trigger, you can handle different events (insert, update, delete, etc.) in a structured way (we'll discuss patterns for that soon). The key is that when an Account record changes, Salesforce will invoke *one* trigger, so you control the flow internally. This approach not only avoids the random execution order problem, but also makes it much easier to locate and update business logic. If a product owner asks, "What happens when an Account is updated?", an architect or developer can point to the single Account trigger and be confident all related logic lives there, instead of hunting through multiple trigger files.

Benefits of Single Trigger per Object: Enforcing a single-trigger-per-object pattern brings several concrete benefits. First, it simplifies debugging. If an Account update isn't working as expected, you only have one trigger to examine rather than wondering which of the many triggers (if multiple were allowed) ran or interfered. Second, it streamlines business logic by consolidating it – you avoid duplicate code spread across triggers and can handle complex logic in a coordinated manner. Salesforce architects often note that this pattern leads to a more organized codebase. All the logic for an object is consolidated in one place, preventing conflicts and making the codebase easier to understand for new team members. Additionally, having one trigger encourages the use of structured patterns (like a handler class) because you naturally end up writing one bigger trigger rather than many small scattered ones. This sets the stage for implementing trigger frameworks or patterns in a consistent way.

From a governance perspective, architects and team leads should **enforce consistency** across the Salesforce environment. Many organizations make the single-trigger-per-object rule a part of their coding standards. Code reviews and static code analysis tools can flag violations. Here's where ChatGPT can help as well: one can use ChatGPT to scan a list of triggers in the org (by name or content) to spot if multiple triggers exist for the same object. For example, you might feed ChatGPT the names of all trigger files and have it report any duplicates per object:

Developer (via ChatGPT file upload): "Here is a list of all Apex trigger files in our org. Can you identify if we have more than one trigger on any object?"

ChatGPT: "After analyzing the trigger list, I found that you have two triggers on the Account object (AccountTriggerHandler and AccountDuplicateChecker). It's a best practice to merge these into a single trigger to avoid uncontrolled execution order. I also see two triggers on Case. It would be wise to consolidate those as well for maintainability."

By using such a prompt (which could be entered by voice or text and accompanied by a file upload of trigger metadata), a developer or architect can quickly confirm whether the single-trigger rule is being followed. This showcases how non-textual interaction – like uploading an exported list of triggers – allows ChatGPT to assist in governance.

Consolidating triggers is not just about moving code around; it requires planning. Suppose you discover multiple triggers on the Contact object performing different tasks – one validates data, another does a roll-up, and another enforces a business rule. You'd want to merge these into one ContactTrigger. A strategic approach is to create a structure within that trigger to call out to different handler methods for each purpose (or even better, use a trigger handler class). The end result is one trigger that might look like:

```
trigger ContactTrigger on Contact (before insert, before update, before delete, after insert, after update) {

        ContactTriggerHandler.handleTrigger(Trigger.isBefore, Trigger.isAfter,

                Trigger.isInsert, Trigger.isUpdate, Trigger.isDelete,

                    Trigger.new, Trigger.old, Trigger.newMap, Trigger.oldMap);

}
```

In this snippet, the single trigger delegates to a handler class method handleTrigger with all the context it needs. This way, the logic that was once split among multiple triggers is unified. The specifics of the handler aren't important yet – the key point is that the consolidation is feasible and usually beneficial. ChatGPT can even assist in this consolidation process. For example, you could provide ChatGPT with the code from two triggers you intend to merge and ask for a combined version:

Developer: "We have two triggers on Case object. One sends an alert on high-priority cases, another updates a field after insert. Can you help merge these into one trigger structure?"

ChatGPT: "Certainly. To merge the triggers, you can create a single CaseTrigger and within it handle both scenarios. Here's an example..."

ChatGPT might then produce a combined trigger code, ensuring that it checks the right context (before/after, insert/update) to perform each piece of logic. This gives the developer a solid starting point, which can then be refined and tested.

Architect Tip: Standardize on naming and structure for single triggers. For instance, name each trigger *<Object>Trigger* to quickly identify it. This makes it obvious in the codebase that there's exactly one trigger per object, reinforcing the convention to all team members. During design reviews, architects should confirm new features integrate into the existing trigger rather than spawning a new one, preserving the one-trigger rule across projects.

By committing to the single-trigger-per-object pattern, teams lay the groundwork for all the following best practices. With one trigger in play, it becomes much easier to manage how that trigger operates, ensure its context-aware, bulkified, and so on. Next, we'll dive into how to make the logic *within* those triggers smart and context-sensitive, so that each trigger behaves correctly depending on how it was invoked.

CONTEXT-AWARE TRIGGER LOGIC

Even with a single trigger per object, the trigger often needs to handle multiple scenarios – inserts, updates, deletions, and even undelete or merge events. **Context-aware trigger logic** means writing your trigger (and its helper code) to respond appropriately based on the context in which it's fired. Salesforce provides a set of trigger context variables (and flags) that let your code know exactly what's happening. Understanding and leveraging these context variables is essential for writing clear, robust trigger logic that doesn't do the wrong thing at the wrong time.

The primary context variables include things like Trigger.isInsert, Trigger.isUpdate, Trigger.isDelete, which return true if the trigger is running for that specific DML event, and Trigger.isBefore / Trigger.isAfter, which indicate if it's the "before" or "after" phase of the trigger. You also have Trigger.new and Trigger.old, which are lists of the new and old versions of the records, respectively (with Trigger.old being non-null only in update or delete contexts). Additionally, Trigger.newMap and Trigger.oldMap provide maps of record IDs to the new and old records, and Trigger.size tells you how many records are in the trigger. These variables collectively give your code a full picture of the state: for example, "We are in a before-update trigger on 20 Account records; here is the list of new Account values, and we can compare them with the old values."

Using these variables makes trigger logic much more readable and less error-prone. For instance, without context checks, a trigger that handles both insert and update might try to access Trigger.old records even during insert (when Trigger.old is null), causing a null-pointer error. A context-aware trigger avoids that by wrapping logic in appropriate if conditions. Consider a scenario: you want to send a welcome email only when a Contact is inserted, not on updates. In your trigger (or handler), you would write:

```
if (Trigger.isInsert && Trigger.isAfter) {

  // send welcome email for each new Contact in Trigger.new

}
```

This ensures the code runs *only* after an insert. If a Contact is updated, that block skips entirely. Such clarity is vital for testers as well – they can see from the code what should happen in each situation and craft tests accordingly.

To illustrate how context awareness simplifies logic, imagine a trigger that maintains a count of related records (say, an Account field tracking the number of its Contacts). Without context awareness, one might write a monolithic block of code that tries to handle insert vs. delete vs. update all together, with interwoven logic. Instead, a context-specific approach would separate these: on Contact after-insert, increment the Account's count; on Contact after-delete, decrement the Account's count; on Contact after-update, maybe adjust the count if the Contact's Account affiliation changed (decrement old account, increment new account). Each of these can be handled in distinct if sections, resulting in cleaner, less error-prone code. It's clear which part runs when, reducing bugs and enhancing readability. In fact, a well-structured trigger handler might even have separate methods for each context, e.g., handleAfterInsertContacts(List<Contact> newRecords) and handleAfterDeleteContacts(List<Contact> oldRecords) – making the code self-documenting.

ChatGPT can assist developers in validating and improving context-aware logic. For example, you can ask ChatGPT to review a trigger to see if all relevant context cases are handled:

Developer: "I have an Opportunity trigger. I handle before insert and before update in it. Can you check if I correctly used context variables?"

ChatGPT: "Sure. In the code you provided, I see you are checking if (Trigger.isUpdate) but inside that block you're still referencing Trigger.new without using Trigger.old or Trigger.oldMap. If you need to compare new and old values (for example, to detect field changes), remember to use Trigger.oldMap for updates. Also, I noticed you don't have any logic for Trigger.isInsert – if that's intentional, that's fine, but if you intended to handle inserts differently, you should add an if (Trigger.isInsert) branch."

This kind of analysis from ChatGPT helps catch common context-related pitfalls. Another usage is simply **asking ChatGPT to explain a context variable** if you're unsure. A tester, for instance, might use voice input to ask, "What does Trigger.old contain and when can I use it?" and ChatGPT would explain that Trigger.old is a list of the old versions of records, available in update and delete triggers. This helps non-developers (like a tester or product owner) understand trigger behavior, fostering better collaboration. A tester who understands context can design tests for each scenario (insert/update/delete) knowing what the trigger should do in each.

Let's consider a concrete example showing context usage. Suppose we have a trigger on Case that should do different things on creation versus closure. We can structure it like so:

```
if (Trigger.isInsert && Trigger.isAfter) {

   // New Case created: send acknowledgement email to customer

   CaseTriggerHandler.handleCaseCreation(Trigger.new);

}
```

```
if (Trigger.isUpdate && Trigger.isAfter) {
  // Case updated: check if it was closed
  for (Case c : Trigger.new) {
    Case oldCase = Trigger.oldMap.get(c.Id);
    if (c.Status == 'Closed' && oldCase.Status != 'Closed') {
      CaseTriggerHandler.handleCaseClosure(c);
    }
  }
}
```

In this snippet, the trigger clearly distinguishes between an insert (new case creation) and an update that marks a case as closed. By comparing c.Status with the old value via Trigger.oldMap, we ensure that we only call the closure handler when the status changed to Closed, not on every update. This prevents running closure logic when it's not needed. Such conditional checks are a form of context-aware logic that also guard against unintended recursion or multiple executions (as we'll explore more in the next section).

Developer Tip: When writing trigger logic, always mentally walk through each DML event type. Ask yourself, "If this trigger fires on a delete, what will happen? On an undelete? On a bulk update of 200 records?" This thought exercise, often done during code review or design, can be enhanced by using ChatGPT as a rubber-duck debugger – explain your trigger flow to ChatGPT and see if it spots a context you missed. For instance, you might describe your trigger's purpose and have ChatGPT list the trigger contexts you should account for, ensuring none are overlooked.

By being diligent with context variables and logic, developers create triggers that behave predictably and correctly no matter how they're invoked. This reduces bugs (e.g., no more null-pointer exceptions because you tried to use Trigger.new in a before-delete trigger where it doesn't exist, or using Trigger.old in insert triggers). It also makes the code easier to maintain, since each logical branch is well-scoped. With context-awareness established, we can now turn to another crucial aspect of trigger design: making sure our triggers handle large volumes of data efficiently. Next, we'll discuss bulkification – the practice of writing triggers that perform well with many records at once, not just one.

ENSURING BULKIFICATION IN TRIGGERS

Salesforce's multitenant architecture means Apex code must be mindful of limits and performance. All triggers are inherently designed to handle bulk operations – in fact, **"all triggers are bulk triggers by default"**, capable of processing up to 200 records at a time. Bulkification is the practice of writing your code to gracefully handle those multiple records in one go, rather than assuming it will only ever process a single record. Ensuring bulkification in triggers isn't just a nice-to-have; it's a necessity for performance and for avoiding governor limit exceptions. A trigger that isn't bulkified might work in a sandbox with one record, but fail in production when an import or mass update triggers it on hundreds of records.

Why is bulkification so important? Consider that a user or an integration could perform an operation affecting many records: a data loader updating 500 Contacts, or a mass reassignment of Accounts. Salesforce will chunk these into batches (up to 200 at a time) and invoke your trigger for each batch. If your trigger code isn't built to handle a list of records, it could do things like query the database or send DML operations inside a loop for each record – quickly exceeding Salesforce's limits on SOQL queries or DML statements. Conversely, a bulkified trigger gathers all needed data in as few queries as possible and performs actions (like updates) in bulk collections rather than one by one. This not only prevents hitting limits but usually runs faster and more efficiently.

Best Practices for Bulkification: The core techniques include:

Avoid SOQL or DML inside loops. If you loop over Trigger.new and perform a SELECT or an UPDATE for each record, you will likely hit the 100 SOQL or 150 DML statements limit in a bulk scenario. Instead, accumulate record IDs or data in collections and query/update outside the loop. For example, collect all Account IDs from Trigger.new records in a Set, do one SOQL query to get needed info for those Accounts, store results in a Map, then loop again to apply the results.

Use collections and bulk operations. Salesforce allows inserting or updating lists of records in one go (e.g., update myListOfContacts;). Always prefer operating on lists or sets instead of single records when possible. If 50 records need a change, add them to a list and do one update, not 50 separate updates.

Leverage Trigger.new as a list. Remember that Trigger.new itself is a list of records. You can pass it directly to methods or use it in a single DML operation. For example, if you have to ensure some field is set on all new records, you can just set it on each record in Trigger.new (looping is fine for simple assignments), and at the end of the trigger, Salesforce will handle updating all those records as part of the transaction.

Use SOQL for loops or stream queries for large data sets. In cases where you truly need to handle thousands of records and query related data, use the SOQL-for-loop construct which batches query results efficiently. This prevents hitting heap limits by retrieving too much data at once.

Think in sets and maps. Bulk design often involves using Sets for unique collection of IDs and Maps for quick lookup of related data by Id. This way, you do things in bulk and have efficient access patterns.

Let's illustrate bulkification with a simple scenario: A trigger on Opportunity needs to ensure that every Opportunity's Account has an updated "Last Deal Closed Date" whenever an Opportunity is closed. A non-bulkified approach might loop through each closed Opportunity, and for each one, perform a SOQL query to get the Account and then do an update. If 100 opportunities closed in one batch, that's 100 queries and 100 updates – which will definitely hit limits (and slow things down). The bulkified approach would be:

Loop through Trigger.new to collect all Account Ids for Opportunities that were closed.

Perform one SOQL query to get all those Account records.

Loop through the Opportunities again, updating the corresponding Account objects in memory (e.g., set each Account's Last Deal Closed Date field).

Perform one DML update for all the Account records that need updating.

This way, regardless of whether 1 or 200 opportunities fire the trigger, it will at most do 1 query and 1 update operation. The logic scales smoothly.

In Apex trigger best practices, it's often emphasized: *"Bulkify your code."* This is so fundamental that it appears in nearly every list of Apex tips. Indeed, not bulkifying is a common development pitfall; many developers initially write triggers assuming a single record context and have to refactor later when issues arise. Avoid that pain by designing bulk handling from the start.

ChatGPT can serve as a useful reviewer for bulkification. You could ask ChatGPT to inspect a trigger for any patterns of SOQL or DML in loops. For example:

Developer: "Here's my trigger code. Do you see any bulkification issues?"

ChatGPT: "I notice that within your for-loop over Trigger.new, you call insert new Task(...) for each record. This is a DML inside a loop. In a bulk context (say 150 records), this will attempt 150 inserts separately, which will likely hit the 150 DML statements limit. A better approach is to accumulate all new Tasks in a list and insert them once after the loop. Similarly, you have a SOQL query inside the loop to fetch a related record – that should be moved outside the loop."

Such feedback can pinpoint exactly where to fix code. Even more interactively, you might use ChatGPT's code generation capabilities to help bulkify code: *upload your non-bulkified trigger as a file* and prompt, "Bulkify this trigger." ChatGPT can then refactor the code, showing how to introduce collections and move operations out of loops. This is an excellent use of the tool's ability to rewrite code while preserving functionality.

Sometimes bulkification involves handling **chunks of data** properly. Remember that if more than 200 records are involved in a single transaction (e.g., a bulk API job of 1000 records), Salesforce will chunk the processing into batches (like 5 batches of 200). Your static variables (if any) will not reset between those batches in the same transaction context, which can affect logic (as we'll discuss in recursion management). For bulkification purposes, just be aware that triggers *will* process records in sets of up to 200, and design your logic as if it will always receive the maximum. A good mental check: ask, "If I had 200 records in Trigger.new, would this code still work? What about 2000 records (which would be 10 trigger invocations of 200 each)?"

Architect Tip: Encourage developers to test triggers with bulk data in mind. During code review, ask for a demonstration or unit test that simulates 100+ records going through the trigger. This not only proves bulk safety but also performance. If the trigger logic uses efficient queries and loops, it should handle 200 records as quickly as 1 record (roughly speaking). If not, that's a sign to refactor. Architects might also integrate static code analysis tools that flag non-bulkified patterns. ChatGPT can complement these by explaining why a pattern is problematic and suggesting fixes, which helps educate the team, not just enforce rules.

Ensuring bulkification at the time of writing the trigger saves considerable time down the road – you avoid emergency rewrites when data volumes grow or when a big data load fails. It's a key part of making triggers scalable. Now that we've covered bulk operations, our next challenge is a subtle one that can arise even in well-bulkified, context-aware triggers: trigger recursion and cascading effects. We'll explore how to manage situations where one trigger might unintentionally fire itself (or another trigger) repeatedly, and what strategies exist to keep that under control.

MANAGING TRIGGER RECURSION AND CASCADING EFFECTS

Apex triggers run as a result of DML events, but what if a trigger itself performs DML? This can lead to **trigger recursion** – a scenario where

Trigger A causes an update that in turn causes Trigger A to run again (or triggers on related objects to cascade in complex ways). If not handled, recursion can lead to infinite loops or "Maximum trigger depth exceeded" errors, and even if it terminates, it might perform the same logic multiple times unexpectedly. Managing recursion and cascading effects is therefore crucial in complex Salesforce orgs, especially when triggers update records that have their own triggers.

Common recursion scenarios: A classic example is a trigger on Contact that, after insert, updates the parent Account (maybe to increment a contact count or update a last modified date). That update to Account could fire the Account trigger (if one exists). If the Account trigger then updates some Contact or another related record, it might end up calling back the Contact trigger again – a cascading chain. Another scenario: a trigger on Opportunity does an update on the Opportunity itself (perhaps adjusting a field) – this will cause the Opportunity trigger to execute again on the same records (since an update inside an update trigger triggers another round, unless it's carefully controlled). Salesforce protects against some forms of immediate recursion (for example, field updates in workflows and flows won't retrigger the same record's save in the same transaction), but Apex triggers have to be carefully coded to prevent self-reinforcing loops.

Techniques to prevent recursion:

Static variables as flags: One common approach is to use a static Boolean flag in an Apex class (often a helper or the trigger handler) to indicate if the trigger has run already in the current transaction. Before executing the trigger logic, you check this flag. If it's false (first run), set it to true and proceed; if it's already true, then you skip the logic to avoid re-entering. This simple mechanism can prevent infinite loops in many cases. However, it has limitations – notably, if you have a bulk operation with more than 200 records causing two chunks in one transaction, the static flag method might skip processing the second chunk entirely (because the flag stays true after the first chunk). So, while static flags are a quick fix for simple recursion, they're not foolproof for all bulk scenarios.

Static sets for record tracking: An improved pattern is to use a static Set<Id> to track which record IDs have been processed by the trigger in the current transaction. When the trigger runs, for each record, check if it's in the set; if not, process it and add its Id to the set. This way, if the trigger fires again (even on another chunk of 200), it will only skip processing if the same record appears again. This prevents double-processing the same record and can handle scenarios where a static boolean would prematurely cut off logic for subsequent batches. It's more granular control than a single flag.

Context-specific condition checks: Sometimes the best way to avoid unnecessary trigger re-invocation is to design your logic such that it doesn't perform a DML unless absolutely needed. For instance, if your trigger is updating a record to "fix" a value, ensure that value is actually wrong before updating. The Nebula Consulting team provides a great example : if a trigger sets a custom Last Modified Date field to today's date, the trigger should first check if that field already equals today's date. If it does, no update is necessary – thus no recursive call. By coding defensive checks like this, you reduce pointless updates that cause recursion. Essentially, *don't update a record if you don't have to*. This naturally curtails recursion and also improves performance.

Using trigger frameworks: Some trigger frameworks provide built-in recursion guards. For example, they might automatically track if a trigger's context has run for a given record to avoid re-entry. We'll discuss frameworks in the next section, but be aware that adopting a framework can save you from writing your own recursion control logic each time.

Let's walk through a scenario to see recursion control in action. Imagine a trigger on Lead that, after insert, wants to update a custom field on the Lead itself (maybe stamping an external ID after creation). If we naïvely update the Lead in an after-insert trigger, that causes an update on the same Lead record – which *will fire the after-update portion of the same trigger again*. To avoid an infinite loop, you could use a static boolean flag LeadTriggerHandler.isFirstRun. The trigger would check if (LeadTriggerHandler.isFirstRun) {...perform update...; LeadTriggerHandler.isFirstRun = false; }. This works for one batch of up to 200 leads – the first run updates them, second run sees the flag false and skips. But as noted, if there are 400 leads inserted in one transaction (split into two batches by Salesforce), the second batch's trigger execution would find the flag already false (set by batch1) and skip processing entirely , leaving those leads un-updated – a bug. A static set approach would do better: use LeadTriggerHandler.processedIds. For each Lead in trigger, check if its Id is in the set. Initially empty, so batch1 processes all 200 and adds their Ids. Batch2 runs, and since its leads have different Ids, none are in the set, so it processes them too. This way, each record is processed once, as intended.

However, even static sets have edge cases – e.g., if a trigger in batch1 and batch2 have overlapping records (which normally shouldn't happen unless the same record somehow comes through twice in one transaction, which can occur with complex inter-object updates). But for most purposes, a static set is a solid guard. It passes the test of preventing true infinite loops and still allowing all intended records to be handled.

ChatGPT can be quite handy in dealing with recursion problems. If you encounter a "maximum trigger depth exceeded" error or notice something odd like a field being updated twice, you can describe the situation to ChatGPT and even share the trigger code. ChatGPT might simulate the logic flow and identify the recursive pattern. For example:

Developer: "My Account trigger updates related Contacts. But when I insert an Account with Contacts, it seems to run multiple times and I get a trigger depth error."

ChatGPT: "It sounds like your Account trigger updates Contacts, and perhaps you have a Contact trigger that in turn updates Account (or the same Account). This could cause a loop. Let's examine: When Account is inserted, trigger runs, updates Contacts. Contact trigger (on update) might update Account's field, which calls Account trigger again. To break the loop, consider using a static flag or checking a condition. For instance, have the Account trigger skip logic if it's running as part of an ongoing update caused by a Contact update. Alternatively, consolidate logic so that only one side does the update, not both."

Through this dialogue, the developer is pointed toward a solution. They might follow up by asking for a specific implementation (e.g., "How do I use a static flag to prevent the Account trigger from running twice in this case?") and get sample code.

For a more direct approach, you could present ChatGPT with two triggers (Account and Contact) and ask how to modify them to avoid ping-pong updates. ChatGPT might suggest adding a static flag or redesigning so that only one trigger does the cross-object update and the other simply doesn't when it detects the change was already handled.

Tester Tip: When testing triggers, include scenarios that could reveal recursion. For example, if a trigger updates a related record, write a test that creates the scenario and assert that the final outcome is correct and that no unintended additional updates happened. One trick is to use the *Limits.getDmlStatements()* in a test to see how many DML operations a single transaction performed – if it's unexpectedly high, there might be a recursive loop. Testers can also check logs for multiple entries of the same debug line which might indicate multiple trigger invocations. ChatGPT can help interpret such logs if you provide them, highlighting where the trigger re-entered.

By proactively managing recursion and cascading effects, you ensure that triggers remain reliable even as they perform complex operations. This often goes hand-in-hand with having a well-structured approach to your trigger logic. In fact, many developers turn to structured **trigger handler patterns** as a way to organize code and implicitly manage issues like recursion and execution order. In the next section, we'll look at implementing trigger handler patterns, which formalize how trigger logic is delegated and organized for clarity and control.

IMPLEMENTING TRIGGER HANDLER PATTERNS

Writing complex logic directly inside a trigger can get unwieldy. As our discussions on context, bulkification, and recursion have shown, there are many concerns to juggle. **Trigger handler patterns** provide a structured way to keep triggers organized, readable, and maintainable by moving the business logic out of the trigger and into dedicated classes. The trigger itself becomes a thin dispatcher – often just a few lines – while the heavy lifting is done in a handler class (or classes). This separation of concerns is a hallmark of clean Apex architecture and is strongly recommended by Salesforce and community experts alike.

What does a trigger handler pattern look like? At its core, it means you have something like MyObjectTrigger (the trigger) which on execution calls methods in, say, MyObjectTriggerHandler (an Apex class). The trigger might pass along context like the Trigger.new list, or it might even call specific methods like beforeInsert, afterUpdate on the handler. The exact structure can vary, but the goal is the same: keep the trigger logic minimal and delegate to a well-defined class structure. This approach yields **logic-less triggers** – a phrase you'll often hear. A logic-less trigger is one whose body doesn't contain business logic, just references to it (like method calls).

For example, without a handler, you might see in a trigger:

```
trigger OrderTrigger on Order__c (before insert, before update) {

  if (Trigger.isBefore && Trigger.isInsert) {

    // 20 lines of logic to initialize some fields

  }

  if (Trigger.isBefore && Trigger.isUpdate) {

    // 30 lines of logic to validate or adjust fields

  }

}
```

With a handler pattern, this would be refactored to:

```
trigger OrderTrigger on Order__c (before insert, before update) {

  OrderTriggerHandler.handleBefore(Trigger.new, Trigger.oldMap);

}
```

And then in OrderTriggerHandler.cls:

```
public class OrderTriggerHandler {

    public static void handleBefore(List<Order__c> newList, Map<Id,
Order__c> oldMap) {

    if (Trigger.isInsert) {
```

```
        doOrderInsertLogic(newList);
      }
      if (Trigger.isUpdate) {
        doOrderUpdateLogic(newList, oldMap);
      }
    }
    private static void doOrderInsertLogic(List<Order__c> newOrders) {
      // formerly 20 lines of logic for insert
    }
     private static void doOrderUpdateLogic(List<Order__c> newOrders,
Map<Id, Order__c> oldOrders) {
      // formerly 30 lines of logic for update
    }
}
```

Now the trigger is just one line, and all the details live in the handler class. This has several advantages:

Clarity: Anyone reading the trigger immediately knows where to find the actual logic (in the handler class). And the handler class is organized by method, often correlating to trigger events, making it easier to follow.

Reusability: The handler methods (being static here) could potentially be called from other places, like maybe a batch job or another class, if needed. Also, they can be unit tested directly by calling them with sample data, which can simplify testing (you're not forced to do DML to invoke the trigger logic).

Consistency: If all objects follow this pattern, the codebase feels consistent. A developer moving from AccountTrigger to ContactTrigger won't be surprised – each will have a handler with similar structure.

Control: It's easier to implement things like recursion flags or specific ordering in code when you have a structured handler. For instance, you could include a check at the top of handleBefore to manage recursion (like our static flags) and that logic is in one place, not scattered in trigger code.

There are various flavors of trigger handler patterns. Some prefer one handler class per object (as shown). Others implement a generic framework where one base TriggerHandler class is extended by object-specific subclasses, possibly using interfaces (we'll touch on frameworks in the next section). Some even break handlers into multiple classes by domain (e.g., one class for validation logic, another for post-save actions, etc.). The right approach depends on the complexity of your org and your team's preferences. However, a common theme is keeping triggers themselves very slim.

To implement a handler pattern in an existing org, you might have to refactor. ChatGPT can assist here by transforming a fat trigger into a handler structure. You could provide ChatGPT the trigger code and ask it to outline a handler class. For instance:

Developer: "Convert this trigger logic into a trigger handler class pattern."

ChatGPT: "Certainly. I will create a handler class with methods corresponding to the trigger events. First, here's your new trigger code... [provides slim trigger]. And here's a draft of the handler class... [provides class code]."

ChatGPT might produce something akin to the example above, saving you from manual boilerplate writing. You'd still test and tweak it, but it accelerates the restructure.

Another aspect of handler patterns is they make it easier to **disable or control triggers** when needed. For example, you might include a static flag in the handler that can be toggled to turn off trigger processing (useful for data loads or tests). With a centralized handler, you can have an if (isTriggerDisabled) return; at the top of your handler execution. If you ever set OrderTriggerHandler.isTriggerDisabled = true; (perhaps via custom setting or during a data migration script), the trigger will effectively skip its logic. This is much cleaner than trying to sprinkle such checks in a messy trigger body.

Let's include an example of a trigger handler pattern in code form to cement the idea. We'll use a simple scenario: an Account trigger that ensures every new Account has a unique external code, and that when an Account is deleted, it logs an entry. Without diving into the logic details, here's how it could look:

```
trigger AccountTrigger on Account (before insert, after delete) {
    AccountTriggerHandler.handleTrigger(Trigger.operationType, Trigger.new, Trigger.old);
}
```

In this case, instead of booleans, we use Trigger.operationType (a newer enum that encapsulates the event) to route logic. Now the handler:

```
public class AccountTriggerHandler {
  public static void handleTrigger(System.TriggerOperation op, List<Account> newList, List<Account> oldList) {
    switch on op {
      when BEFORE_INSERT {
        assignExternalCodes(newList);
      }
      when AFTER_DELETE {
        logDeletions(oldList);
      }
      when else {
        // no action for other events
      }
    }
  }
  private static void assignExternalCodes(List<Account> newAccounts) {
```

```
    // generate and set unique codes for each account
  }
  private static void logDeletions(List<Account> oldAccounts) {
    // create log entries for deleted accounts
  }
}
```

This example demonstrates a pattern using a single handler method with a switch on the operation type (available in newer Apex versions). It's yet another style to achieve the same goal: logic is compartmentalized and the trigger just delegates.

Architect & Developer Tip: Start new projects with a trigger template that includes a handler from the get-go. Many teams even have a snippet or a SFDX generator for "create new object with trigger & handler." By ingraining this pattern, you avoid the temptation to cram logic into triggers. Architects should also document this pattern in the project's development guide – possibly illustrating it with a diagram. For instance, a simple diagram could show: ***Object → Trigger → TriggerHandler → Business Logic Methods***, indicating how data flows from the trigger to the handler to various methods. This visual aid can be generated with ChatGPT's help as well, or even using an image generation tool if you want an illustration in documentation. The consistency pays off when multiple developers work on triggers across different objects, as everything will follow a familiar structure.

Diagram: The trigger handler pattern delegates logic from a single trigger into a handler class. The trigger (one per sObject) acts as a ***delegator***, calling the handler which then executes multiple pieces of business logic. This results in cleaner separation of concerns, with all business logic housed in handler methods instead of the trigger body.

By implementing trigger handler patterns, you not only make the current trigger easier to manage, but you set yourself up to adopt broader trigger frameworks more easily. In fact, many full-fledged trigger frameworks are built on the concept of a single trigger delegating to structured handler logic. Up next, we'll explore how to choose and use Apex trigger frameworks – standardized approaches that can further improve consistency and add powerful features to your trigger management strategy.

CHOOSING AND USING APEX TRIGGER FRAMEWORKS

As Salesforce orgs grow, teams often gravitate from simple patterns to more structured frameworks for triggers. A trigger framework is essentially a predefined architecture or library that standardizes how triggers delegate work, enforce best practices (like one trigger per object, logic-less triggers, etc.), and sometimes provide extra capabilities like easier control of trigger order, bulk state management, or even the ability to turn off triggers dynamically. Using a trigger framework can accelerate development by providing a blueprint that all developers follow, and it can simplify troubleshooting because the framework imposes a known structure. However, choosing the right framework (or deciding to build a lightweight one) requires understanding the options and the needs of your organization.

Benefits of a Trigger Framework:

Standardization: Every trigger looks and behaves similarly, because the framework dictates the pattern. For example, frameworks often require triggers to call a specific interface or base class method. This uniformity means any developer looking at any trigger can quickly understand it.

Reduced code duplication: Frameworks typically handle common tasks (like preventing recursion, handling bulk batch iterations, etc.) internally, so individual triggers don't each implement their own version of these. This leads to less repeated code and fewer inconsistencies.

Advanced control: Some frameworks allow you to control trigger behavior with configuration. For instance, a framework might allow administrators to disable certain pieces of logic via Custom Metadata (commonly done in complex orgs for flexibility during deployments or data loads). Others can manage the order of execution for different chunks of logic within a single trigger context (something vanilla Apex doesn't provide across multiple triggers).

Ease of onboarding: New developers can be trained on the framework once, and then they can write triggers for any object following that pattern without reinventing the wheel each time.

Popular Framework Styles: According to Salesforce and community discussions, there are a few common ways to implement a trigger framework :

Trigger Handler Pattern (Simple): We discussed this – one trigger, one handler class per object. This is sometimes just called a pattern rather than a full framework, but it's the basis of many frameworks. It's simple and works well for a lot of orgs.

Framework using a Base Class (Virtual Class): In this style, you might have an abstract class like TriggerHandler with methods for beforeInsert, beforeUpdate, etc. Each object gets its own class extending TriggerHandler and implementing those methods as needed. The trigger then simply instantiates its handler class and calls a common execute method (which in turn calls the appropriate overridden methods). Kevin O'Hara's famous trigger framework (from around 2014) uses a base class approach, where triggers register actions and a central coordinator executes them. The advantage is you have structured overrideable behavior and possibly some shared logic in the base class.

Framework using an Interface: This approach defines an interface (say ITriggerHandler) with methods for various contexts. Each object's handler class implements that interface. The trigger then might call something like ITriggerHandler handler = new MyObjectHandler(); handler.beforeInsert(triggerNew, triggerOldMap); depending on context. It's similar to the base class approach but using interface to enforce the contract.

Metadata-Driven or Advanced Framework (Trigger Orchestration): Some frameworks go further, using custom metadata to define what actions to run in triggers, so the trigger code doesn't even need editing to add new logic – you just create a new handler class and update metadata. The "Trigger Actions Framework" by Salesforce is an example, which lets admins configure the order of operations via metadata rather than code. This is powerful in large orgs where multiple teams might contribute business logic – they can plug into the framework rather than touching the single trigger class.

When selecting a framework, consider your org's complexity and your team's comfort. If you have a small to medium org with a handful of triggers, the single trigger + handler pattern (perhaps with a simple base class to share a recursion flag or utility methods) might suffice and be easiest to maintain. If you have a very large org or multiple packages with triggers, a more robust framework that can coordinate triggers across packages or manage turn-on/turn-off via config might be beneficial.

Criteria for Choosing:

Organization size and team: A larger dev team with many moving parts may benefit from the strict standards a framework imposes. Also, if you foresee dozens of triggers, a framework helps manage that scale.

Need for advanced features: Do you need the ability to easily disable triggers (for migrations)? Do you want to control execution order of different logic pieces? If yes, lean towards a more feature-rich framework.

Learning curve: Some frameworks are more complex to grasp. If your team is relatively new to Salesforce, a simpler pattern might be better initially. You can evolve into a framework as skills grow.

Community support: Using a well-known framework (like fflib Apex Common's Domain layer or Kevin O'Hara's framework) means you can find community examples and perhaps even plug into open-source code that's maintained. A custom homegrown framework can be tuned to your needs but requires you to maintain it.

ChatGPT can help evaluate framework options by summarizing pros/cons. For example:

Architect: "Give me a comparison of a simple trigger handler pattern vs using the Apex Trigger Actions Framework."

ChatGPT: "A simple trigger handler pattern (one trigger & one handler class per object) is easy to implement and understand, but relies on developers to consistently follow the pattern and handle things like recursion on their own. It's lightweight but doesn't offer dynamic control. The Apex Trigger Actions Framework, on the other hand, is a metadata-driven approach where you can configure multiple actions for triggers and their order. It requires more setup and understanding of the framework's abstractions, but it provides out-of-the-box solutions for ordering and toggling actions. If you need non-developers to control trigger behavior or expect frequent changes in business logic that you want to configure, the Trigger Actions Framework is beneficial. Otherwise, for simplicity and low overhead, the basic pattern might be sufficient."

This kind of advice can guide decision-making. You could also ask ChatGPT specific questions like "How do I implement a trigger framework using virtual classes?" and it could outline the steps or provide sample code structure. If you have requirements (e.g., "I need to sometimes bypass certain trigger logic during data loads"), you can inquire how different frameworks support that.

Let's say the team decides to adopt a particular framework – how to proceed? Often, it means refactoring existing triggers to fit the mold. Suppose you choose a base class framework:

You'd create an abstract base class with methods for each trigger context.

Each trigger gets a handler class extending that base.

Triggers become one-liners that call into a central runner, often something like TriggerHandler.run(MyObjectHandler.class); – a static method in the base class that handles instantiating the handler, setting context, and calling the appropriate methods.

Add utility for recursion control in base class so all handlers benefit.

The specifics aside, once set up, adding new logic means adding code in the handler methods rather than new triggers.

Architect Callout: Ensure alignment of the chosen framework with business objectives and technical standards. For example, if the business requires that certain operations can be toggled without a deployment (a product owner might say "we need to be able to turn off auto-notifications triggers during holiday season"), then a framework that supports configuration (via Custom Metadata or Custom Settings) is worth the investment. Communicate to product owners how a framework might require a bit more initial development effort but will pay off in faster changes down the line (this helps justify the work in terms they understand – flexibility and reliability). From the developer perspective, once a framework is selected, provide training or examples so everyone uses it correctly. If the framework is complex, perhaps have a mini workshop or pair programming session with the team. Use ChatGPT to generate a ***cheat sheet*** for the framework – e.g., ask it to list the steps to create a new trigger using the framework, and share that in your internal wiki.

Product Owner Tip: Although the technical implementation of a trigger framework might seem deep in the weeds, it has business impacts. A well-chosen framework can reduce bugs (meaning fewer firefights that disrupt business) and speed up feature development (new logic can be added in a consistent way). If your technical team advocates for implementing a trigger framework, understand that it's an investment in quality and agility. It may not produce a shiny new feature by itself, but it creates a foundation for future features to be delivered more smoothly. Support such internal improvements in planning and recognize them in timelines.

To illustrate framework usage, consider that after adopting a framework, adding a new piece of logic is like "registering" a new handler rather than writing a brand new trigger. For example, using a hypothetical metadata-driven framework, an admin could add a new Custom Metadata record saying: "On Contact after insert, run class ContactAutoWelcomeEmail with order 1." The trigger (already existing in code) will dynamically pick this up and execute the new class's logic when applicable. This level of decoupling is powerful.

In summary, trigger frameworks range from simple to sophisticated. Choosing the right one means balancing complexity with control. Whether you stick to a straightforward single-trigger pattern or leverage a full-fledged framework, the key is that your triggers are no longer ad-hoc – they follow a plan. With the triggers structured and managed through patterns or frameworks, the final piece of the puzzle is making sure they work correctly through thorough testing. In the next section, we will discuss strategies for testing Apex triggers to ensure reliability and performance in real-world conditions.

TESTING STRATEGIES FOR APEX TRIGGERS

Quality assurance for triggers is paramount. Apex triggers often execute critical business rules, and a mistake can lead to data corruption or business process failures. Therefore, robust testing strategies are needed to ensure triggers behave as expected under various conditions. In Salesforce development, testing isn't just a best practice – it's required (with code coverage mandates). But beyond hitting coverage numbers, we want our tests to truly validate trigger logic, including bulk behavior and interactions with other automation.

Key considerations for testing triggers:

Cover all trigger contexts: If your trigger (or its handler) has branches for before insert, after insert, before update, after update, etc., your test classes should have test methods that cause the trigger to execute in each of those modes. For example, to test an after-insert scenario, you perform an insert in the test; for after-update, you perform update in the test, perhaps changing specific fields to trigger certain logic.

Bulk testing: Simulate bulk operations in tests. If a trigger is bulkified, test it by inserting, updating, or deleting, say, 50 or 100 records in one go (you generally don't need to hit 200 exactly, just more than 1 to see bulk behavior). Then verify that the outcomes are as expected for all records. For instance, if a trigger is supposed to set a field on all new records, insert 50, then query those 50 and assert the field is set on each. This ensures no record "fell through the cracks" in bulk.

Boundary and negative cases: Consider edge conditions. For example, test that your trigger doesn't perform actions when it shouldn't (like ensure that your after update logic that sends an email doesn't send when an irrelevant field changes). If your trigger has a recursion guard, test a scenario that would cause recursion and assert that it indeed prevented double-processing (one way could be to count how many times a certain record was updated or how many log entries were created).

Test interactions with other triggers/automation: If possible, write tests that include interactions. For example, if you have both a trigger and a Flow on the same object, you might want a test that runs to see if combined they produce the desired result (though flows can be tricky to unit test in Apex). However, in pure Apex context, if an Account trigger creates a Contact, and you have a Contact trigger, your test should verify the end-to-end outcome (Account insert -> Contact created -> Contact trigger ran).

Use of test utilities: Use Test.startTest() and Test.stopTest() around sections of your test to simulate separate transactions, especially if testing asynchronous behavior (like if your trigger enqueues a job or sends an email, which is processed after stopTest).

One challenge in trigger testing is that triggers run in system mode, so things like field-level security are not enforced by default. This means your tests might bypass certain issues that occur in production if not careful. If your trigger code queries or updates fields that a normal user can't, consider using runAs with a low-privilege user in tests to simulate real conditions (though in system mode, some differences remain).

Another aspect: triggers often use static variables (for recursion control or to communicate between trigger contexts). Those static states do not automatically reset between test methods (since each test method runs in a fresh context) – that's good, but within a single test method, static state persists unless you manually clear it or hit stopTest (which acts like a commit). Ensure your tests don't inadvertently let static flags from one operation affect another in the same method; often its simplest to do one trigger execution per test method to keep things isolated.

ChatGPT can help generate test code if you provide it with the trigger logic. For instance:

Developer: "Here is my trigger handler code for Opportunity. Can you create a sample Apex test class to cover its functionality?"

ChatGPT: "Sure. I will write a test class TestOpportunityTrigger with methods to cover insert, update scenarios..." and it will produce a skeletal test code with inserts/updates and some assertions.

While you'd need to fill in exact assertions (since GPT doesn't know your exact expected outcomes without description), it can scaffold the test methods, which is a big time saver. It might even remind you to test bulk behavior:

ChatGPT: "I included a test method testBulkOpportunityInsert that inserts 10 opportunities to ensure bulk behavior is handled. Adjust the number as needed. This will validate that your trigger logic works when multiple records are inserted at once."

Testers (or developers in the testing phase) can also use ChatGPT to interpret test failures. If a test fails because "expected X, actual Y", you can feed the relevant piece of trigger code and the failing test scenario to ChatGPT and ask for possible reasons. Because triggers can have side effects, sometimes the cause of a failure isn't obvious. ChatGPT can help trace through the logic, acting like a debugging buddy:

Tester: "Our test expected an Account's field to be 5 after inserting contacts (trigger supposed to count contacts), but it was 4. What could cause this off-by-one?"

ChatGPT: "Perhaps the trigger counting contacts is not handling one of the cases – for example, if one of the contacts in the batch failed to insert or was filtered out in the trigger logic. Or the trigger might be excluding something like inactive contacts. Check if your trigger uses Trigger.isInsert vs. isUpdate correctly for that field. Also, ensure that the test is committing all contacts (Test.stopTest() if any async processing is involved). The off-by-one suggests one contact wasn't counted – maybe a duplicate or some condition prevented it from incrementing. Reviewing the trigger code for any conditional logic around counting will help pinpoint it."

This analysis might lead you to discover, say, the trigger skipped counting contacts missing an email (just a hypothetical condition in logic), and your test data had one such contact, hence count 4 instead of 5. In this way, ChatGPT helps diagnose and thereby improve tests and code.

Continuous integration (CI) should run all trigger tests on every deployment. It's wise to integrate code coverage checks and perhaps static analysis for triggers. ChatGPT, while not a CI tool, can still help maintain test quality by suggesting additional test cases. For example, after writing basic tests, you might ask:

Developer: "Can you suggest more scenarios to test for this trigger?"

ChatGPT: "Consider testing an update where no relevant fields change, to ensure the trigger doesn't perform unnecessary actions. Also test the scenario where an exception might be thrown (if your trigger does addError or such) – for instance, if your trigger prevents deletion under certain conditions, write a test that tries to delete a record violating that condition and assert that the error is thrown. Additionally, test the maximum bulk size (like 200 records) if feasible, to ensure performance is acceptable."

Such suggestions can enhance your test coverage beyond the obvious paths.

Tester & Developer Tip: Treat trigger testing like black-box testing of a feature. Start from what the trigger is supposed to do (the requirement) and ensure the tests verify those outcomes, not just the code execution. For example, if a trigger ensures "whenever an Opportunity is closed, a Task is created," then write a test that closes an Opportunity and then queries for the Task, asserting it exists and has correct fields. This is an end-to-end verification. Test not only the "happy path" but also intentional failure cases: if the trigger is supposed to block deletion of an Account with active projects (via *addError*), your test should attempt that deletion and expect an exception. By verifying these business rules in tests, you protect against regressions – if someone modifies the trigger later and accidentally removes or alters that logic, a test will catch it.

Product Owner Tip: While product owners may not write tests, they define the scenarios that need to work. It can be useful for a product owner or business analyst to review the test case scenarios (in plain language) to ensure all acceptance criteria are covered. For instance, if a trigger is part of a user story, the PO can list the cases: "create X, expect Y; update Z, expect W," which developers can translate into test methods. This ensures the development team's understanding aligns with the business expectation. Involving POs in this review, even informally, increases confidence that the trigger's behavior is well understood and validated.

To conclude the testing section, remember that testing triggers is not just about achieving 75% code coverage to deploy – it's about ensuring reliability. Triggers often have far-reaching effects. Rigorous tests act as a safety net as the system evolves. When new features are added, if they inadvertently affect existing triggers (or vice versa), a solid test suite will flag it immediately. Incorporating ChatGPT in the testing process – for generating tests, explaining failures, or brainstorming scenarios – can enhance the thoroughness of your tests.

Having covered design, patterns, frameworks, and testing of triggers, we've assembled a comprehensive toolset for strategic trigger use and control. By following these practices, teams can leverage the full power of Apex triggers while keeping systems stable and maintainable.

SUMMARY

Apex triggers are a powerful mechanism in the Salesforce platform, and this chapter has explored how to use that power in a disciplined, strategic way. We began by emphasizing the critical role triggers play and why uncontrolled trigger logic can lead to significant problems if left unchecked. By adopting a mindset focused on clarity, scalability, and maintainability, we can avoid common pitfalls like random execution order, performance bottlenecks, and tangled logic.

We then reviewed seven key areas of Apex trigger best practices:

Single-trigger-per-object pattern: Ensuring only one trigger exists per object to maintain an organized codebase and eliminate unpredictable execution order. This pattern simplifies debugging and fosters clean architecture by consolidating all related logic in one place.

Context-aware trigger logic: Using trigger context variables (isInsert, isUpdate, Trigger.new, Trigger.old, etc.) to write logic that responds correctly to different events. We saw how context awareness leads to more robust and readable code, preventing errors like null pointer dereferences and making the trigger's behavior explicit for each scenario.

Bulkification: Always designing triggers to handle batches of records efficiently, thereby respecting Salesforce governor limits. We covered techniques to avoid SOQL/DML in loops and to use collections for operations, supported by Salesforce's own guidance that triggers must support bulk operations. Bulkifying triggers ensures that your automation scales and performs well as data volumes grow.

Recursion and cascading control: Recognizing when trigger logic can inadvertently cause itself (or other triggers) to fire repeatedly, and applying strategies to prevent infinite loops or double-processing. Whether via static flags, static sets, or conditional checks, we make sure our triggers execute only as intended, even in complex, multi-object interactions.

Trigger handler patterns: Structuring triggers to be logic-less and delegating work to handler classes. This leads to cleaner, modular code where triggers act as dispatchers and all business logic resides in well-defined methods or classes. The handler pattern not only improves readability but also eases testing and future maintenance.

Trigger frameworks: Taking patterns a step further by possibly adopting a formal framework for triggers. Frameworks bring consistency and added capabilities (like easier control of execution flow and toggle-able trigger logic) across the entire org. We discussed different framework approaches and how to align the choice with an organization's needs, ensuring that architects, developers, and even product owners are on board with the benefits and trade-offs.

Testing strategies for triggers: Finally, we underscored the importance of testing triggers thoroughly. From covering all contexts and bulk scenarios to validating integration points and using tools like ChatGPT to generate and debug test cases, we established that robust tests are the guardian of trigger quality. They verify that our strategic designs actually hold up in practice and continue to do so as requirements evolve.

Throughout the chapter, we demonstrated how ChatGPT can serve as a valuable ally in each of these areas. It can analyze code for best practice adherence, suggest improvements, generate templates (for example, merging triggers or creating handler classes), and assist in troubleshooting issues like recursive behavior or failed tests. We also highlighted using ChatGPT beyond plain text – you can verbally ask for explanations, upload code files for review, or even have it help craft an image or diagram to communicate a trigger architecture to your team. By integrating an AI assistant into the development workflow, teams can iterate faster and with more confidence, effectively having a smart pair programmer or reviewer available at all times.

In sum, disciplined and strategic trigger management significantly enhances a Salesforce org's stability and maintainability. By following the best practices outlined – from design patterns to frameworks and testing – architects and developers can ensure that triggers serve the business without unexpected side effects. Testers and product owners also benefit: the system behaves predictably, and new features can be delivered with less risk because the underlying automation is reliable.

As we move forward in this journey of Salesforce Enterprise Thinking, the focus will shift to other critical aspects of Apex and platform development. Having tamed the complexity of triggers, the next logical step is to explore how our code interacts with Salesforce's data layer. In the following chapter, we'll dive into efficient SOQL and database operations, learning how to optimize queries and transactions so that our applications perform well and scale – a perfect complement to the robust trigger foundations we've established here.

4

EFFICIENT SOQL AND DATABASE OPERATIONS

Efficient SOQL queries and optimized database operations are the backbone of high-performing Salesforce applications. Every time an Apex class retrieves or modifies data, it interacts with the Salesforce database layer via SOQL (Salesforce Object Query Language) queries or DML (Data Manipulation Language) statements. If these interactions are not carefully designed, they can become bottlenecks that degrade user experience and limit your system's scalability. Poorly optimized queries might hit Salesforce's governor limits, cause slower page loads for end users, or even result in runtime exceptions that break transactions. For an enterprise Salesforce implementation, such issues can translate into frustrated users, incomplete processes, and increased maintenance costs.

This chapter focuses on how to write Apex code that handles database interactions efficiently, ensuring your Salesforce apps remain responsive and scalable even as data volumes grow. We will explore techniques for crafting selective and efficient SOQL queries, minimizing unnecessary database calls, and handling bulk data operations gracefully. Inefficient SOQL or DML can cripple an otherwise well-designed feature – for instance, a naive query that scans an entire object table can time out or throw a "Non-selective query" error when data is large. By learning best practices, you'll prevent such problems and keep your applications running smoothly under heavy workloads.

Crucially, these best practices benefit every role involved in Salesforce projects. **Architects** will see that applying rigorous query and DML standards leads to systems that scale without costly refactoring. **Developers** gain the ability to build features without constantly fighting limits, resulting in cleaner code and faster development cycles. **Testers** can more easily validate performance and catch potential issues early when code follows known patterns of efficiency. **Product owners and managers** ultimately benefit through reliable application performance – users get faster response times, and the business experiences fewer disruptions from governor limit errors or data processing delays. In an enterprise setting, efficient database operations aren't just a developer concern; they improve stakeholder confidence that the Salesforce implementation can handle growth.

Throughout the chapter, we will also highlight how to leverage ChatGPT as a virtual assistant to improve your Apex database logic. Modern AI tools can analyze SOQL queries for optimization opportunities, suggest more selective filters, or review Apex code to spot inefficiencies. For example, a developer could **ask ChatGPT (even using voice input)** to review a query that seems slow; ChatGPT might point out that the filter isn't indexed or that a suboptimal pattern is used, providing suggestions to fix it. By iteratively refining queries and DML usage with ChatGPT's help, teams can quickly converge on best practice solutions. ChatGPT's support isn't limited to text – you might upload an Apex class file for analysis (with appropriate file privacy settings enabled) or use a plugin to visualize query execution plans, all of which can streamline the optimization process.

Efficient SOQL and Database Operations is organized into seven sections, each tackling a key aspect of database performance in Apex. First, we will cover the fundamentals of writing clear and effective SOQL queries, laying the groundwork for more advanced techniques. Next, we'll dive into query performance and selectivity – understanding how Salesforce indexes data and how to ensure your queries remain fast as data grows. With the basics and performance considerations in hand, we'll then address **bulkification**: designing queries that handle large volumes of records without hitting limits. Following that, we explore how to manage related data through parent-child relationship queries, retrieving complex data in a single round-trip. We will then discuss dynamic SOQL, which offers flexibility at runtime but comes with its own best practices to ensure safety and efficiency. The sixth section shifts focus to write operations, showing how to perform inserts, updates, and other DML optimally (for example, batching records and handling partial successes). Finally, we conclude with advanced SOQL techniques like aggregations and grouping, which allow you to get summary insights from the database with minimal Apex logic.

By the end of this chapter, you will understand how to design Apex code that interacts with the database in a robust and scalable manner. More importantly, you will have seen examples of how ChatGPT can assist at each step – from analyzing a simple query for clarity to generating a complex aggregate report – making it an invaluable partner in applying these best practices. Let's begin our journey with the foundation of all database interactions: mastering SOQL fundamentals and best practices for query writing, which is the first crucial step toward efficient Salesforce data operations.

MASTERING SOQL FUNDAMENTALS AND BEST PRACTICES

To build efficient database interactions in Apex, one must start with mastering the fundamentals of SOQL query design. SOQL (Salesforce Object Query Language) is Salesforce's query language for retrieving records, and its basic syntax is similar to SQL: you specify **SELECT** fields, **FROM** an object, and often a **WHERE** clause to filter results. For example, a simple SOQL query might be:

SELECT Id, Name, Industry

FROM Account

WHERE BillingCountry = 'USA'

LIMIT 100

This query would retrieve the Id, Name, and Industry of up to 100 accounts that have BillingCountry equal to 'USA'. Writing clear and purposeful queries like this is the first step in optimizing data access. A well-formed SOQL query should only select the fields you truly need and only filter the records you actually require. By avoiding the retrieval of superfluous data, you reduce memory usage and execution time.

Key SOQL fundamentals to remember:

Select only required fields: Don't query more fields than necessary. This reduces data transfer and processing overhead.

Filter results with WHERE: Always constrain queries by appropriate conditions to avoid pulling entire tables. Even if you need all records now, adding specific filters or at least a LIMIT makes your intent clear and guards against growth.

Use LIMIT to cap volumes: If you only need the first few records (or any sample of records), use a LIMIT clause. This prevents unexpectedly large result sets from consuming resources.

Avoid redundant queries: Re-use data you've already queried instead of querying again. And never put a SOQL query inside a loop; if you need to query multiple records, collect the criteria first and query once.

Another key aspect of SOQL fundamentals is readability and maintainability. Just as with Apex code, queries should be easy for others (and your future self) to understand. This means using descriptive variable names for query results in Apex and possibly formatting complex queries across multiple lines for clarity. For instance, if a query has multiple conditions, you might split each condition onto a new line and indent them in your Apex code. While Salesforce's query syntax is straightforward, complex business logic can lead to long queries – formatting them neatly helps avoid mistakes and makes code reviews smoother. It's a good practice to document the intent of non-obvious queries with a comment, so that testers or other developers know why certain filters are in place. Clarity in query writing also aids maintainability: when requirements change, you can more easily modify a well-structured query without introducing bugs.

Fundamental best practices in SOQL also improve reliability by reducing the risk of hitting governor limits. By selecting only necessary fields and filtering results properly, you implicitly conserve two precious resources: SOQL query limits and heap memory. Every Salesforce transaction can execute a limited number of SOQL queries (100 in a synchronous transaction) and can retrieve at most 50,000 records per query. If you follow the basics—no unnecessary queries, no unfiltered queries that retrieve massive datasets—you're far less likely to run into the infamous "Too many SOQL queries: 101" error or to chew through heap space with unneeded data. For example, a query that fetches every Contact in a dev sandbox might work fine with 500 records, but the same unfiltered query in production with 5 million records would likely hit limits or cause timeouts. By adhering to these basic principles early, such issues can be prevented long before deployment.

In short, simple tweaks like adding a WHERE clause or reducing fields can be the difference between a successful operation and a governor limit exception. Adopting these habits from the start makes your code inherently more robust.

To ensure you're applying these fundamentals, you can use ChatGPT as a sanity check for your queries. For example, you might have written a query and wonder if it's as clear and efficient as it could be. You could ask ChatGPT to review it:

User: "I have an Apex method that runs the following SOQL query:

SELECT Id, Name, Email, Phone, Address, Birthdate

FROM Contact

It fetches all contacts without any filter. Is this a good practice for production code?"

ChatGPT: "Fetching all Contact records without a filter is generally not a good practice. Even though Salesforce limits a single query to 50,000 records, querying a large object without constraints can lead to performance issues or governor limit errors if the data set is big. In production, it's rare that you truly need every record. It would be better to add a **WHERE** clause to narrow down results (for example, by region or date) or use a **LIMIT** clause if you only need a sample. Also, select only the fields you need – if you don't actually use the Phone or Address in your code, omit them. By refining the query to target just the needed data, your code will run faster and be less likely to hit limits."

In this example, ChatGPT identified the lack of a filter as a problem and reminded us to limit fields. These are the kinds of fundamental checks that can catch issues early. You could also use ChatGPT to format a long query or to explain what a complex query is doing. ChatGPT can even help non-developers: for instance, a product owner could simply describe a need (e.g. 'list all high-priority open cases for customers in Europe') and ChatGPT would generate the corresponding SOQL query. This capability helps ensure that even initial drafts of queries include proper filters (status, priority, region in this case) according to best practices, bridging the gap between business requirements and implementation. Such AI assistance is helpful for onboarding testers or new developers who need to understand your code or data requirements. Even using ChatGPT's voice interface, a developer might quickly ask, "What does this query return?" and get an explanation, ensuring they understand the code's intent.

Developer Tip: Establish internal coding standards that require every SOQL query to be purposeful. Before writing a query, clarify what data you need and why. During code reviews, check that each query has appropriate filters and doesn't retrieve more fields or records than necessary. This discipline not only prevents inefficiencies but also makes the codebase more readable.

Tester Tip: When reviewing user stories or test cases, pay attention to the data scope of each feature. If a requirement says "show customers in the U.S. with recent purchases," ensure the Apex queries reflect that (e.g., there's a WHERE Country__c = 'US' filter). Test with both small and large data sets – this way you confirm that the query returns correct results and observe its performance. Clear query logic will usually mean fewer surprises in testing, whereas an overly broad query might work in a dev sandbox with 100 records but fail in UAT with millions.

By mastering these SOQL fundamentals, you create a strong foundation for all database interactions. Clean, focused queries are easier to optimize and less likely to cause problems. Next, with these basics in hand, we can move on to examining how to make queries not just correct, but truly performant at scale – starting with understanding SOQL selectivity and performance considerations.

OPTIMIZING SOQL PERFORMANCE AND SELECTIVITY

Writing correct queries is not enough – they also need to be *selective* and efficient, especially as your data grows. Query selectivity refers to how well a query's filters target a small subset of records. Salesforce's query optimizer will use database indexes to speed up queries that filter on certain fields, but only if the filter is selective. A query is considered selective when it uses at least one indexed field in the WHERE clause and the filter reduces the result set significantly (generally below a threshold percentage of the total data). If a filter isn't selective – for example, querying a field with no index or one that matches a very large portion of the data – Salesforce may perform a full scan of the object, which is much slower. In extreme cases, such a query can even be rejected with a runtime error for being "non-selective" if used in a context like a large data trigger. In short, selective queries run fast and scale well; unselective queries do not.

Indexed fields in Salesforce include commonly used ones like record IDs, foreign keys (lookup/master-detail fields), audit dates (such as CreatedDate), and any custom field marked as Unique or External ID. Salesforce automatically indexes these standard fields and any custom External ID fields. If your query filters on one of them (e.g. WHERE AccountId =... or WHERE CreatedDate >...), the platform can quickly narrow the search using the index. However, if you filter on a field without an index – say, a text field like Industry or a formula field – the database has to check every record, which is inefficient. Often the fix is to add an index: for a custom field that you frequently query by, you can ask Salesforce Support to create a custom index (or mark it as an External ID if possible) so that queries on it become selective.

Beyond indexing, you should also avoid certain filtering patterns that prevent selectivity. Some query conditions inherently can't use indexes even if the field is indexed:

Negative filters: Avoid using operators like != or NOT in WHERE clauses. For example, WHERE Status__c != 'Closed' is non-selective because the database must find all records not matching a value, which is effectively a full scan.

Null or empty comparisons: Avoid checking if a field *is null* or equals an empty string. Indexes generally don't track null values, so WHERE Phone = '' won't be selective.

Leading wildcards in LIKE: A WHERE Name LIKE '%ABC' clause cannot use an index because the wildcard at the start means the database must examine every value for a match. (If possible, use a trailing wildcard like ABC% which can still use indexing, or consider using SOSL for such scenarios.)

Uneconomical operators on text fields: Avoid using greater-than/less-than on text fields or applying functions to indexed fields (for instance, WHERE YEAR(CreatedDate) = 2023). These patterns prevent index usage. It's better to compare ranges on an indexed date directly (e.g. CreatedDate >= 2023-01-01) rather than wrap it in a function.

By adhering to these guidelines, you ensure the query optimizer can do its job. Selective filters dramatically improve performance by letting the database skip over vast swathes of records up front. Salesforce's own documentation notes that as a rule of thumb, a filter that targets under about 10% of an object's rows (up to a maximum of around 333,000 records) is considered selective. Exceeding that, especially on large objects, can cause slow performance or a non-selective error. As an example, imagine an Account object with 5 million records: a query filtering by an unindexed field like Industry might return 2 million rows – that's 40%, far above the selectivity threshold, likely triggering a warning or error. But a query filtering by CreatedDate = Last 30 Days might return only 50,000 rows (1%), which the optimizer would happily run using an index on CreatedDate.

Developers can use Salesforce's Query Plan tool to evaluate query selectivity in the Developer Console. This tool shows the query's cost and which indexes, if any, are used. A cost threshold of 1.0 or lower indicates a selective query; anything above 1.0 means the query is unselective and might not use indexes efficiently. If you see a high cost, consider restructuring your query or adding indexes. Sometimes the solution is to limit the scope (e.g. split one huge query into multiple smaller selective queries by region or date range).

ChatGPT can assist in analyzing and improving query performance as well. For example, a developer might ask why a particular query is slow:

User: "This SOQL query is timing out: SELECT Name, Email FROM Contact WHERE Description__c = 'High Value Client'. The Description__c field isn't indexed. How can I improve it?"

ChatGPT: "The performance issue is because Description__c is not indexed, and the filter value 'High Value Client' likely matches a large number of Contact records. Salesforce has to scan all Contact records to find matches, which is slow for a big data set. To improve this query, try to make the filter more selective. For instance, if there's another criterion you can add that uses an indexed field (like a checkbox or picklist for high-value clients, or a LastModifiedDate range), include that in the WHERE clause. You could also consider marking Description__c as an External ID or Unique field if appropriate – that would index it. Otherwise, you may need to use a different approach (such as a scheduled batch to mark high-value clients with an indexed flag) to avoid querying on Description__c directly. Essentially, we want to reduce the result set that Salesforce has to sift through, ideally leveraging an index."

In this dialogue, ChatGPT identified the core issue (filtering on a non-indexed field that is too broad) and suggested making the query more selective by either adding an indexed filter or indexing that field. You could similarly input an error message like "System.QueryException: Non-selective query against large object type" into ChatGPT and get an explanation of which part of your SOQL is problematic and how to fix it. This can save developers time in pinpointing why a query isn't selective.

Architect Tip: When designing the data model, think about reporting and querying needs. If certain fields (e.g. Account Number, Order Status) will be used frequently in filters or reports, define them as **External ID** or add custom indexes early. Planning for indexes in advance avoids major refactoring when data volumes grow.

Developer Tip: Use the Query Plan tool on queries that will run on large datasets, and monitor for any non-selective warnings. If a query plan shows a cost above 1, don't ignore it—optimize the query or reach out to a Salesforce admin to create an index. Also, include realistic data volumes in performance testing. A query that runs fine with 1,000 records might fail with 1,000,000 records if not selective, so testing with scale data (or using ChatGPT to hypothesize performance) ensures your solution remains fast in production.

By optimizing for selectivity, you make your SOQL queries scalable and efficient. Next, we'll address another crucial aspect of performance: ensuring that queries and their surrounding code handle large volumes of data in bulk without breaking Salesforce's limits.

BULKIFYING SOQL QUERIES FOR SCALABILITY

Scalability in Salesforce isn't just about single query performance – it's about handling bulk data without hitting limits. Apex code often runs in contexts (like triggers or batch jobs) where it processes many records at once. Bulkifying your SOQL queries means making sure that no matter if you're handling 1 record or 200, you perform a fixed minimal number of queries. The key rule is simple: never put a SOQL query inside a loop. If you do, you risk executing that query once per record, which can quickly exceed the governor limit of 100 queries per transaction. This anti-pattern is known as the "N+1 query problem" – one query for each of N records – and it's a common source of runtime errors and slowdowns.

To bulkify such code, follow these general steps:

Collect identifiers: In your loop, collect the record identifiers or keys you need to query by (e.g., all AccountIds from a list of Contacts) into a Set or List.

Query in one go: Outside the loop, perform a single SOQL query using those collected IDs with an IN clause (or perform an aggregate query if counting or summarizing is needed). This retrieves all required data at once.

Store results: Store the query results in a Map (keyed by the record Id) or another structure so you can easily find the data for each original record.

Process using the results: Loop through your original records again and use the Map/structure to get the related data instead of querying again. Now your loop contains no SOQL calls, and you've eliminated the risk of query limits.

Let's illustrate the difference. Imagine code that looks like this (bad practice):

```
// BAD PATTERN: SOQL inside a loop

for (Contact c : contactList) {

  Account parent = [SELECT Name FROM Account WHERE Id = :c.AccountId];

  //...do something with parent.Name

}
```

If contactList has 150 contacts, this code executes 150 SOQL queries (one per contact) – blowing past the limit of 100 and throwing a governor exception. The bulkified version below does only one query, regardless of list size:

```
// GOOD PATTERN: Bulkified query outside the loop

Set<Id> accountIds = new Set<Id>();

for (Contact c : contactList) {

  accountIds.add(c.AccountId);

}

Map<Id, Account> accountMap = new Map<Id, Account>(

  [SELECT Id, Name FROM Account WHERE Id IN :accountIds]

);

for (Contact c : contactList) {

  Account parent = accountMap.get(c.AccountId);
```

//...do something with parent.Name

}

In the bulkified approach, we moved the SOQL query out of the loop. We collect all Account Ids we need, then retrieve all those Account records in one go. We use a map to efficiently find which Account belongs to each Contact. This pattern uses slightly more memory (to store the map) but far fewer database calls – it's much more scalable. In fact, bulkifying like this is not only about avoiding errors; it also improves performance. One query that returns 150 records is generally faster than 150 queries returning 1 record each, due to reduced query overhead.

Salesforce also provides **SOQL for loops** (using a query directly in a for-loop) to efficiently handle very large query results by internally batching them, and **Batch Apex** for processing records in chunks when one transaction isn't enough. These advanced techniques still follow the same principle: perform as few queries as possible and handle many records per query.

ChatGPT can help identify and fix unbulkified code. For instance, you could provide a snippet and ask for optimization:

User: "How can I improve this Apex code? It's hitting the SOQL 101 limits:

for (Opportunity opp : oppList) {

opp.totalContacts__c = [SELECT COUNT() FROM Contact WHERE AccountId = :opp.AccountId];

}

ChatGPT: "The code is encountering a governor limit because there's a SOQL query inside the for loop, running once per opportunity. To bulkify this, you should query all needed Contacts in one go, outside the loop. For example, gather all AccountIds from `oppList` into a set, then perform a single aggregate query on Contact like `SELECT AccountId, COUNT(Id) FROM Contact WHERE AccountId IN :acctIdSet GROUP BY AccountId`. This will return one row per Account with the count of Contacts. You can then loop through the query results and store the counts in a map keyed by AccountId. Finally, iterate over `oppList` and set `opp.totalContacts__c` from that map. This way, you replace potentially hundreds of small queries with one big query."

In this answer, ChatGPT spotted the N+1 query pattern and outlined a bulk solution (even using a clever aggregate query to count related records in one go). You could follow such guidance to rewrite the code, perhaps even asking ChatGPT to produce the sample bulkified code for you.

Bulkifying queries goes hand-in-hand with bulkifying Apex triggers and other operations. Salesforce architects and lead developers often enforce the rule that any loop must not contain a SOQL (or DML) call. It's wise to use code scanning tools or even manual code reviews to catch these issues early. The same goes for DML statements in loops.

Developer Tip: Always think in sets and collections. If you're writing a loop that needs data from another object, pause and refactor to do one SOQL query outside the loop. Leverage collections (sets and maps) to gather IDs and relate data back. This ensures your code runs within limits even as data volumes grow.

Tester Tip: Include bulk data scenarios in your testing. For a given feature, create test cases that pass multiple records (50, 100, 200) into the Apex method or trigger at once. The code should handle them in one go without errors. This simulates real scenarios like data imports or mass updates where many records hit the system at once. If a governor limit is hit during a bulk test, that's a red flag indicating an unbulkified query or operation. By catching this in testing (or using ChatGPT to analyze stack traces), you can ensure the issue is fixed before production.

By systematically bulkifying your SOQL queries, you make your code capable of handling large batches of records efficiently. This sets the stage for the next consideration: even when querying in bulk, you may need to retrieve complex related data. In the following section, we'll look at managing parent-child relationships in queries to further streamline data retrieval.

MANAGING RELATIONSHIPS AND PARENT-CHILD QUERIES

Many Salesforce data needs involve multiple related objects. Apex developers frequently need to retrieve a record along with its related records – for example, getting an Account and all of its Contacts, or a list of Cases with their associated Account information. Instead of writing separate queries and then manually linking the data in Apex, you can often use **relationship queries** in SOQL to pull related data in a single round trip. Salesforce supports two types of relationship queries:

Child-to-parent: You can select fields from a parent object directly in a query on the child object, using dot notation. For instance, if you are querying contacts, you can retrieve each contact's Account Name in the same query: SELECT Id, Name, Account.Name FROM Contact WHERE Account.Industry = 'Banking'. Here Account.Name is a parent field reference; this query returns Contact records, but each Contact result will include its parent Account's Name. Under the hood, Salesforce performs a join, so you don't have to query Account separately.

Parent-to-child: You can select a parent record and a collection of its child records by using a subquery. For example, to get each Account and its related Contacts, you could write: SELECT Id, Name, (SELECT FirstName, LastName FROM Contacts) FROM Account WHERE BillingState = 'NY'. The subquery in parentheses after the Account fields retrieves child Contact records for each Account. This single SOQL will return a list of Account objects, and each Account in the result has a Contacts field containing a list of its Contact children that met the criteria.

Using these relationship query patterns can dramatically reduce the number of queries you need. In the example above, one SOQL replaced what might otherwise require one query for Accounts and a loop of queries for Contacts for each account. Not only is this more convenient, it also conserves query limits and improves performance by retrieving everything at once.

There are some considerations when working with parent-child queries. First, the normal governor limits still apply: the total number of rows returned (parent plus children combined) can't exceed 50,000 in a single query. If each parent has many children, a relationship query could hit this limit. For instance, if you query 1,000 Account records and each has 100 Contacts, that's 100,000 Contact rows – too many for one query. In such cases, you may need to filter the children or break the query into smaller chunks (or use batch processing). Salesforce won't fetch more than 2,000 child records per parent in a single query by default either, so very large one-to-many relationships might need an alternative strategy (or multiple queries) to retrieve all data. Another consideration: you generally can't filter on child fields in the parent query's WHERE clause. In other words, WHERE Id IN (SELECT AccountId FROM Contact WHERE...) is possible (a semi-join), but you can't directly put child conditions in the parent query's WHERE. If you need to filter by child criteria, you might use a subquery with a semi-join or do the filtering in Apex after querying.

Despite these caveats, relationship queries are powerful. They leverage defined lookup relationships (including custom lookups and master-detail relationships) via the **relationship names.** When writing a parent-to-child subquery, you must use the child relationship name as defined in the object schema (often the plural label, like Contacts for the child relationship from Account to Contact). ChatGPT can help here if you're unsure of the relationship name – you can ask something like "What's the child relationship name for contacts on Account?" and it can tell you that it's "Contacts".

To see relationship queries in action, consider a scenario: initially, a developer wrote separate queries to get Accounts and then looped through them to query Contacts. This can be refactored into one SOQL. Let's use ChatGPT to assist:

User: "I have Apex code that first queries Accounts, then within a loop for each Account, queries its Contacts. Can I combine that into one SOQL query?"

ChatGPT: "Yes! You can use a parent-to-child subquery to get Accounts with their Contacts in one go. For example:

SELECT Id, Name, (SELECT FirstName, LastName FROM Contacts)

FROM Account

WHERE Name LIKE 'Acme%'

This single query retrieves all Accounts whose name starts with 'Acme' and includes each Account's related Contacts. In your Apex, you'll get a list of Account objects, and for each Account, you can iterate over account.Contacts to access its Contact records. This is much more efficient than running a query per account."

In this response, ChatGPT provided the exact SOQL syntax for a relationship query, showing how to replace multiple queries with one. You could paste such a query into your code (or refine the filters as needed) and remove the extra loop logic, simplifying your codebase.

Relationship queries can also simplify reporting logic. For example, if an application needs to display all Cases with the name of the Account and Contact on each case, a single SOQL like SELECT Id, Subject, Account.-Name, Contact.Name FROM Case... will retrieve everything needed. The developer doesn't have to later fetch Accounts or Contacts separately – the data is ready to use, which is a huge win for clarity and efficiency.

Architect Tip: Design your data model and queries to leverage relationships instead of manual joins. If you notice a pattern of querying an object and then querying its children in code, consider rewriting that with a SOQL subquery. However, be mindful of extremely large hierarchies. For objects with huge numbers of children, it may be wiser to use roll-up summary fields or a scheduled job to aggregate data, rather than trying to pull thousands of child records at once in a real-time query.

Developer Tip: Get familiar with relationship names (the ones used in subqueries and dot notation) by checking Salesforce schema or using tools like Workbench. Using these names in SOQL will allow you to fetch related data easily. Always test your relationship queries with realistic data volumes – for example, an Account with hundreds of children – to ensure performance remains good. If a relationship query returns too much data, try adding filters on the child query (e.g., only active child records) or retrieve children in batches. And as always, keep an eye on the query row count to stay within limits.

By tapping into Salesforce's relational querying capabilities, you minimize the need for iterative processing and make your Apex code cleaner and faster. Next, we'll explore a different kind of query flexibility: dynamic SOQL, which allows you to construct queries at runtime – and how to use it safely and effectively.

DYNAMIC SOQL – USAGE AND BEST PRACTICES

Sometimes you won't know exactly what to query at compile time. Maybe you're building a search feature with user-defined filters, or you need to query different fields based on context. In such cases, **dynamic SOQL** – constructing the query as a string at runtime and using Database.query() – offers the flexibility you need. Dynamic SOQL allows you to decide fields and conditions on the fly, something not possible with static (hard-coded) SOQL. However, this flexibility comes with trade-offs and risks, so it should be used carefully.

The primary concern with dynamic queries is **SOQL injection** – if you concatenate raw user input into a query string, malicious input could alter the query to retrieve data the user shouldn't see (or simply break the query). The best defense is to avoid putting unchecked user input directly into the query. Salesforce's recommendation is to use static SOQL with bind variables whenever possible (which automatically handles input safely). If you must build a query string, *always* sanitize user inputs by escaping quotes. In fact, the Apex String.escapeSingleQuotes() method is provided for this exact purpose. For example:

// Unsafe dynamic SOQL (userInput comes from a form)

String soql = 'SELECT Id, Name FROM Account WHERE Name = \'' + userInput + '\'';

List<Account> results = Database.query(soql);

If userInput contained something like **a' OR Name != 'b**, the query would become...WHERE Name = 'a' OR Name != 'b'..., effectively ignoring the Name filter entirely. To prevent this, use escapeSingleQuotes:

String soqlSafe = 'SELECT Id, Name FROM Account WHERE Name = \''
+

String.escapeSingleQuotes(userInput) + '\'';

```
List<Account> results = Database.query(soqlSafe);
```

This ensures any rogue quotes in the input are neutralized and the query remains syntactically correct and secure. A Salesforce developer once asked if using escapeSingleQuotes everywhere in dynamic SOQL is sufficient to prevent injection – and the consensus was yes, it's the essential step.

Instead of Database.query, you can use Database.queryWithBinds, which allows you to safely bind variables into your dynamic SOQL queries without manual string concatenation. This method helps mitigate SOQL injection risks by automatically handling user input, much like bind variables in static SOQL. Using queryWithBinds also enhances code readability and maintainability, especially when dealing with multiple parameters.

Another issue with dynamic SOQL is the loss of compile-time checking. With static SOQL, the platform verifies your field names and syntax at deployment. With dynamic SOQL, a mistake in the field name or query only surfaces at runtime (perhaps as a QueryException). For this reason, **use static SOQL when the query structure is fixed**. If a field name used in a dynamic query is changed or removed in the org, you won't get a compile-time error – the failure will happen at runtime, possibly in production. With static SOQL, such issues are caught during deployment or by tests. Dynamic SOQL thus demands vigilant maintenance as your data model evolves. Reserve dynamic SOQL for scenarios where the query truly needs to vary. As one expert put it, inline (static) queries are preferable whenever possible, since the compiler can catch typos in field names. On the other hand, if you have a highly flexible search with many optional criteria, writing dozens of static query variations is impractical – dynamic SOQL may be the cleaner solution. For instance, if there are 8 independent filters a user can toggle, theoretically that's 256 possible query combinations; constructing a single dynamic query that includes filters only when needed is far more maintainable.

Best practices for dynamic SOQL include:

Whitelisting fields: If your dynamic query chooses fields or sorts based on user input, allow only expected field names or use a mapping. Never directly trust a user-provided field name to avoid security holes.

Sanitizing values: As discussed, escape quotes in text inputs, and ensure numeric inputs are actually numeric (or use Integer.valueOf etc. to throw errors on bad input).

Use bind variables where possible: You can sometimes use bind variables even in dynamic SOQL (by constructing a string like... where Name = :someVar and letting the Database.query evaluate the bind). This isn't always straightforward, but if feasible, it adds safety.

Test thoroughly: Dynamic queries should be covered by unit tests for various scenarios – valid input, unexpected input (like special characters), and boundary cases – to ensure they behave as expected.

Let's see how ChatGPT can help in a dynamic SOQL scenario. Imagine a developer wrote code like the unsafe example above and wants to verify its security:

User: "I build a SOQL string in Apex by concatenating user input. It works, but is it secure against injections?"

ChatGPT: "Directly concatenating user input into a SOQL query is dangerous because of SOQL injection risks. A user could input a value that tricks the query into changing its meaning. To make it secure, you should never insert raw user input into the query without cleaning it. The best practice is to use bind variables in a static query whenever possible, as Salesforce will handle quoting for you. If you must build the query string dynamically, make sure to escape user input using String.escapeSingleQuotes(). This will handle quotes in the input by prefixing them with a backslash, preventing malicious termination of your string literal. Also consider whitelisting: only allow expected filter values or field names. In summary, yes, you need to actively defend against injection – simply concatenating strings is not safe."

Here, ChatGPT correctly warns the developer and reinforces the best practices we discussed. You could follow up by asking ChatGPT to review your dynamic query code for any other issues (like missing commas or typos in field names).

To summarize, dynamic SOQL is a powerful tool when used in the right situations. It provides flexibility for complex search and filtering use cases, but with that power comes responsibility: you must guard against security vulnerabilities and runtime errors. By validating inputs and leveraging Salesforce's built-in escaping and binding mechanisms, you can safely use dynamic queries. And whenever you find yourself reaching for dynamic SOQL, consider if it's truly necessary – often, a well-structured static query or a slight schema change (like a formula or roll-up field) can eliminate the need for dynamic construction altogether.

Next, we will turn our attention to optimizing database write operations (DML). Just as we bulkify and secure our queries, we need to ensure our inserts, updates, and deletes are handled efficiently and safely in Apex.

EFFICIENT DATABASE OPERATIONS – DML OPTIMIZATION

Just as we optimize SOQL queries for bulk operations, we must also handle data modifications (DML) in bulk. **DML** stands for Data Manipulation Language – operations like insert, update, upsert, delete, and undelete in Apex. The same golden rule applies: don't perform DML inside a loop. Instead, batch your changes and commit them in one (or a few) statements. Salesforce allows at most 150 DML statements per transaction, so doing a separate insert for each record in a loop will quickly hit that limit. Moreover, bulk DML is more efficient: inserting 200 records with a single insert call is much faster than 200 individual inserts.

Bulk DML pattern: Collect the records you need to create or change into a list, and call insert/update/etc. once on that list. For example, if you're processing opportunities and need to create a related task for each, don't call insert new Task(...) in each iteration. Instead, build a List<Task> of tasks in the loop, then call one insert tasksList after the loop. This way, whether there's 1 record or 200, you stay within limits and use resources efficiently.

However, bulk DML introduces a question: what if one record in the batch fails? By default, Salesforce applies an "all-or-none" behavior – if any record triggers an error (say a validation rule or required field missing), the entire insert/update will fail and rollback. That ensures data integrity (no partial transaction commit), but sometimes you want to save the valid records even if a few have issues. To support this, Apex offers partial success processing via the **Database.* methods.** For instance, Database.insert(records, false) will attempt to insert all records and allow some to succeed even if others fail. No exception is thrown for the failures; instead, you get back a list of SaveResult objects to check which records succeeded or had errors.

Here's how you might handle an insert with partial success:

```
Database.SaveResult[] results = Database.insert(recordList, false);

for (Integer i = 0; i < results.size(); i++) {

  if (!results[i].isSuccess()) {

    // Handle the error for recordList[i]

    Database.Error err = results[i].getErrors()[0];

     System.debug('Record '+ recordList[i].Id +' failed to insert: ' + err.get-
Message());

  }

}
```

In this snippet, we insert a list of records with allOrNone set to false. We then loop through the results to log or address each error. Successful inserts will have isSuccess() == true, and failed ones provide an error message. This pattern is very useful in data loading or integration scenarios where you don't want one bad record to prevent all others from saving.

If you use the default all-or-none behavior (for example, calling insert recordsList; without the Database class), you should wrap it in try-catch to handle any DmlException. That exception's message will indicate the first error encountered and abort the operation. Often in user-facing logic (like a batch update from a UI button), product owners might prefer a partial save: save what can be saved, and report the ones that failed. It's important to make that decision explicitly and implement accordingly.

Beyond basic inserts and updates, consider **transaction control**. Apex allows you to start a **savepoint** and perform a series of operations, then roll back to the savepoint if something goes wrong. This is useful if you have multiple DML operations that must either all succeed or all fail as a unit (for example, insert an Order and related OrderLine records together). By using Savepoint sp = Database.setSavepoint(); and Database.rollback(sp) in a catch block, you can undo partial changes if later steps fail. Keep in mind that rolling back will undo all DML up to that point, so it should be used carefully. In most Apex code, you won't manually control transactions, relying instead on the platform's automatic all-or-none behavior or partial success flags.

Let's use ChatGPT in a scenario involving bulk DML and error handling. Suppose a developer has a loop doing individual updates and wants to improve it:

User: "My code is updating contacts one by one in a loop. How can I make it bulk and handle errors better?"

ChatGPT: "You should accumulate those Contact records in a list and perform a single update on the entire list outside the loop. For example, gather all the contacts you want to update in contactsToUpdate, then call update contactsToUpdate once. This avoids the 'Too many DML statements' limit. For error handling, consider using Database.update(contactsToUpdate, false). By setting the second parameter to false, the update will continue even if some contacts fail to update. You can then check the Database.SaveResult[] returned to see which records failed and why. This way, one problematic record (say, missing a required field) won't stop all the others from being updated. Remember to remove or fix the loop that was doing one update at a time – after these changes, your code will handle all records in bulk."

The assistant's advice covers both bulkification and partial error handling. After implementing it, the developer would have cleaner code that respects governor limits and a strategy for dealing with failures.

Developer Tip: Use the Database.insert/update methods with the allOrNone flag strategically. For back-end processes like data migrations or nightly jobs, partial success (false) is often preferred so that one error doesn't halt the whole job. But for front-end transactions where atomicity is important (e.g., creating an Invoice and its Line Items together), you might stick to all-or-none and let the transaction roll back on any error. In all cases, bulkify the calls and handle errors explicitly rather than ignoring them.

Tester Tip: When testing bulk DML logic, include scenarios where one or more records are set up to fail (e.g., put invalid data in one record among a batch of valid ones). Verify that the code handles it as expected: if partial success is enabled, the failing records should be reported and the others committed. If all-or-none, the whole transaction should roll back. Also test that the system doesn't exceed DML limits when processing large batches.

Product Owner Tip: Decide on the desired behavior for bulk operations in your requirements: should the operation fail entirely if any record is invalid, or process the valids and skip the invalids? Communicate this to developers so they can use the appropriate approach. Bulk operations can impact data integrity, so it's a balance between strictness and throughput. Also, ensure that if partial failures occur, there is a mechanism (like an error log or user message) to surface those issues so they can be corrected.

By treating DML operations with the same bulk mindset as SOQL, and taking advantage of Apex features for partial success and error handling, you make your data modifications robust and scalable. For extremely large datasets (millions of records), consider using Apex Batch jobs or the Salesforce Bulk API to process data in segments – these tools follow the same bulk principles under the hood. Our final section will look at advanced SOQL techniques for aggregation and grouping, rounding out our performance toolkit.

ADVANCED TECHNIQUES – SOQL AGGREGATIONS AND GROUPING

Our last topic deals with letting the database do more of the work: using SOQL's aggregate functions to summarize data and using **GROUP BY** for categorization. These features allow you to get counts, sums, averages, and other metrics directly from Salesforce in a single query, rather than retrieving rows and computing values in Apex. This can greatly improve efficiency for reporting and analytics use cases.

SOQL supports aggregate functions like **COUNT()**, **SUM()**, **MIN()**, **MAX()**, and **AVG()**, as well as **GROUP BY** clauses (and an optional HAVING clause to filter aggregate results). For example, suppose you want the number of contacts associated with each Account. You could write:

```
SELECT AccountId, COUNT(Id)

FROM Contact

WHERE AccountId IN :acctIds

GROUP BY AccountId
```

This query will return one result per Account (for the Accounts specified in acctIds if used as a bind variable), along with the count of Contact records for each. In Apex, the result comes back as a list of **Aggregate-Result** objects, where you can call get('expro') or get by alias to retrieve each aggregate field. In this case, you might do:

```
for (AggregateResult ar : results) {

  Id acctId = (Id) ar.get('AccountId');

  Decimal contactCount = (Decimal) ar.get('expro');

  //... use acctId and contactCount...

}
```

It may look a bit cumbersome, but it's far more efficient than running a separate count query for each Account or pulling all contacts and counting in memory.

You can also group by other fields. For instance, to get the total pipeline value by Stage for opportunities, you might use:

```
SELECT StageName, SUM(Amount)

FROM Opportunity

WHERE IsClosed = false
```

GROUP BY StageName

HAVING SUM(Amount) > 1000000

This will give you one row per stage (for open opportunities), with the total Amount for that stage, and only including stages where the sum exceeds $1,000,000. A query like this could replace many lines of Apex code that would otherwise loop over opportunities to calculate these totals.

Using aggregate queries has some limitations to keep in mind. Firstly, the results of a GROUP BY query cannot exceed 2,000 rows. This means if you group by a field with a very high number of distinct values, the query may fail or truncate results. For example, grouping all Contacts by City could hit this limit if you have contacts in over 2,000 cities. In such cases, you might need to refine the query (perhaps filter by region or use multiple queries). Also, aggregate queries still count against the total SOQL query governor limit (each aggregate query is one SOQL call) and must obey the overall 50,000 row retrieval limit. However, note that when you use an aggregate function like COUNT(), the query doesn't count each record toward limits in the same way as retrieving those records would – it's the number of aggregate result rows that matters.

Another consideration: not every type of field can be grouped. You can GROUP BY text and picklist fields, dates, and so on, but you cannot group by long text areas or rich text, for example. Also, when using aggregate functions, you can't directly retrieve non-grouped fields in the same query (every selected field must be either aggregated or grouped). This is different from SQL in some environments and is just something to be aware of in SOQL.

Despite these limitations, aggregate queries are incredibly powerful for Salesforce analytics within Apex. They push computations to the database level (which is optimized for such work) and reduce data transfer. Instead of querying 10,000 records and then iterating to compute totals or counts, you can get just the 10 summary results you need. This is a common pattern in enterprise applications – let the database do heavy lifting whenever possible.

ChatGPT can assist with writing these queries if you're not familiar with the syntax. For instance, you might ask: "How can I write a SOQL query to count contacts per account?" and it would show a query similar to our example. Or: "How do I use HAVING in SOQL?" and it could explain the syntax and give an example. If you have a specific scenario (like "find the max ClosedDate per Account"), ChatGPT can often produce the correct aggregate query.

As a best practice, use aggregate SOQL when it significantly simplifies your code or improves performance. If you find yourself writing nested loops to compute statistics, that's a good sign an aggregate query could replace that logic. That said, if you need very complex multi-level aggregations or more than 2,000 group results, you may need to use a different approach (such as breaking the problem into multiple queries or using an external analytics tool). Also, consider whether a **Roll-Up Summary** field could meet the requirement if it's a straightforward "per-parent" aggregation. Roll-up summaries (available in master-detail relationships) let Salesforce maintain counts or sums automatically, which can be leveraged in Apex without any query at all.

Architect Tip: Encourage the use of aggregate queries for reporting data that doesn't need to be real-time per record. They are great for generating summaries for dashboards or batch processes. Ensure that developers understand the governor limit of 2,000 aggregate results – if a use case might hit that, plan to partition the query or adjust the data model (maybe categorize data so that group counts are manageable).

Developer Tip: When writing aggregate queries, give aliases to your aggregate fields for clarity (e.g., SELECT AccountId, COUNT(Id) cnt...). This makes it easier to retrieve via ar.get('cnt') instead of expro. Always test the SOQL in Query Editor or with a small Apex script to make sure you understand the results structure before coding with AggregateResult. And include unit tests for your aggregate logic – for example, verify that a COUNT query returns the expected numbers for a given test dataset.

Tester Tip: Validate that the aggregate queries are correct by comparing their output to a known dataset or to manual Admin reports. For instance, if the code is supposed to sum up all Invoice Amounts by

Region, create sample records and manually calculate the expected sums to see if the Apex aggregate query returns matching results. Also test boundary conditions – what if there are no records for a certain group? Does the code handle a missing group gracefully? – since AggregateResult will simply omit groups that have no data.

Product Owner Tip: Be aware that some aggregate queries might not run instantaneously if they have to crunch a lot of data (especially without proper filters or indexes). When defining requirements for metrics or summaries, specify if real-time calculation is needed or if a nightly batch is acceptable. Often, heavy aggregation can be done asynchronously to offload processing. Work with the technical team to decide if a roll-up summary field could deliver the metric continuously or if a query is needed. This collaboration will ensure you get accurate results within acceptable performance.

Using aggregation and grouping in SOQL is like having a built-in reporting engine in your Apex code. By smartly applying these techniques, you can obtain insights and summaries with minimal code and without exhausting system resources. This concludes our exploration of efficient SOQL and database operations.

SUMMARY

In this chapter, we explored how to make Salesforce database operations as efficient and scalable as possible. We began with SOQL query **fundamentals**, emphasizing clear intent, selective field retrieval, and proper filtering to return only the data you need. We then examined **query performance and selectivity**, learning how indexed filters and selective queries keep things speedy and avoid the dreaded non-selective query errors. With that foundation, we moved on to **bulkification** – designing queries to handle many records at once (never in a loop!) so that our code works just as well for 200 records as for 2. We applied the same bulk mindset to relationships, using **parent-child SOQL queries** to fetch related data in a single swoop instead of making separate calls. Next, we discussed **dynamic SOQL**, which offers runtime flexibility for complex search scenarios, and we established best practices to use it safely (like escaping inputs and favoring static SOQL when feasible). We didn't forget DML: we covered **efficient DML operations**, demonstrating how batched inserts/updates and partial success options can make data modifications robust and limit-friendly. Finally, we unlocked **advanced SOQL techniques** with aggregations and grouping, showing how to get summary insights from the database without extra Apex processing.

A core theme across all these topics is that disciplined design – writing queries and DML logic with bulk operations, selectivity, and governor limits in mind – greatly benefits your Salesforce org in the long run. Applications built with these principles are more **scalable** (able to handle growing data volumes) and more **performant** (running within CPU and memory limits, providing fast responses). They also tend to be easier to maintain, because following best practices often leads to cleaner and more declarative code.

Throughout the chapter, we also saw how **ChatGPT** can serve as a valuable ally in implementing these best practices. From identifying unindexed filters in a SOQL query to suggesting how to bulkify a piece of Apex code, an AI assistant can accelerate the debugging and optimization process. It can explain error messages (like governor limit exceptions) and offer code snippets or query examples to solve the problem. By incorporating ChatGPT into your development workflow – as a code reviewer, generator, or even a tutor – you can reinforce the habits and techniques we discussed.

By now, you should have a firm grasp on writing Apex that interacts with the database efficiently. You know how to query selectively, process data in bulk, secure dynamic queries, handle transactions smartly, and leverage aggregate queries for analytics. In the next chapter, we will extend our focus beyond SOQL to Salesforce's search language, SOSL, exploring how to perform efficient text searches across objects while applying the same performance-first mindset. The journey to mastering Salesforce data operations continues there, building on the foundation we've established in this chapter.

5

EFFICIENT SOSL PRACTICES

Salesforce Object Search Language (SOSL) is a powerful tool in the Salesforce platform that allows you to perform text-based searches across multiple objects. Unlike SOQL, which retrieves records from one object at a time based on structured field conditions, **SOSL enables broad keyword searches** across many objects and fields in one go. This means you can find data by content – for example, searching all accounts, contacts, and leads for the word "Acme" – without knowing exactly which object or field it lives in. SOSL is what powers the global search box you use in Salesforce's UI, and it brings that flexible search capability into your Apex code.

In this chapter, we will learn how to use SOSL efficiently and effectively. We'll see typical use cases where SOSL excels: performing a global search in a custom application, implementing a single search feature that spans multiple Salesforce objects, or enabling end-users to explore data dynamically through keywords. SOSL shines in scenarios like a customer 360 lookup (where you enter a customer name and retrieve accounts, contacts, cases, and opportunities related to that name all at once) or any time you need to provide a "search bar" experience in your Salesforce solution. By understanding when to use SOSL vs. when a targeted SOQL might suffice, you can build features that are both powerful and performant.

Throughout this chapter, our goals will be to learn how to write **efficient and scalable SOSL queries**, understand how indexing and search algorithms affect performance, manage Salesforce governor limits related to SOSL, and avoid common pitfalls that can trip up developers. We'll cover how to construct SOSL queries that return the results you want without waste, how to limit and handle the results so you stay within limits and deliver a good user experience, and how to secure SOSL queries especially when constructing them dynamically. In short, by chapter's end you should be comfortable incorporating SOSL into your Apex arsenal with confidence in both its capabilities and its constraints.

We will also highlight how ChatGPT can assist you in mastering SOSL as you learn. ChatGPT can be a handy *pair programmer* for Salesforce development – you can ask it to help form a SOSL query, to explain why a certain search isn't working, or to suggest optimizations. For example, if you're unsure about your SOSL syntax, you might prompt ChatGPT and get immediate feedback or corrections. If you have a large result set, ChatGPT might suggest ways to narrow the search. We'll give prompt examples where ChatGPT provides query validation, optimization tips, and troubleshooting guidance for SOSL problems. Using such AI assistance can accelerate your learning and help you catch mistakes early.

This chapter's insights are valuable to every role on a Salesforce project team. **Architects** will gain patterns for designing search features that are both user-friendly and technically sound. **Developers** will get concrete best practices and examples for coding SOSL queries efficiently. **Testers** will learn what to watch out for when validating search functionality (like making sure security filtering is correct and limits aren't exceeded). **Product owners** and project sponsors will see how thoughtful use of SOSL can improve application usability and performance, justifying investment in these best practices. By the end, all stakeholders should appreciate how a well-implemented search can elevate the user experience of a Salesforce app and how to achieve that through best practices.

We've structured the chapter into seven sections, each building on the previous. We'll start with SOSL fundamentals and syntax, ensuring we have a solid foundation. Next, we'll explore multi-object searches to harness SOSL's cross-object powers. Then we'll dive into performance tuning and indexing, followed by strategies for managing results within limits. We'll discuss dynamic SOSL – constructing queries at runtime – and the security measures that go with that. After that, we'll look at integrating SOSL calls in Apex logic smoothly, and finally, we'll walk through common SOSL issues and how to troubleshoot them. By progressing through these sections, you'll gradually build a comprehensive understanding of SOSL from basics to advanced usage.

Let's jump in. We'll begin by reviewing the fundamentals of SOSL and its core syntax, to make sure we're speaking the same language (literally) and to set the stage for more advanced techniques.

SOSL FUNDAMENTALS AND CORE SYNTAX

To use SOSL effectively, we first need a solid grasp of its basic syntax and how it compares to the more familiar SOQL. At its core, a **SOSL query** uses a structure like:

FIND {search_query} [IN search_group]

RETURNING ObjectA(fields...) , ObjectB(fields...) ,... [LIMIT n]

This might look abstract, so let's break it down:

FIND {search_query}: This is the heart of the SOSL query. You put the word or phrase you're searching for inside curly braces. For example, FIND {Acme} will search for the term "Acme". SOSL is case-insensitive, so Acme and acme are treated the same. You can include multiple words (e.g. {John Doe}) to search for records containing both "John" and "Doe". If you want to match partial words, SOSL supports wildcards: an asterisk * matches zero or more characters and a question mark ? matches exactly one character. For instance, FIND {Mi*} could find "Mike" and "Michael", while FIND {Jo?n} might match "John" or "Joan". (Be careful: SOSL requires at least two characters in the search term, so a lone {A} will throw an error for being too short.)

IN search_group: This optional clause lets you narrow down which fields are searched. By default, SOSL searches all searchable text fields (IN ALL FIELDS). You can restrict the scope to make searches faster or more precise. For example, IN Name Fields means "only look at Name fields on records" (such as Account Name, Contact Name), whereas IN Email Fields would search only email address fields, and IN Phone Fields only phone number fields. Using a specific group focuses the search. If you know the user is searching for a name, searching just name fields is efficient. If you're doing a broad search where the keyword could be anywhere (notes, descriptions, etc.), then use ALL FIELDS. *(For instance, FIND {Bob} IN Name Fields will not match a record that has "Bob" only in an address field, but FIND {Bob} IN ALL FIELDS would.)* If in doubt about which to use, you could even ask ChatGPT *"What does IN ALL FIELDS vs IN NAME FIELDS do?"* and it would explain that one searches everywhere and the other limits to names.

RETURNING Object(...): The RETURNING clause specifies which objects to search and what data to return. You list one or more object types and the fields you want from each, in parentheses. For example: RETURNING Account(Id, Name), Contact(Id, FirstName, LastName, Email). This query would search Accounts and Contacts; for any matching Accounts it returns the Id and Name, and for Contacts it returns Id, FirstName, LastName, Email. You can list multiple objects separated by commas. This is where SOSL's multi-object capability shines: a single SOSL can search across objects that aren't directly related. (In SOQL you'd need separate queries for each object, since SOQL can only query one object type at a time.) If you omit the RETURNING clause entirely, SOSL will by default return the IDs of matching records (and their object types), but no other field data. In Apex, it's almost always better to specify the fields you need so you get useful data back.

WHERE filters: Within each object's parentheses, you can add an optional WHERE clause to filter the results of that object. This works like a post-filter. SOSL will find all records that match the search term, then apply your WHERE condition to those records to decide which ones to include. For example, FIND {Acme} RETURNING Opportunity(Id, Name WHERE StageName='Closed Won') will find all Opportunities with "Acme" in any searchable field, then return only those opportunities whose StageName is "Closed Won". It's important to know that the filter is applied *after* the text search, not before. In practice, that means the search might match more records behind the scenes, but you only see the ones that pass the filter. This is fine for most uses (like the example, which finds "Acme" opportunities but only shows won deals). Just remember that if your search term is very broad, Salesforce will only consider the first several hundred or thousand matches for filtering. We'll discuss SOSL limits and performance in a later section.

ORDER BY and LIMIT: You can also specify sorting and limits for each object's results. For example, you might use RETURNING Contact(Name, Email ORDER BY LastName DESC LIMIT 10) to get only the first 10 contacts found, sorted by name descending. If you're searching multiple objects, you can even put a limit per object (as in the example) to ensure one object doesn't dominate the results. There is also a global LIMIT you can put at the end of the query to cap the total number of records returned across all objects (the default maximum is 2,000 records total). In practice, using individual limits in each RETURNING clause gives you more control. We'll revisit result limits when we talk about managing SOSL output.

One fundamental difference from SOQL is how SOSL returns its results in Apex. A SOSL query returns a **list of lists** of sObjects. Each sub-list corresponds to one object type in the RETURNING clause (in the same order). For example, if you search Accounts and Contacts, the result might be a list with two sub-lists: the first is a list of Account records that matched, the second is a list of Contact records. If an object has no matches, its sub-list will just be empty. You'll typically need to cast these sub-lists to the appropriate types in Apex. For instance, List<List<SObject>> results = [FIND 'Acme' IN Name Fields RETURNING Account(Id,Name), Contact(Id,Name)]; List<Account> accts = (List<Account>) results[0]; List<Contact> cons = (List<Contact>) results[1];. This structure is unique to SOSL; by contrast, a SOQL query on a single object just returns a single list.

Quick comparison to SOQL: Both SOSL and SOQL are used to retrieve data in Apex, but they serve different needs. Use **SOQL** when you know exactly which object and fields you need (e.g. "find all Contacts where State = NY"). Use **SOSL** when you have a search term that could be in multiple objects or fields (e.g. "find all records with 'Salesforce' anywhere in them"). They do have some things in common: both support **bind variables** (you can use an Apex variable in place of a literal by prefixing with : in the query) and both enforce user permissions and sharing rules on the returned data. They also each have governor limits (e.g. you can do at most 20 SOSL queries in a transaction, similar to SOQL limits). We'll dive more into when to choose SOSL or SOQL in context as we proceed.

To reinforce these basics, here are a few example prompts you could try with ChatGPT:

"What's the difference between SOSL and SOQL?" – ChatGPT might explain that SOSL is for multi-object keyword searches (great for a global search bar) while SOQL is for structured queries on one object at a time, and give examples of each.

"Give me an example of a SOSL query to find 'Acme' in accounts and opportunities." – ChatGPT could produce something like FIND {Acme} IN ALL FIELDS RETURNING Account(Name), Opportunity(Name) and explain that it searches both objects for the keyword.

"What does using 'IN EMAIL FIELDS' do in a SOSL query?" – ChatGPT would clarify that it limits the search scope to email fields only (ignoring names, etc.), and might note you'd use it when you specifically want to find a keyword in email addresses.

Finally, a practical tip: always validate your SOSL queries. If you're a **developer**, you can use the Salesforce Developer Console's Query Editor to run SOSL and see if you get the expected results. You can also enlist ChatGPT as a second set of eyes – for example, ask *"Is this SOSL query syntax correct?"* before you run it, especially if you're trying something new with the format. If you're a **tester**, design test scenarios for both hits and misses: ensure that a search term that should find a record indeed finds it, and that an irrelevant term returns no results. Also verify edge cases (like very short search terms, or searches that should return a lot of data) to see that the system handles them gracefully. Catching an unexpected empty result or an overly broad result is easier early on, and tools like ChatGPT can help brainstorm such scenarios (e.g., *"Give me ideas for testing a search feature that finds contacts by name"*).

By mastering these fundamentals, you've built a foundation for all the SOSL techniques to come. Now that we know how to construct a SOSL query and retrieve its results, we can explore how to leverage SOSL's multi-object searching power even further. In the next section, we'll look at strategies for using SOSL to search across multiple objects efficiently and how to structure those searches for best performance and relevance.

LEVERAGING SOSL FOR EFFECTIVE MULTI-OBJECT SEARCHES

One of SOSL's greatest strengths is the ability to search across multiple objects at the same time. This "one query, many objects" capability is a major advantage in terms of speed and flexibility. Instead of writing separate queries for Accounts, Contacts, Cases, etc., and then merging the results, a single SOSL query can look through all of them in one go. Under the hood, the Salesforce search engine is optimized for this – it indexes text across objects and can retrieve matches very quickly. In practical terms, this means a well-crafted SOSL query can function like a mini "global search" for your application, returning a consolidated set of results from a variety of sources with minimal coding and minimal wait.

To illustrate, imagine a customer support scenario: a support agent enters a customer's name. With SOSL, a single query could return that customer's Contact record, any open Cases related to them, and even relevant Knowledge articles that mention their company – all at once. The agent gets a 360-degree view from one search action. Achieving the same with SOQL would require three separate queries (one for Contacts, one for Cases, one for Knowledge), and you'd have to manually combine the outputs. SOSL handles the multi-object retrieval and gives you structured lists for each object.

However, with great power comes great responsibility – just because you *can* search many objects at once doesn't mean you always *should*. Here are some best practices for structuring multi-object SOSL queries effectively:

Keep the scope focused: Include only the objects that make sense for the user's context. Every additional object in a SOSL query adds potential noise (irrelevant results) and could slightly increase search time. If users are primarily interested in Accounts and Contacts, there's no need to also search Opportunities and Cases in the same query (you could provide a separate search option for those, or include them only when appropriate). By narrowing the object list, you improve relevance and potentially performance. An architect should work with the business to decide, for each search feature, which objects are in scope. It can be tempting to "just search everything," but that often isn't the best experience for users or the system. Product owners might initially request an all-object search for completeness; architects can guide them by explaining the trade-offs and suggesting a curated set of objects that cover 90% of use cases with far less clutter.

Be aware of result distribution: Salesforce applies a limit of 2,000 records total per SOSL query (by default). When multiple objects are searched, that limit is essentially shared among them. In a multi-object search, the platform will allocate slices of the 2,000 record pool to each object (not necessarily evenly, but enough to give each object some representation). For example, if you search five objects, you might get at most a few hundred from each, even if one of those objects had 1,000 matches by itself. On the other hand, if you limit the search to just one object, that object alone can return the full 2,000 records. This means that extremely broad searches (many objects, very common term) could potentially omit some matches from an object simply because the quota for that object was filled. The practical take-away: don't overload a single SOSL query with too many objects, especially if some of those objects have high chances of large result sets. If you truly need to search a very large set of objects, consider whether the search should be split into two different queries or user-initiated options. For instance, you might let users toggle between searching "Sales data" (Accounts, Opportunities, Leads) and "Support data" (Cases, Knowledge, Contacts) separately, rather than one giant search over all six objects. This way, each query has a narrower focus and is less likely to hit distribution limits.

Separate concerns when necessary: If two objects have very different purposes or audiences, it might be better to search them separately. For example, you might have a unified search for customers (which looks at accounts, contacts, leads) and a separate search feature for internal records (like an employee directory or internal project objects). Combining too many disparate areas into one query could confuse users

("Why did my customer search return internal project records?") and make the results harder to parse. From a developer standpoint, maintaining several focused SOSL queries can be easier than one mega-query that tries to do everything. It also allows you to handle the results differently — maybe you display account/contact results in one part of the UI and cases in another, rather than one long mixed list.

When planning these multi-object search patterns, **architects** should also consider the future. Ask questions like: "If we add a new object type next year (say, a custom object for Contracts), should it be part of this search?" Designing with configurability in mind can save time later. One strategy is to store the list of objects to search in a custom setting or configuration, so that adding or removing an object doesn't require code changes. For example, you could have a config that says "Global Search Objects = Account,Contact,Case,Knowledge" and your code builds the SOSL query dynamically from that. Then an update to include "Contract__c" is just a config change. ChatGPT can help here by generating dynamic Apex code snippets if you describe this requirement. You might prompt: *"How can I dynamically build a SOSL query from a list of object API names in Apex?"* and use the suggestions to implement a flexible design.

ChatGPT can be a useful assistant when refining multi-object searches. Here are a few ways you might use it:

Validating object choices: "We plan to have a search feature for support reps that searches Cases, Contacts, and Accounts. Should we include the Knowledge object as well?" ChatGPT might reason about the scenario and suggest whether adding Knowledge articles is beneficial or if it could overwhelm the results, helping you make a decision.

Optimizing query structure: *"My SOSL query searches 8 objects and seems slow – how can I improve it?"* ChatGPT might recommend splitting the query, or confirm whether performance issues are likely due to too many objects vs. other factors (perhaps it will remind you of the 2,000 record distribution, and suggest narrowing the search criteria).

Generating the query syntax: If you give ChatGPT a simple description like "search accounts (name, industry) and contacts (name, email) for 'Acme'", it can draft the SOSL query for you. This is handy to double-check that you haven't made a syntax mistake when listing many objects and fields. An architect or developer could even prepare a list of objects and their key fields in a text file and upload it to ChatGPT, asking it to produce a well-formed SOSL query. *(If you do upload any file, remember to omit sensitive data and follow your organization's privacy guidelines.)*

Let's not forget governance: Salesforce limits the number of SOSL queries (max 20 per transaction), so you generally want to use one SOSL to cover multiple objects rather than firing many separate SOSLs. In other words, combine objects into one query when it makes sense to do so, but don't combine just for the sake of it if the use case doesn't require it. It's a balance between breadth and specificity.

Architect Tip: When designing multi-object search, consider the user interface as well. How will results be displayed? Often, it's wise to group results by object (e.g., show contacts in one section, cases in another) so that users can quickly scan what they found. If your SOSL query returns multiple object types intermixed, plan to separate them in the UI. This not only improves user experience but also helps you internally to manage the results (since you know which list corresponds to which object). ChatGPT can even assist in mocking up how a multi-object search output might look — you could ask, for example, *"What's a good way to present search results from accounts, contacts, and cases on one page?"* and use those ideas in your design discussions.

In summary, multi-object searches are a powerful feature of SOSL that can significantly streamline user workflows. By thoughtfully selecting which objects to include, being mindful of the underlying limits, and structuring queries in a maintainable way, you ensure that this power is used effectively. The ability to search across entities simultaneously can feel almost magical to users, but it works best when it's scoped and tuned to the task at hand. With these practices, you can leverage SOSL to deliver fast, relevant results from across your Salesforce org, all in one convenient search.

As we continue, we'll shift focus from *what* we're searching to *how* to make those searches as efficient as possible. The next section dives into performance optimization and indexing – understanding how Salesforce indexes your data for SOSL and how to write queries that make the most of those indexes for speed.

OPTIMIZING SOSL QUERY PERFORMANCE AND INDEXING

To get the best performance from SOSL, it helps to understand how Salesforce indexes data for search and how to craft queries that work with those indexes. The reason SOSL is so fast is that it doesn't scan database records directly at query time – instead, it searches through a pre-built **search index** of terms. Salesforce automatically indexes most text-based fields in each object for SOSL use. This includes standard fields like Name, Email, Phone, and the contents of Text or Text Area fields (up to a certain size), as well as custom text fields. Because SOSL leverages these indexes, it can swiftly sift through large volumes of data and return matches without performing full table scans. In fact, SOSL uses indexed searches under the hood, making it much faster than SOQL when dealing with big data sets.

What is indexed (and what isn't): As a rule of thumb, any field you would consider a "text" or "string" field is likely searchable by SOSL. This includes names, addresses, phone numbers, picklist values (treated as text), and even long text areas (though very long fields may be truncated in the index). On the other hand, fields like numbers, dates, and checkbox values are not directly searchable using SOSL. You also can't use SOSL to look for values in formula fields or roll-up summary fields if those formulas result in non-text values, because those aren't indexed in the full-text search engine. (If a formula returns a text string, it might be searchable, but it's not guaranteed unless that formula's result is stored in a text field.) This means that if you have a specific data point you need to search that isn't a text field (say, an order number or a birthdate), you might need to either convert it into a text representation for search purposes or use a SOQL query instead.

Salesforce's indexing strategy for SOSL is to tokenize text into searchable terms. It will break up strings by spaces, punctuation, etc., and index those pieces (including variants, like lowercasing words, maybe removing plural suffixes, etc.). For example, a field value "Acme Corporation" would be indexed as "Acme" and "Corporation" as separate tokens. If a user searches for "Acme", that record will be found based on the first token. If they search for "Corp", they might not find it unless the query uses a wildcard (since "Corporation" doesn't start with "Corp"). Understanding this helps in crafting efficient queries:

Use Selective Search Terms: The more specific your search terms, the fewer records Salesforce has to consider, and the faster your query will run. For instance, searching for FIND {John Smith} is going to be more selective than FIND {John} alone. With John Smith, the search engine will look for records containing both "John" and "Smith", drastically reducing the result set and work to do. If you only search John, especially across common fields like Name, you might hit many records, possibly thousands. That isn't inherently "slow" (because of indexing), but it could approach the 2,000 record output limit and result in a lot of data transfer. In extreme cases, very broad searches can lead to what Salesforce calls "search crowding" – relevant results beyond 2,000 get cut off because too many things matched. To optimize, encourage users (or design your application) to include multiple keywords or more distinctive terms when possible. You can also leverage SOSL's support for logical operators: for example, FIND {John AND Smith AND California} will only return items where all three terms appear, which is far more targeted than just John. ChatGPT can help brainstorm or refine search phrases. If you told ChatGPT, *"My SOSL for 'John' is too slow/returns too many hits"*, it might suggest adding another term (like a last name or filter by some field using a WHERE clause) to narrow it down.

Leverage Search Groups and Where Clauses: This ties in with content from earlier sections, but it's worth repeating from a performance angle. By using IN Name FIELDS or IN Email FIELDS etc., you limit the scope of the index that SOSL has to scan. Searching "all fields" is convenient but means the engine has to consider every indexed term in every field. If you know the user is searching for a name, restricting to Name fields not only increases relevance of results, it also cuts out a lot of unnecessary index lookup, making the search slightly faster. Similarly, adding a WHERE clause (post-filter) to your SOSL can reduce the volume of data returned. For example, FIND {Acme} RETURNING Opportunity(Name WHERE IsClosed=false) will search all opportunities for "Acme" but then only return the open ones. Even though the filtering happens after finding matches, it means less data sent back to your code, and potentially less post-processing. The takeaway: narrow the search target as much as you reasonably can; the system will appreciate it.

Be smart with wildcards: Wildcards can be a double-edged sword for search performance. SOSL allows wildcards at the end or middle of search terms (but not at the very beginning of a term). For example, FIND {Joh*} will match "John", "Johnson", "Johansson", etc.. This is useful if users might input partial names. Using a wildcard at the end of a term (suffix wildcard) still leverages the index effectively, as the search engine can jump to the "Joh" section of the index and then scan forward. However, using a wildcard at the beginning of a term (prefix wildcard, like *son) is not allowed in SOSL and would force a full scan if it were allowed. Similarly, putting wildcards on both ends of a very short term (like *oh*) isn't meaningful in SOSL (you'd just use a single term without wildcards in that case). The guidance is: use wildcards sparingly and only when needed to catch variations. If you can avoid a wildcard by searching the exact token, do so. For instance, if you know your data always starts with certain prefixes (account numbers like "ACC-1001"), searching for ACC-100* is fine, but searching for 100* alone might not match anything unless those digits appear as a separate token in some field. Often, users don't need wildcards at all if they provide distinct keywords; the search engine automatically finds tokens that *start* with the given term. (Searching for "John" will match "Johnathan" because the token "Johnathan" starts with "John". You'd only need a wildcard if you wanted to match something like "AJones" by searching "*Jones", which SOSL won't do.)

Understand indexing delays: Salesforce's search index is updated on a schedule – it's near real-time, but not instantaneous. This means there's a slight delay between a record being created or updated and that record being searchable via SOSL. In a typical user scenario, this delay might be a few seconds and not noticeable. But in an automated scenario (or a unit test), if you insert a record and immediately perform a SOSL query, you might not find the record. From a performance perspective, this isn't about query speed but about query completeness. For developers and testers, it's important to know that SOSL might "miss" very new data. The index update is an asynchronous process. So, if a search immediately after an insert doesn't find something, perhaps retry after a short wait or fall back to a targeted SOQL query if absolutely necessary. ChatGPT can help explain odd cases like this – for example, if a tester asks *"Why does my SOSL query not find a record that I just inserted?"*, the explanation about indexing delays would come up, and they could plan around it.

Monitor and tune: As with any query, you should monitor the execution time of your SOSL searches, especially those that run in bulk or are critical to user workflow. Use debug logs or the Salesforce Developer Console to see how long SOSL queries take and how many records they return. If you consistently hit the 2,000 record limit or see searches taking longer than a second or two, consider it a sign to refine the query. Perhaps add another search term, or break one very broad search into a couple of narrower ones as discussed earlier. Also, ensure that the fields you expect to be searchable are indeed populated. A search on a field that is often blank won't perform well not because of speed, but because of lack of results – users might then repeatedly try broad queries to find what they need. From an architect perspective, if you know a certain field will be heavily searched (e.g., a SKU or product code), make sure it's a text field and not, say, buried in an attachment or a child record. You might even denormalize or copy data into a searchable field to facilitate fast searches. A little upfront schema planning can go a long way for search performance.

ChatGPT prompt examples (performance and indexing):

"Our SOSL search for 'Jones' in Contacts is returning too many results and timing out. How can we make it more selective?" – ChatGPT might suggest searching full names (first and last together) or adding an additional term like city or company to narrow down "Jones" results.

"Are picklist fields included in SOSL searches?" – ChatGPT could confirm that yes, picklist values are treated as text and indexed (so searching for a picklist value is possible via SOSL).

*"Does using * in a SOSL search slow it down?"* – Expect an explanation: a wildcard at the end of a term (like "Jen*") is fine and still uses the index, but a wildcard at the beginning isn't allowed and would be bad for performance if it were. Essentially, ChatGPT would advise to avoid unnecessary wildcards.

"Why can't my test see SOSL results for data I just created?" – ChatGPT would explain the indexing delay and the nature of SOSL in tests, perhaps suggesting using SeeAllData or adjusting the test approach.

In summary, optimizing SOSL performance is mostly about guiding the search engine to do less work: use specific terms, limit the scope when feasible, and avoid patterns that prevent index use (like leading wildcards or extremely generic queries). The good news is that out-of-the-box, SOSL is already quite fast thanks to Salesforce's indexing – you often won't need to micro-optimize. Just keep an eye on your queries in a production-like environment; if a particular search is a hotspot, apply the principles above to keep it snappy. By understanding how indexing works, you as a **developer or architect** can structure both your data and your queries to play to SOSL's strengths, ensuring that even as your org's data grows, your search features remain responsive and reliable.

We've now looked at how to make our SOSL searches functionally and performance-wise efficient. Next, we'll turn to managing the results that come back from these searches. In the following section, we'll discuss strategies for handling SOSL results, limiting them, and working within Salesforce's governor limits to avoid running into issues when searches yield a lot of data.

SOSL RESULT MANAGEMENT AND LIMITING

Efficiently managing SOSL results is as important as crafting the query itself. Even a well-optimized search can return a large number of records, and you need to handle that data without hitting Salesforce's limits or overwhelming your users. Salesforce imposes some constraints on SOSL results by default: a SOSL query will return at most 2,000 records total, no matter how many more might match. Additionally, each SOSL query counts toward governor limits (you can do at most 20 SOSL queries per transaction). In this section, we'll look at best practices to keep within these bounds and to make sure the results you do get are manageable.

Use LIMIT clauses to cap results: The simplest way to prevent retrieving too many records is to use the LIMIT keyword in your SOSL query. You can apply LIMIT to each object's results in the RETURNING clause. For example, FIND {Acme} RETURNING Account(Name **LIMIT 100**), Contact(FirstName, LastName **LIMIT 50**) will return at most 100 accounts and 50 contacts matching "Acme". This is a smart practice when you expect that a search term could match a lot of records. By putting an upper bound, you ensure that even if thousands of items technically match, you're only pulling a reasonable subset into memory. Often, the user is interested in the "top" matches (which Salesforce determines by relevance) rather than every single match. In fact, the platform by default is already doing something similar by cutting off at 2,000, but you might not even need that many. Ask yourself: how many results would a user realistically sift through? If the answer is, say, 100, then limit it to 100. This reduces data transfer and processing. Using LIMIT per object also guarantees a bit of balance; it prevents one object with tons of matches from crowding out others. Without explicit limits, SOSL might return, for instance, 1,990 Contact records and only 10 Accounts, even if there were more Account matches, simply because Contacts had more hits up front. By limiting to 100 each, you ensure you get a mix of both.

Implement sensible pagination or user prompts: Out-of-the-box, Apex SOSL doesn't provide a way to get "records 2001-4000" – once you hit 2,000, that's the cap for that query. If your use case truly requires navigating through more than 2,000 results, you'll need to design for it. One approach is to guide the user to refine their search (which is usually the better UX anyway). For example, if a search term is very broad, you might display the first X results and then a message like "Please refine your search for more specific results." Another approach is to perform multiple SOSL queries with additional filtering to partition the result space. This is an advanced technique and can get complicated. For instance, you might split by alphabet: search A-M and then N-Z separately if you were searching names – but this requires knowing something about the data distribution and still may not guarantee no overlap or gaps. In general, if you find yourself needing to page through huge result sets in SOSL, it might be a sign that the search criteria should be tightened. In contrast to SOQL (which has cursors and queryMore for large results in some contexts), SOSL's philosophy is to return the most relevant hits quickly, not every possible match. So, design your UI and logic with that in mind. If a user truly needs to see

5,000 items, perhaps a report or SOQL query is more appropriate than a free-text search.

Beware of post-processing limits: Getting SOSL results is just step one – what your code does with those results can also run into limits. For example, if you loop through 1,000 results and perform a DML update on each, you could hit DML governor limits or CPU time limits. Always try to bulk-handle the results. If you need to, say, send an email to each result record, collect their IDs and do it in a single email invocation or batch the operation. Or if you must loop, ensure you aren't doing something inefficient inside that loop (like running additional queries – the classic mistake would be doing a SOQL or another SOSL for each SOSL result, which could easily breach the 20 or 100 query limit). Since SOSL already returns a list of sObjects, you can often feed that directly into another operation (for example, maybe you got a bunch of Accounts and now you want to update a field on all of them – you can just do update accountsList without further querying). A developer should pay attention to what happens when the search returns a lot of data: run a test with a simulated large result set and ensure no limits are hit in processing. This is where a tester can also contribute by writing test methods or using automation to simulate heavy search usage and confirming the system handles it gracefully (for example, it doesn't time out or throw a too many DML statements error).

Handle no-results and full-results scenarios: Good result management also means dealing with edge cases. If SOSL returns zero records, your code should handle that elegantly (maybe show "No results found" on the UI, and certainly avoid null-pointer exceptions by checking that each sublist exists and has content before accessing it). If SOSL returns the maximum 2,000 records, be mindful that there might have been more matches that were not returned. In some situations, you might want to inform the user like "Showing 2,000 of many results, please refine your search." This is not something the platform does for you – you have to decide if it's relevant to your users. For internal admin tools, maybe it's fine; for customer-facing search, it might improve the experience to let them know there could be more. From a testing perspective, testers should validate both ends of the spectrum: a query that returns nothing (to ensure the system handles the empty state), and a query that returns a very large set (to see that performance is acceptable and any truncation messaging or limits are working as expected).

(At this point, recall that SOSL automatically sorts results by relevance. You cannot directly control ordering in SOSL beyond that, aside from grouping by object in the RETURNING clause or limiting per object. If needed, you can always sort the combined results in Apex after retrieval or present them grouped by object to avoid confusion. Typically, though, trusting SOSL's relevance sorting is fine – it usually brings the most likely matches to the top.)

With these practices, you can manage SOSL output cleanly. The user sees a concise, relevant list of results, and your code stays well within platform limits. Next, we'll tackle dynamic SOSL – how to construct SOSL queries on the fly – and the security considerations to keep in mind when you do so.

DYNAMIC SOSL – TECHNIQUES AND SECURITY PRACTICES

In previous sections, we used static SOSL queries – the search terms and fields were hardcoded in our Apex. But there are scenarios where you might need to construct a SOSL query dynamically at runtime. For example, you might take a keyword input from a user and plug it into a SOSL search, or you might build the RETURNING clause based on user-selected objects. Dynamic SOSL gives your application flexibility (users can search for what they want), but it also introduces some risks and challenges. In this section, we'll outline how to implement dynamic SOSL safely and efficiently, with a strong emphasis on security to prevent SOSL injection.

When and why use dynamic SOSL: Suppose you're building a global search bar in a custom app – you won't know the search term ahead of time; it comes from the user at runtime. Or imagine an admin tool where the administrator can choose which object to search and what fields to display; your code might need to construct a SOSL query string on the fly based on those choices. These are cases for dynamic SOSL. Essentially, whenever the search criteria aren't fixed in your code, you are dealing with dynamic SOSL. The value is clear: more flexibility and a more interactive experience. A **product owner** might love the idea that a user can type any keyword and search across multiple objects freely. However, with that freedom comes the responsibility to handle arbitrary input correctly.

Security risk – SOSL Injection: Anytime you take user input and inject it into a query, there's a risk of what's known as injection. In the context of SOSL, "SOSL injection" refers to malicious or accidental input that breaks the query syntax or alters the query logic in unintended ways. For instance, if your code naively does String userInput = ApexPages.currentPage().getParameters().get('q'); query = 'FIND {' + userInput + '} IN ALL FIELDS RETURNING Account(Name)';, what if the user input is '} RETURNING Contact(Name) OR Account.Name LIKE '%' or some such attempt to manipulate the query? They could prematurely close the braces or add logical operators to the search string. In a worst-case scenario, they might try to retrieve data they shouldn't by altering the query (though Salesforce's security like field-level security and sharing still apply to results, so they can't access records they have no rights to – but they could, for example, broaden a search that was meant to be restricted). At the very least, a malformed input could break your code by causing a query parse error.

To prevent SOSL injection, always **sanitize and bind user inputs**. The best practice is to use bind variables in SOSL rather than string concatenation. In Apex, you can do this by writing your SOSL like a formula: FIND :searchTerm IN ALL FIELDS RETURNING.... Here, :searchTerm is an Apex variable that holds the user's input. When you use this syntax, the platform handles escaping special characters and quoting for you. It's similar to using prepared statements in other languages – you aren't directly splicing the raw text into the query, so it can't bust out of the query format. If for some reason you cannot use a bind (binds are allowed for the search term and for values in the WHERE clause of a SOSL, but you can't bind, say, an object name or field list), then you need to carefully escape the input. At minimum, call the String.escapeSingleQuotes() method on any user-provided string that you include in a SOSL query. This will add a backslash before any single quote in the input, preventing a quote from accidentally terminating a string literal in your SOSL. It's also wise to validate the characters of the input – for example, you might reject or strip out curly braces {} since those have special meaning in SOSL syntax, or logical operators like AND, OR to prevent the user from injecting their own conditions. In many applications, users only need to input alphanumeric characters and maybe spaces for a search term. You can enforce that with a simple regex check. By constraining what the user can do, you both improve security and ensure the query won't blow up with weird input.

Techniques for dynamic SOSL construction: Aside from handling the search text safely, dynamic SOSL might involve varying which objects or fields to search. Let's say an admin can select which objects to include (Accounts, Contacts, or both). One approach is to write multiple SOSL queries and choose one based on input (not very flexible, but straightforward). A more dynamic approach is to build the RETURNING clause as a string. For example:

List<String> selectedObjects = getSelectedObjectsFromUI(); // e.g. ["Account","Contact"]

String returningClause = '';

if(selectedObjects.contains('Account')) returningClause += 'Account(Name, Industry), ';

if(selectedObjects.contains('Contact')) returningClause += 'Contact(FirstName, LastName), ';

// remove trailing comma

if(returningClause.endsWith(', ')) returningClause = returningClause.removeEnd(', ');

String soslQuery = 'FIND {:searchText} IN ALL FIELDS RETURNING ' + returningClause;

List<List<SObject>> results = Database.search(soslQuery);

In this pseudo-code, we dynamically built the RETURNING clause and then used Database.search() (which is equivalent to Search.query()) to execute the SOSL, binding the search text safely (note the {:searchText} placeholder for bind variable – in a dynamic string, one way is to use

Database.search with a binding context or you'd have to escape the search text manually and insert it).

Important: If you go this route, ensure that the object names and field names you concatenate are not coming from user input directly, but from controlled logic or options. In the above example, selectedObjects might come from a multi-select picklist on a Visualforce page – the user isn't typing "Account" or "Contact", they're selecting from predefined options that we trust. This avoids a scenario where a user could input an arbitrary object API name (which could lead to errors or attempts to search objects they shouldn't). Always whitelist dynamic parts of your SOSL. For instance, have a set of allowed object names or field names and confirm that any "dynamic" value is in that set. If not, you default or reject. This is basic defensive coding that prevents misuse.

Testing dynamic SOSL: As a **tester**, pay special attention to how the system behaves with unusual inputs. Try inputs with quotes ('), braces ({ }), backslashes (\), etc., and confirm that the SOSL either still works (because the characters were escaped) or that the input is rejected with a user-friendly error. Also test some attempted injections: for example, type in a search term like '} OR name != '' and see if the code handles it (it should ideally either find literal records matching that strange string or more likely just return no results – but it should not error out or expose data). If the application provides an advanced search feature where users can add filters (like a dropdown to choose a field and a textbox for a value), ensure that mechanism is implemented with proper binding as well (likely it's using SOQL in that case, but same principle – use bind variables and input validation techniques).

ChatGPT can assist in security reviews: If you have a piece of code and you're not sure if it's safe, you can ask ChatGPT to analyze it. For example: *"Here's my Apex code that builds a SOSL query: <paste code>. Do you see any security issues?"* ChatGPT might point out that you are concatenating user input without escaping and remind you to use a bind or escape method. You could also ask *"How do I prevent SOSL injection in this scenario?"* and it would likely advise the use of escapeSingleQuotes and other input validation techniques. Another helpful use is generating test cases: *"Give me examples of malicious input for a search box to test for injec-*

tion vulnerabilities." You might get a list of patterns to try in your testing, which you can then use to ensure your code is robust.

In summary, dynamic SOSL is a powerful feature that, when used carefully, can greatly enhance the flexibility of your Salesforce apps. The key is to **never trust raw user input**: always sanitize and/or bind it so that it cannot alter the structure of your query. Treat dynamic queries with the same caution you would treat constructing a SQL query in a traditional app – the concepts of injection and escaping are very similar. By following Salesforce's best practices (use bind variables whenever possible, or escape user strings), you can allow users to search freely while keeping your application secure and stable. As an **architect or lead developer**, establish guidelines in your team for any code that constructs SOQL or SOSL from user input, and consider using automated code scanning tools (Salesforce's security scanner or PMD rules) which can catch obvious injection risks. With these precautions in place, you can confidently implement dynamic search features that are both powerful and safe.

Now that we've covered dynamic query construction and security, our next step is to see how SOSL fits into the broader Apex logic of an application. In the following section, we'll examine patterns for integrating SOSL queries within Apex code flows, including error handling and ensuring a good user experience once the search results are obtained.

INTEGRATING SOSL WITH APEX LOGIC

Writing an efficient SOSL query is one thing; using it effectively within your Apex code and application flow is another. SOSL doesn't live in isolation – you'll use it inside Apex classes, triggers, Lightning controllers, or batch processes. In this section, we'll discuss how to integrate SOSL into your codebase cleanly, manage exceptions properly, and ensure users get a smooth experience when performing searches.

Calling SOSL from Apex methods: Typically, you'll call SOSL in response to some event – for example, a user clicks a "Search" button (Apex controller calls a method that executes SOSL) or an Apex method is invoked via REST API to perform a search. Make sure to keep the SOSL call itself outside of any loops (as with SOQL, we want to avoid running multiple queries in a loop). Often, you will gather any inputs you need, then have a single line like List<List<SObject>> results = [FIND :searchText IN Name Fields RETURNING...];. After that, you usually process the results (maybe format them for display, or take some action). If multiple parts of your code need to search for the same thing, consider wrapping SOSL calls in a utility or service class. For instance, you might have a SearchService.searchContactsAndLeads(String name) that internally does the SOSL and returns a structured result. This promotes reuse and centralizes any special handling (like input sanitization or logging of search queries).

Error handling and SearchException: In Apex, if something goes wrong with a SOSL query at runtime, Salesforce will throw a SearchException (a system exception). The most common cause is a bad search syntax or violating a rule (for example, the infamous "search term must be longer than one character" error). This happens if you accidentally try to search for a single character or if the search string after cleaning becomes empty. To handle this, you should always validate the search term before calling SOSL. A simple check: ensure searchText.length() >= 2 (after trimming spaces). If not, you can either skip running the SOSL and perhaps inform the user ("Please enter at least 2 characters to search.") or you can catch the exception. Wrapping your SOSL call in a try-catch for SearchException is a good safeguard:

```
List<List<SObject>> results;

try {

        results = [FIND :searchText IN ALL FIELDS RETURNING
Account(Name)];

} catch (SearchException e) {

  results = new List<List<SObject>>();

  // Maybe log the error or set a message indicating a bad search input.

}
```

However, proactively validating input (and possibly adjusting it) is better than relying on exceptions. For example, if a user enters "A" or even something like "??", you know SOSL will complain. You can catch that before calling SOSL. Perhaps append a wildcard in such cases or simply block it. Some implementations automatically append * to single-character inputs (turning "A" into "A*"), which can then search for all terms starting with "A". But use that strategy with caution; it can lead to very broad searches. It might be better to enforce a minimum length or suggest the user enter more.

Integrating into UI logic: When a SOSL query is triggered by user action, think about the user experience. If the search might take a moment (say it spans many objects or large data), provide some loading indication on the UI. Thankfully, SOSL is usually very quick due to indexing, but in a complex org it could take a bit. In a Lightning Component or LWC, you'd call an @AuraEnabled or @Invocable Apex that runs the SOSL and returns results. Ensure that method is optimized (as we've covered) and doesn't do unnecessary extra work. Once the results are returned, format them for display. Often you'll loop through the sublists and perhaps create a unified list of some custom wrapper class instances (e.g., a wrapper that has fields like recordName, recordType, recordLink, so you can show a mixed list of accounts, contacts, etc.). This is fine – just be mindful of not hitting view state limits in Visualforce or similar issues when returning a lot of data to the front end. Sometimes, if results are large, you might truncate them before sending to the UI for performance. For example, maybe only show the first 200 records to the user even if 500 matched, to keep the page snappy.

Using SOSL in triggers or backend logic: Most often, SOSL is used for user-driven searches, but there are use cases for using SOSL behind the scenes. For instance, a trigger after insert could use SOSL to find related records by a keyword (maybe to link a new case to existing knowledge articles by a keyword in the case description). If you do this, bulkify it – i.e., if multiple records fire the trigger, perhaps combine their keywords into one search if possible (though SOSL doesn't accept multiple separate search terms unless you combine them with OR in the same FIND). Alternatively, limit such logic to contexts where it's safe. You wouldn't want to run a heavy SOSL for every record in a 200-record batch. One approach could be to put SOSL logic into a @future method or Queueable if it's not time-sensitive, so that it runs asynchronously and outside of the main transaction (thus not adding to the trigger's governor limits beyond the async invocation). As an **architect**, question whether a SOSL in a trigger is truly needed or if a SOQL or different approach would suffice – SOSL is great, but using it in automation requires careful thought about execution limits and consistency (remember the slight indexing delay; a trigger that does a SOSL on the record that just triggered might not find it because it's not indexed yet, for example).

Ensuring a good user experience: From a user's perspective, a search function should feel responsive and reliable. That means:

- Validating input quickly (maybe even on the client side) to avoid obvious errors (like the one-character issue).
- Handling the "no results" case gracefully (display a friendly message, maybe suggestions like "Check spelling or try a different term").
- If partial results are shown (like we only show top 100), make that clear to manage expectations.
- Possibly offering a follow-up action if needed, such as "Too many results? Try adding more keywords."

ChatGPT for integration guidance: You can use ChatGPT to review your integration approach. For example: *"I plan to call a SOSL in a trigger to find similar records. Is this a good idea?"* ChatGPT might remind you about bulk triggers and indexing delays, suggesting perhaps using after-commit logic. Or *"How can I combine results from SOSL with results from another query?"* – maybe you have to merge SOSL results with SOQL results (ChatGPT could propose how to unify or iterate them properly). If you encounter an error like SearchException and aren't sure why, you could paste the error and your code to ChatGPT: *"Why am I getting System.SearchException: search term must be longer than one character?"*, and it would explain the cause (likely an empty or too-short term) and how to fix it. This can save time debugging.

Testing the integration: We mentioned testers verifying certain scenarios in previous sections. When SOSL is part of a larger process, test the whole flow. For example, if it's in a trigger, create records in a test that should invoke it and assert the expected outcomes (maybe the trigger populates a field or creates a link if a match is found). Use the

Test.setFixedSearchResults in test methods to simulate SOSL returning specific records without relying on actual data – this Salesforce feature lets you set what a SOSL query will return (by ID) so you can test logic deterministically. Also test failure modes: simulate a SearchException by calling the search with something like just "{ }" (which will trigger an exception) inside a System.runAs that doesn't have try-catch, to see if your code properly handles it by catch or validation.

By thoughtfully integrating SOSL into your Apex logic, you make sure that all the best practices in writing the query translate into real benefits in your application. The goal is to use SOSL as a seamless part of your app's functionality – users shouldn't have to think about the underlying query language at all; they just experience a fast and helpful search feature. Meanwhile, you as a developer make sure it runs within limits and doesn't break under odd conditions. With these integration patterns in mind, you can confidently include powerful search capabilities in your Salesforce solutions.

As we near the end of our deep dive into SOSL, we've covered query writing, performance, results, security, and integration. But no journey is complete without addressing the bumps along the road. In the final section of this chapter, we will tackle common SOSL issues and troubleshooting techniques, so you know what to do when SOSL doesn't behave as expected.

TROUBLESHOOTING COMMON SOSL ISSUES

Even with best practices in place, you may encounter situations where SOSL doesn't behave as expected. Perhaps a search isn't returning a record you think it should, or you run into an error or limit. This section covers common SOSL-related issues and how to diagnose and fix them.

Issue 1: Expected records not found – One frequent puzzle is when you perform a SOSL query and certain records that you *know* contain the search term aren't appearing. There are a few reasons this can happen:

Field not searchable: Not every field is searchable by SOSL. For example, formula fields that output text are generally not searchable unless the output is stored in an indexed field. Also, an administrator can mark custom fields as "Not Searchable" (via field settings) – though this is uncommon, it can happen for large text fields to save indexing space. If you're missing records, first verify that the field where the term resides is indeed among those SOSL considers. You might try a broader search (e.g. IN ALL FIELDS) to see if the record comes up; if it doesn't, the field might not be indexed for search. Check Salesforce documentation or object settings for whether a particular field is searchable. Standard fields like Name, Email, Phone, etc., are searchable by default. But if your term is hiding in a long Description or a Notes field, there could be nuances (for instance, some very large text areas might only index the first X thousand characters).

Recently added or updated records: If a record was created or modified just moments ago, the SOSL index might not have caught up yet. Salesforce's search indexing is fast but not instantaneous. This is noticeable in tests or in rapid data entry scenarios. If you suspect this, wait a bit or rerun the search after a minute. For automated processes, design them to be tolerant – e.g., if you insert a record and then immediately try a SOSL search for it, consider adding a slight delay or using a SOQL query as a backup if immediate consistency is crucial.

Term mismatch or stemming: SOSL does some normalization (like lowercasing terms). But if your search term is a partial word that doesn't match from the beginning of a token, it might miss. For example, searching for "corp" might not find "Acme Corporation" because "Corporation" doesn't start with "corp" (unless you use a wildcard). Or if there's a typo or different spelling, SOSL won't magically correct it. If you suspect a term mismatch, try using wildcards or different forms of the word to see if the record appears. Also be aware that certain very common words might be treated as "stop words" and ignored. For instance, searching for "the" might return nothing because "the" is too generic. If a user insists on searching such terms, you'll need to explain the limitation or implement a different strategy.

User permissions and sharing: Remember that SOSL respects object and field level security and sharing rules. If you run a SOSL as a user who doesn't have access to certain records, those records won't appear in results (they're filtered out after the search). This might cause confusion if an admin tests a search and sees a record, but an end user doesn't. The solution is to verify the permissions: does the user have at least Read on that object and field? If not, that's by design. The fix could be adjusting sharing or instructing the user.

Issue 2: Too many results (search crowding) – On the flip side, you might have a search that returns so many results that it hits limits or seems to drop data. As we covered, SOSL will only ever return 2,000 records maximum. If your query logically matches more, Salesforce will truncate the results to 2,000 (or to your specified LIMIT). In such cases, you might observe that the results feel incomplete. This is sometimes called "search crowding" – not all relevant records are returned because the limit was reached. The fix here is not technical (you can't lift the 2,000 cap), but rather to refine the search. Encourage or enforce more specific search input (multiple terms, etc.) or split the query by some criteria. Also consider using the per-object LIMIT trick: for example, if one object is flooding the results, limit it so others can surface. If a user truly needs to see beyond 2,000, you may need a custom workaround (like narrowing by alphabet or date range and doing a second query), but these can get complex. Most often, the answer is to prevent the need for so many results at once.

Issue 3: Search query errors (SearchException) – We've touched on the SearchException for short terms. Another scenario is if your dynamic SOSL construction has a bug, you might get a runtime exception. For instance, if you forget a quote or brace in your SOSL string, it will throw a SearchException complaining about bad syntax. These can be tricky to debug because the error might just say "Malformed search" or something similar. If you hit a SearchException and the cause isn't obvious, log the query string that was attempted (if you're building it dynamically, output it to debug logs). Often, seeing the full query text will make the issue clear (maybe two braces in a row or an extra quotation mark). In a production scenario, consider wrapping SOSL calls in try-catch and capturing e.getMessage() to report it somewhere, because you won't have a debugger there.

Issue 4: Performance issues over time – If a SOSL query that used to be fast becomes slow, consider:

Has the data grown significantly? Perhaps the org now has millions more records, and while SOSL should scale via indexing, more data can still mean slightly slower queries or more results to handle.

Is the search term too broad? Maybe users started entering just "a" or "test" and causing heavy loads. Check logs to see what search terms are being used (you could instrument your code to log search terms frequency, though be mindful of not logging sensitive info).

Did the underlying index get affected? (This is rare; Salesforce maintains indexes internally. But if you added many new custom text fields and marked them searchable, the index is now larger and the search might consider more fields.)

To troubleshoot, you can raise a case with Salesforce support if you suspect an internal indexing issue, but typically, the solution is at the application level: narrow the search or archive some old data if it's truly not needed (Salesforce search omits records in the recycle bin or archived activities by default, so those usually aren't a factor).

Issue 5: Sorting and relevancy confusion – Sometimes users may wonder why results are ordered a certain way. Salesforce automatically sorts SOSL results by relevance, not strictly alphabetically or by record type. This means a Contact might appear before an Account if the Contact's fields match the search term more closely. This isn't an error, but it's good to be aware of. If needed, you can post-process results to order them as you prefer (e.g., group by object type) or simply explain that the search engine ranks results by how well they match the query.

Also, note that Salesforce's **global search** (the search bar in the standard UI) uses the same SOSL engine. If you're debugging, try replicating the search via global search – it can help verify if the issue is with your Apex code or with the search index itself. For example, if global search also fails to find a record, it confirms an indexing or data issue rather than an Apex bug. On the other hand, if global search finds something your SOSL didn't, double-check your SOSL syntax and context.

In general, approach SOSL troubleshooting methodically: isolate the query (e.g., run it in the Developer Console's SOSL query editor), check for exceptions or error messages in debug logs, verify the user's permissions, and confirm the field and data are meant to be searchable. Often these steps reveal the root cause of the issue, whether it's a permissions problem, a query syntax mistake, or a limitation being hit. Most SOSL quirks boil down to the considerations we've already discussed in this chapter.

By methodically analyzing the symptoms, you can usually pinpoint the cause of SOSL issues. Most often, the problems aren't bugs in Salesforce but nuances of how search operates (security, indexing, limits, etc.). The good news is that once you identify the cause, the solutions are usually straightforward – adjust the query, clean the data, or guide the user. A quick checklist (field searchable? term length? user access? number of matches?) can often save you a lot of time. And remember, tools like ChatGPT or Salesforce's community forums can be great resources if you're stuck: describing your issue and getting an outside perspective can illuminate something you overlooked.

With robust troubleshooting skills in your toolkit, you'll be able to tackle any SOSL hiccups that come your way. And that brings us to the end of our journey through efficient SOSL practices. Let's recap the key lessons from this chapter and prepare to apply these insights in our Salesforce projects.

SUMMARY

In this chapter, we explored the ins and outs of **Efficient SOSL Practices.** We began by understanding what SOSL is and how it differs from SOQL, learning that SOSL's strength lies in full-text searches across multiple objects. We practiced the core SOSL syntax – using FIND with search terms, optional field scope, and the powerful RETURNING clause to fetch records from various objects in one query. With ChatGPT's help, we saw how to verify SOSL queries and clarify the differences between SOSL and SOQL for our teams.

We then leveraged SOSL's ability to search multiple objects at once. By applying strategies to focus the scope and limit results, we ensured that multi-object searches remain fast and relevant. We discussed how architects can plan which objects to include and how to handle the distribution of results when many objects are involved, using techniques like per-object limits to avoid one type of result crowding out others. Throughout, we kept in mind the user perspective – providing comprehensive search without overwhelming them – and used ChatGPT prompts to refine query structures for optimal outcomes.

Performance was a key theme: we reviewed how Salesforce indexes data for SOSL, which fields are indexed, and how to write selective queries that cooperate with those indexes. We learned to use precise search terms and judicious wildcards to get faster results and avoid hitting the 2,000-record limit unnecessarily. We also touched on how to design data models and queries to make the most of indexing, ensuring our SOSL queries remain efficient even as data grows.

Managing SOSL results is as important as writing the query. We covered best practices for handling result sets – from using the LIMIT clause to cap the number of records, to designing user interfaces that page or prompt for refinement when searches are too broad. We emphasized testing for both "no results" and "very many results" scenarios so that our application behaves gracefully at both extremes. Additionally, we kept governor limits in mind (like the 20 SOSL per transaction limit), structuring our code to use as few queries as needed and never in a loop.

When it came to dynamic SOSL, we balanced flexibility with security. We saw how dynamic construction of SOSL queries can empower user-driven search experiences, but we made sure to include defenses against SOSL injection. By using bind variables and input escaping, and by whitelisting which fields or objects can be searched dynamically, we protect our applications from malicious or accidental query tampering. Testers were encouraged to validate these protections with creative inputs, and ChatGPT proved useful for spotting potential security holes in our approach.

Integrating SOSL into Apex logic rounded out our discussion. We learned patterns for calling SOSL from Apex controllers and triggers, with an eye on proper error handling (like catching SearchException for bad queries) and maintaining a responsive user experience. We stressed the importance of input validation (e.g., requiring a minimum of two characters to avoid runtime errors) and considered the impact of SOSL in various contexts (synchronous searches vs. asynchronous or trigger-based searches). By following these integration patterns, we ensure that SOSL becomes a reliable component of our system's functionality, delivering search results when and where they're needed without surprises.

Finally, we equipped ourselves with troubleshooting techniques for common SOSL issues. Whether it's missing results (due to permissions or non-indexed fields), overly broad queries hitting limits, or unexpected query errors, we now have a toolkit to diagnose and resolve problems. We learned to double-check field searchability, compare behavior with Salesforce's global search, and use tools like debug logs (and even ChatGPT) to figure out why a query might not be doing what we expect. This proactive debugging mindset will save time and keep our search features running smoothly.

By mastering these SOSL best practices, you as a Salesforce professional – whether an architect, developer, tester, or product owner – can build search capabilities that are powerful, fast, and user-friendly. Effective SOSL usage contributes to more responsive applications and happier users who can find the information they need quickly. We've seen how ChatGPT can assist at every step, from brainstorming query optimizations to double-checking security, making the development process more collaborative and insightful.

With a firm handle on efficient SOSL practices, you're ready to apply these lessons to your own Salesforce projects. As we move forward to the next chapter, we will shift our focus from search queries to **robust exception handling**. Just as a solid search function improves user experience, a solid error-handling strategy ensures application stability and maintainability. In the upcoming chapter, we'll explore how to gracefully handle errors in Apex, create resilient code with proper try-catch logic, and how ChatGPT can aid in refining our approach to Apex exception handling. This will further equip you to build Salesforce applications that are not only powerful and efficient but also reliable and fault-tolerant.

6

EXCEPTION HANDLING AND ROBUST ERROR MANAGEMENT

Errors are inevitable in software, but how we handle them makes all the difference between a resilient Salesforce application and one that frustrates users and developers alike. In Salesforce Apex—a server-side language that powers complex business logic—robust exception handling is a cornerstone of stability and reliability. A poorly handled exception can roll back an entire transaction, souring end-user experience and potentially causing data loss. In enterprise environments with thousands of users and intricate integrations, even a single unhandled error can cascade into significant user dissatisfaction, prolonged troubleshooting, and elevated maintenance costs. Small and mid-size businesses (SMBs) are not immune either; a seemingly minor error can consume disproportionate time to diagnose if proper practices aren't in place. This chapter dives into how systematic error management in Apex keeps applications stable and maintainable across organizations of all sizes.

At its core, an *exception* in Apex is an event that disrupts the normal flow of execution. Exceptions can arise from many sources: a null pointer when code tries to use an uninitialized object, a failed database operation, a callout to an external service that times out, or a governor limit violation unique to Salesforce's multitenant environment. No matter the cause, unhandled exceptions will abort the operation and roll back any database changes, which is Salesforce's way of preserving data integrity. This protective behavior is useful, but it means that if we don't anticipate and catch exceptions, users might see their work vanish due to a trivial issue and get greeted with a generic *"internal server error"*. In this chapter, we'll learn how to prevent such scenarios by implementing structured exception handling and robust error management in Apex code. We'll explore the **core principles** of handling exceptions, how to write clear and effective **try-catch-finally blocks**, the use of **custom exception classes** for more granular control, and strategies for **logging** and monitoring errors so that nothing falls through the cracks. We'll also tackle specialized contexts: dealing with errors in **asynchronous processes** (like batch jobs and future calls) and in **integration scenarios** where Apex interacts with external systems. Rounding out the chapter, we'll look at **proactive techniques** to prevent exceptions before they happen, through defensive coding and validation.

Throughout each section, we will highlight how different stakeholders benefit from diligent error handling. **Salesforce architects** will see patterns that improve system reliability at scale. **Developers** will gain techniques to write cleaner, safer code that is easier to debug. **Testers** will learn what to look for when validating error scenarios and how to ensure coverage of edge cases. **Product owners and project managers** will appreciate how robust error management reduces downtime, support tickets, and unscheduled firefighting, ultimately protecting the project's ROI and user satisfaction. Keep an eye out for special callouts with tips for each of these roles as we progress.

Importantly, this chapter demonstrates how ChatGPT can be a valuable companion in mastering and implementing these practices. We'll incorporate extensive ChatGPT prompt examples to illustrate how you can use AI assistance for tasks like analyzing stack traces, generating example code, and even brainstorming recovery strategies. You'll see prompts that leverage **non-text interactions** as well—using voice queries for quick Q&A, uploading code files or log files for analysis (with due attention to privacy settings), trying out the new ChatGPT "Operator" to fetch documentation or automate debugging steps, and even using image generation to visualize error flows or create diagrams for your team's understanding. For example, you might use voice on a mobile device to ask, *"What does a DmlException mean in Salesforce?"*, and get an explanation on the fly, or you could upload an Apex log file for ChatGPT to parse and summarize the root cause of an error (after ensuring your file is shared in a safe, private manner).

By the end of this chapter, you will have a 360-degree view of Apex error handling. We start from foundational principles and build up to enterprise-level strategies for resilience. Each section is designed to deepen your understanding and provide practical techniques (with ChatGPT as a helpful aide) to make your Apex code not only handle exceptions gracefully but also emerge stronger from them. Let's begin our journey by establishing the core principles of exception handling in Apex and why a thoughtful approach to errors is so critical in any Salesforce project, big or small.

CORE PRINCIPLES OF EXCEPTION HANDLING IN APEX

What exactly is an exception in Salesforce Apex, and why is handling exceptions systematically so crucial? In simple terms, an exception is a runtime error that disrupts the normal flow of code execution. Apex, like most programming languages, uses exceptions to signal that something unexpected happened: maybe a required field was missing during a database save, or an operation tried to divide by zero. If not caught, an exception in Apex will cause the entire transaction to roll back, meaning all the database changes made in that execution context are undone. This behavior protects data integrity but also means that one unanticipated error can wipe out a lot of work. For example, if a trigger processing 200 records encounters a null pointer on the 150th record and nothing handles that exception, none of those records will save — a frustrating outcome for users and a nightmare for developers to debug after the fact.

Robust exception handling aims to prevent such scenarios by catching errors *before* they cause a wholesale failure, allowing the system to recover or at least fail gracefully. A core principle is **fail early, but handle gracefully**. This implies that our code should catch exceptions at appropriate boundaries and take some action: log the error, clean up partial data, inform the user with a friendly message, or all of the above. In contrast, failing to catch exceptions (or catching them incorrectly) leads to what we call *ghost errors* that appear as sudden crashes or silent rollbacks with cryptic messages. For enterprise applications, this can erode user trust and lead to significant maintenance overhead. As one guide succinctly puts it, "*a well-handled exception not only prevents application crashes but also provides valuable insights for continuous improvement*". In other words, every exception is telling you a story about what went wrong, and robust handling ensures you hear that story and learn from it, rather than sweeping it under the rug.

Key Best Practices at a Glance: At the heart of Apex error management are a few golden rules :

Catch specific exceptions whenever possible. It's tempting to write one broad catch (Exception e) to handle everything, but specificity is key. If you expect a DmlException from a database operation or a CalloutException from an HTTP call, catch those types explicitly. Being specific helps handle each error type more effectively. We'll explore this in depth in the next section.

Never swallow exceptions without action. A common mistake is catching an exception and then doing nothing or just printing a debug log. This "swallowing" means the error is essentially ignored, which can lead to inconsistent data or harder-to-diagnose issues later. Always handle it meaningfully: log it, fix the state, or rethrow it as a different exception that higher layers will handle.

Maintain system stability. Your error handling strategy should aim to keep the application running or data safe, even when things go wrong. This might mean designing fallback logic or ensuring partial failures don't break the entire process. For example, if part of a batch of records fails, the system should continue processing the rest if possible. Stability also means rolling back changes when necessary to avoid corrupt data, but doing so deliberately and informing someone (a user or admin) about the problem.

Keep the end-user experience in mind. Users should ideally never see a raw Apex exception message or, worse, an unhandled exception screen. Design your error messages for clarity and helpfulness. For instance, if a payment process fails because an external service is down, an end user message might say, "We're sorry, we couldn't process the payment at this time. Please try again later," rather than "CalloutException: HTTP 503". We'll discuss user-facing error messages and techniques like using custom exceptions or error results to convey issues to the UI later in this chapter.

Log and learn. Every exception is an opportunity to improve. Implement logging for exceptions so that your team can diagnose issues quickly. Good logs capture what happened and when, and sometimes even auto-notify the team. For example, in an enterprise scenario, you might have a logging framework that records exceptions to a custom object and triggers an email or Slack notification to the support team for critical errors. Even for an SMB with fewer resources, simply using system debug logs or sending an email to an admin when a critical operation fails can dramatically cut down troubleshooting time.

Plan for recovery and prevention. Handling an exception is good; preventing it next time is even better. Incorporate fixes or checks once you know what caused an error. In Apex, this might mean adding a validation to prevent bad data from reaching that point, or adding retry logic for a transient failure. Always remember the cycle: catch -> log -> fix (if possible). A minor example: if you catch a NullPointerException in a method and realize it's because a map wasn't initialized, the long-term fix is to initialize that map or add a guard (if (map == null)) before using it, so the exception won't happen at all.

Let's consider a quick scenario that highlights these principles and how ChatGPT can assist. Imagine you have an Apex method that performs several steps: it creates an Opportunity, calls an external pricing API, and then updates the Opportunity with the returned price. You've been getting reports that occasionally the process "does nothing" and the Opportunity isn't created, but no error is shown to the user. This is a red flag that an exception might be occurring silently. You review the code and see something like:

```
public void createOpportunityWithPrice(Opportunity opp) {

  try {
```

```
    Decimal price = PricingAPI.getPrice(opp); // callout to get price
    opp.Price__c = price;
    insert opp; // insert Opportunity
  } catch (Exception e) {
    System.debug('Error in process: ' + e.getMessage());
  }
}
```

At first glance, it looks okay – it's catching any exception. But on closer inspection, this code is *swallowing* exceptions by catching the generic Exception and merely writing a debug log. If the pricing API call fails (perhaps a callout timeout), the catch will execute, log the error, and then the method ends normally. The user who invoked this (say via a Lightning component) sees no error message, but the Opportunity wasn't updated with a price (and in fact, since the catch handles it, the originally inserted Opportunity remains in the system without a price). This is confusing behavior: the operation partially succeeded (Opportunity created) but the user wasn't informed that the pricing step failed.

A better approach is to catch specific exceptions and handle each appropriately. For instance, catch CalloutException to handle pricing API errors: maybe set a default price or mark the Opportunity for manual pricing later, and inform the user that pricing service is unavailable. Catch DmlException around the database operations to handle any Salesforce-specific errors (like validation rules preventing save) and perhaps roll back the Opportunity if it can't be priced. And importantly, present a message back to the user or the calling context. In a Lightning context, one could throw a AuraHandledException with a friendly message so the UI can display it. The key is not to simply absorb the error.

To refactor this, you might use ChatGPT by uploading the code file or snippet and asking for suggestions on improving its exception handling (remember to exclude or anonymize sensitive data and consider turning off chat history for privacy when sharing actual code). For example, you could prompt:

User (File Upload & Prompt): *Uploading OpportunityService.cls* – "ChatGPT, this Apex method is meant to create an opportunity and get a price from an API, but it sometimes fails silently. How can I improve its exception handling?"

ChatGPT might analyze the code and respond:

ChatGPT: "It looks like the method catches all exceptions generally, which can hide specific issues. I recommend splitting the try-catch and handling callout exceptions separately. For example, use one catch (CalloutException e) to handle pricing service errors – perhaps set a flag or default price and inform the user – and another catch (DmlException e) for Salesforce data errors – maybe roll back the record or alert an admin. Also, consider using finally if any cleanup is needed after these operations. You should avoid catching Exception unless you rethrow or handle it meaningfully. You can propagate exceptions up if you want the caller (e.g., a UI component) to handle them. In Lightning, throwing a System.AuraHandledException with a user-friendly message would let the component show that message to the user."

Through this process, we apply the core principles: be specific, don't swallow, and keep the user and system in mind. Early validation can also help here: for instance, ensure the opp isn't null and has required fields before insertion to avoid a DmlException in the first place. This brings us to a callout tip:

Developer Tip: Always do a quick audit of your Apex methods to see what exceptions they might throw and ensure you have a plan for each. If you find a catch-all Exception as a placeholder, don't leave it that way. Decide how to handle each likely scenario or at least log it and rethrow. If you're unsure, ask ChatGPT (via text or voice) something like, "What exceptions could this code throw?" to get a list of potential issues, then handle each accordingly.

Having established the importance of catching exceptions properly and not letting them fail silently, let's dive deeper into how to structure those try-catch blocks in Apex for maximum clarity and effectiveness.

STRUCTURING TRY-CATCH-FINALLY BLOCKS FOR MAXIMUM CLARITY

The try-catch-finally construct is the primary tool for handling exceptions in Apex. Writing a try-catch is straightforward, but writing an *effective* try-catch requires careful thought. In this section, we'll explore best practices for structuring try-catch-finally blocks to ensure your code remains clear, maintainable, and does the right thing when exceptions occur. We'll also highlight common pitfalls, like overly broad catches and misusing the finally block, and show how to avoid them.

The Anatomy of Try-Catch in Apex: A quick refresher: you wrap the code that might throw an exception in a try {... } block. Immediately after, you provide one or more catch blocks to handle specific exception types. For example:

```
try {
  // risky operations
} catch (DmlException e) {
  // handle database errors
} catch (CalloutException e) {
  // handle callout errors
} catch (Exception e) {
  // handle any other exceptions
} finally {
  // cleanup actions
}
```

Apex allows multiple catch blocks, just like Java, and you should order them from most specific to most general. This means catch the subclass exceptions (e.g., DmlException, NullPointerException, etc.) before a general Exception catch. If an exception is thrown, Apex will run through the catches top-down and execute the first one that matches the exception type. One important Apex nuance: unlike some languages,

Apex does not support *checked exceptions* (all exceptions in Apex are unchecked), so you are not forced to catch any particular exception. It's up to you to anticipate and catch what you can handle.

Be Specific and Intentional: As noted earlier, specificity in catches is crucial. For example, if you have database operations that might fail due to record locking or validation rules, catch DmlException specifically. This not only allows you to handle that scenario (maybe by informing the user which record failed or by retrying later), but also serves as documentation in code — anyone reading it sees that you expected and planned for a DML issue. A generic Exception catch, on the other hand, could mean anything went wrong. It's a "catch-all" that obscures the nature of failures. Salesforce provides around 20+ built-in exception types in the System namespace (such as DmlException, QueryException, NullPointerException, CalloutException, LimitException, and so on). Knowing these types helps you write tighter error handling. As one Salesforce expert advises: *"catch only what you can handle... catching the generic Exception and calling it a day is like putting a band-aid on a broken arm"*. In practice, this means if you catch an Exception broadly, you should either be rethrowing it as something more specific or at least logging it with full context, because otherwise you might mask a serious issue.

Let's illustrate with a quick fix on a sample code. Suppose we have a snippet:

```
try {
  update accountsList;
  makeSomeCallout(); // could throw CalloutException
} catch (Exception e) {
  // logging the error
  System.debug('Error: ' + e);
}
```

This code catches any exception, but it doesn't distinguish between a database error from the update and a callout failure. Merging those paths might be inappropriate. A better version:

```
try {
  update accountsList;
  makeSomeCallout();
} catch (DmlException dmlEx) {
  // Handle database error (e.g., notify about failed records)
  System.debug('Database error: ' + dmlEx.getMessage());
    // perhaps take specific action like marking records or reversing
changes
} catch (CalloutException callEx) {
  // Handle external callout error (e.g., set a flag to retry later)
  System.debug('External call failed: ' + callEx.getMessage());
  // maybe send an alert or fallback to offline processing
} catch (Exception e) {
  // Handle any other unexpected type of exception
  System.debug('Unexpected error: ' + e.getMessage());
  throw e; // rethrow to ensure it's not silently ignored
}
```

In the revised version, each expected problem has its own catch. We even added a generic catch at the end to capture truly unforeseen exceptions, but importantly, we rethrow it (throw e) after logging. Rethrowing is often better than swallowing when you *don't know how to handle an exception* — it allows higher-level code or the platform to deal with it, rather than continuing in a possibly inconsistent state. Notice that we didn't include a finally here; we'll discuss finally in a moment.

Avoid Swallowing Exceptions: A swallowed exception is one that is caught but not properly acted on. Logging it with System.debug (which goes to debug logs) is better than nothing, but remember that debug logs in Salesforce are only retrievable if you proactively go looking (and have debug logging turned on for that user). In production, unless a developer is monitoring, a debug log message might go unnoticed. So swallowing with a debug statement is effectively silencing the error in many cases. If you don't intend to surface an error to a user, at least consider surfacing it to an admin or support system. For example, within a catch you might increment a custom "error count" field, or add an entry to a custom Log__c object, or send an email using Messaging.sendEmail to the support address. Those actions make sure the exception isn't forgotten. A Trailblazer Community best-practice post puts it succinctly: *"You must do something with every handled exception. Either you are notified about it and can investigate or you correct the issue in code."*. In other words, catching an exception obligates you to handle it responsibly.

Tester Tip: When reviewing user stories or test cases, ensure there are scenarios covering exceptions. If a feature involves a callout to an external system, there should be a test (or at least a plan) for what happens if that callout fails. As a tester, you can ask the developer, "What should happen if this external service is down?" If the answer is "the user sees an error message saying try again later," then verify that's implemented. If the answer is "I hadn't thought of that," that signals a gap. ChatGPT can help testers generate a list of "what-if" questions for a given feature description – consider using voice to quickly query: *"What possible errors should I test for in a feature that does X?"* and use the answer to guide your testing plan.

Using finally Wisely: The finally block in a try-catch-finally structure is executed after the try and catch blocks, regardless of whether an exception was thrown or not. It's commonly used for cleanup operations. In Apex, you might use finally to do things like: reset static variables, clear out collections, or release resources. However, heavy resource management is less common in Apex than in, say, Java on a server, because Apex doesn't deal with open file streams or manual memory management (the platform handles most resources). A practical use of finally in Apex is in test code or utility code; for example, if you start an asynchronous job in a try, you might have a finally that always logs the end of the process or always unlocks a record if you had some locking mechanism. The important thing is not to put essential logic that changes the outcome of the operation in finally—because finally runs even if you had an error. For instance, do not decide success/failure in finally; that should be determined in catch. One common pattern is to use finally to **perform actions that must happen no matter what**, like clearing a temporary record you created or sending a notification that a process completed (success or fail). Always be cautious: if your finally block itself throws an exception, it could mask an exception that was being handled. Apex will throw an unhandled exception if something in finally throws (since there's no further catch for it). So keep finally blocks simple and guard them with their own try-catch if necessary.

Consider this scenario: you have an Apex method that sets a static "maintenanceMode" flag to true at the start of a critical update (to prevent other processes from running concurrently), and you need to set that flag back to false when done. You'd use finally to ensure the flag resets:

```
SystemController.maintenanceMode = true;

try {

  performCriticalUpdates(); // may throw exceptions

} catch (Exception e) {

  // handle or log exception

  throw e;

} finally {

  SystemController.maintenanceMode = false; // always reset

}
```

Here, no matter what happens in try (success or error), the system leaves maintenance mode. Without the finally, an unhandled exception could leave the flag stuck in true, impacting other operations.

Consistent Error Propagation: Decide at what level of your application an exception will be handled and at what level it will be allowed to bubble up. For example, in a multi-layer application, you might decide that all database-related exceptions are caught and transformed in the service layer, but anything that reaches the top of the call stack (say an Apex controller for a Lightning page) triggers a user-facing message. In Apex, uncaught exceptions in a Visualforce or Lightning context bubble up to the user as an error page or aura message. That's rarely user-friendly, so often the top controller will catch any remaining Exception and handle it. One pattern is to rethrow exceptions as needed to move them up. For instance, if a low-level method catches a QueryException (no records found) but that just means the record wasn't there, it might not consider that an "exceptional" case and simply return null or a status. Conversely, if a mid-level method catches something it can't resolve, it might wrap it in a custom exception (e.g., new MyServiceException("Failed to process X", e)) and throw that upward, so the higher layer knows a specific service failed but doesn't need the gory details. The higher layer (e.g., a batch finish method or a controller) can then decide final actions like logging or notifying.

If you are unsure whether to handle an exception or let it propagate, ask yourself: "Does this code block know what to do about this error, or would a caller have more context to handle it?" Often, a lower-level method should not pop up a user alert—that's the job of a higher-level context. Instead, it might throw a custom exception describing the failure, and the caller (which could be a UI controller) catches that to show an appropriate message. This separation keeps your logic modular.

Leverage ChatGPT for Code Reviews: It can be challenging to spot ineffective try-catch usage in a big codebase. A useful trick is to use ChatGPT's file analysis capabilities. For example, you could ask ChatGPT to review an Apex class for exception handling issues by providing the code (ensuring you've anonymized data and disabled history for privacy). A prompt might be:

User: "Here is an Apex class file (attached) containing multiple try-catch blocks. Can you identify any places where exceptions are caught too generally or potential issues where an exception might be swallowed?" *[Attach Apex class file]*

ChatGPT could then highlight sections like, "In method updateOrders, there is a catch (Exception) that only does System.debug. This might swallow exceptions. Consider handling specific exceptions or rethrowing." This gives you a checklist of fixes to make. Using voice commands for quick checks is also possible; for example, you might use a voice prompt while coding: *"ChatGPT, what's a best practice for using finally in Apex?"* and get a concise reminder. This kind of just-in-time assistance can reinforce good habits.

Architect Tip: Consider establishing an *organizational standard* for exception handling in Apex. For instance, mandate that no method should catch Exception without also handling it or rethrowing; or define that all service classes should use custom exceptions for business logic errors. You can even use static code analysis tools (like PMD) with custom rules to flag broad catches or empty catches. ChatGPT can help write a draft of these coding standards. You might ask, "Help me write a policy document section on how developers should use try-catch in Apex." The AI can generate a nicely formatted list of dos and don'ts, which you can refine and share with the dev team.

By structuring try-catch-finally blocks with clarity and intention, you make your Apex code much more resilient and understandable. Catches become part of the documented behavior of your code rather than afterthoughts. Next, we'll extend our toolkit by looking at custom exceptions – a way to create your own exception types that can make error handling even more expressive and fine-tuned.

CUSTOM EXCEPTIONS AND EXCEPTION CLASSES

Built-in exceptions like DmlException and NullPointerException cover many scenarios, but as your application logic grows, you may encounter situations where defining your own exception class provides clarity and control. Custom exceptions in Apex are user-defined classes that extend the base Exception class (or another exception subclass). They allow you to throw and catch errors that are specific to your domain or business logic, making your code more readable and easier to maintain.

Why Create Custom Exceptions? Think of a custom exception as a way to give a distinct *identity* to an error condition. Instead of catching a generic Exception and then inspecting its message to figure out what went wrong, you could throw (and catch) a InvalidOrderStateException or PaymentProcessingException that immediately tells you the nature of the problem. As one author described, using only generic exceptions is like chasing a mystery with vague clues, whereas custom exceptions are "detailed, specific clues that lead you straight to the heart of the problem". When a PaymentProcessingException is thrown, any developer reading the logs or catch block will know the issue was related to payment handling, not just a random error. Custom exceptions thus act as self-documenting signals in your code.

For example, imagine a situation in an order management system: you have a method that finalizes an order. Several things could go wrong: the order might already be closed, the payment might fail, or inventory might be insufficient. You could handle all these with one generic exception or maybe reuse something like ApplicationException (if you had one), but it may be clearer to define exceptions like OrderAlreadyClosedException, PaymentFailedException, and InventoryShortageException. That way, each distinct cause of failure is handled in its own catch, or logged with its own error type. This granularity is especially useful in enterprise settings where errors might trigger different workflows. For instance, a PaymentFailedException might trigger an automatic email to the billing department, whereas an InventoryShortageException might alert the supply chain team. By categorizing via custom exceptions, you separate concerns cleanly.

Defining a Custom Exception: In Apex, any top-level class (one that is not inside another class) can extend Exception. By convention, we name custom exceptions with an "Exception" suffix (e.g., OrderAlreadyClosedException). A simple example:

```
public class OrderAlreadyClosedException extends Exception {}
```

This is the most basic form – it inherits everything from Exception. You can add constructors to pass messages or other data. For example:

```
public class PaymentFailedException extends Exception {

  public PaymentFailedException(String message) {

    super(message);

  }
```

}

Now you can throw this exception with a specific message:

```
if(paymentService.process(cardInfo) == false) {
    throw new PaymentFailedException('Payment gateway declined the card.');
}
```

And catch it specifically:

```
catch (PaymentFailedException pf) {
    // handle payment failure, maybe roll back order or mark as pending
  System.debug('Order not completed: ' + pf.getMessage());
}
```

Custom exceptions can even have member variables or methods if you want to attach additional info. For instance, you might include an error code, or a reference Id. They can also implement interfaces if needed. Salesforce's docs remind us that exceptions can be complex classes: they "can have member variables, methods, and constructors". However, keep them lightweight – usually they just carry a message and maybe an error code or context info.

When to Use Custom Exceptions: Not every error needs a custom class. Use them when you need to handle different error scenarios differently or when propagating an error up requires more context than a generic exception provides. Some common use cases:

Business rule violations: e.g., throwing BusinessRuleException when a certain business condition isn't met (like "customer credit limit exceeded"). This could be caught at a higher level to inform the user in a specific way. In one example, a business logic method might throw BusinessLogicException if an invariant is violated.

Multi-layer applications: If you have service classes, they might catch lower-level exceptions (like DML or callouts) and throw a custom exception like ServiceException or more specific ones. This way, the controller layer only has to catch ServiceException rather than every possible low-level exception.

Integrations: Wrapping external system errors. For instance, if you call out to a shipping API and get an error, you might throw a ShippingServiceException within your Apex. That exception can include details like the endpoint or payload. Later, when catching it, you could decide to retry or log those details. In our earlier example, throwing a PaymentGatewayException gave a clear signal that an external payment callout failed.

Complex processes: In a batch job, if you have logic that processes records and you want to signal a special type of failure for certain records, throwing a custom exception could help differentiate a "data error" from a "system error," for example. However, note: in batch Apex, unhandled exceptions in execute will cause that batch chunk to roll back and possibly be retried or reported – batch handling is special.

One thing to remember: a custom exception thrown is still an exception – it will behave like any other in terms of rollbacks. So use them to improve communication and handling, but they won't magically bypass the transaction rollback rules. If you throw a custom exception and don't catch it, the transaction aborts just as with a System exception.

Example – Using Custom Exceptions in Practice: Let's say we implement an order finalization method:

```
public void finalizeOrder(Id orderId) {

    Order__c order = [SELECT Status__c, Total__c FROM Order__c
WHERE Id = :orderId];

  if(order.Status__c == 'Closed') {

      throw new OrderAlreadyClosedException('Order ' + orderId + ' is
already closed.');

  }

  if(!InventoryService.checkInventory(order)) {

      throw new InventoryShortageException('Insufficient inventory for
order ' + orderId);
```

```
  }
  try {
    PaymentService.charge(order.Total__c);
  } catch (PaymentGatewayTimeoutException pgte) {
    // maybe a specific exception from PaymentService
    // Wrap into our own exception to throw upward
    throw new PaymentFailedException('Payment timed out for order ' +
orderId);
  }
  //... continue order finalization
}
```

Here we throw OrderAlreadyClosedException and InventoryShortageException for business conditions. We also catch a hypothetical PaymentGatewayTimeoutException (maybe the PaymentService threw that) and wrap it into our PaymentFailedException – perhaps we consider any payment exception as the same category of failure for the caller. On the calling side, we could then do:

```
try {
  OrderService.finalizeOrder(someOrderId);
  // if we reach here, success
} catch (OrderAlreadyClosedException | InventoryShortageException e)
{
  // handle known business rule issues
    ApexPages.addMessage(new ApexPages.Message(ApexPages.Severity.ERROR, e.getMessage()));
} catch (PaymentFailedException e) {
    ApexPages.addMessage(new ApexPages.Message(ApexPages.Severity.WARNING,
```

```
    'Payment could not be processed. Please try again later.'));
  // maybe also log this for finance team
} catch (Exception e) {
  // catch any other unexpected error
    ApexPages.addMessage(new ApexPages.Message(ApexPages.Severity.ERROR, 'Unexpected error finalizing order.'));
  // log e for dev team
}
```

In Visualforce, using ApexPages.addMessage would show the user a page message. In Lightning, we might instead throw an AuraHandledException with a message so it surfaces to the user without a full stack trace (Lightning framework catches AuraHandledException and passes the message to the client). The key is that our catches cleanly separate the three scenarios: order already closed or inventory issues (likely user correctable or data issues), payment failure (external issue, maybe temporary), and anything else (unexpected). Without custom exceptions, we would be stuck parsing an error message or using flags to differentiate these cases, which is messy.

Architect Tip: Align custom exceptions with your application's error model. Define a clear hierarchy if needed. For example, you might have a base ApplicationException (extending Exception) and derive more specific ones like ValidationException, IntegrationException, etc., under it. This way, you can catch a broad category (say all integration-related exceptions) in one go if needed. Apex doesn't have a formal exception hierarchy beyond what you build, so if it serves the design, create one. Just avoid over-engineering; two or three well-placed custom exceptions often suffice for a given module. Create what adds clarity. An added benefit in enterprise contexts: if you throw a custom exception all the way up, it will appear by class name in logs. Seeing "ApplicationException: ..." vs "Exception: ..." in a stack trace can help teams quickly identify that it was a known, handled scenario vs a surprise.

Validating and Reviewing Exception Strategy with ChatGPT: If you're introducing custom exceptions into an existing codebase, you might want to review where they could be useful. ChatGPT can assist by analyzing code for repeated patterns of error handling that could be refactored into custom exceptions. For example, if multiple classes all throw generic exceptions with messages containing the word "Inventory" or an error code "INV_", that's a candidate for consolidating into an InventoryException. You could prompt:

User: "Scan the following Apex code for error handling. Identify if there are cases where I should use a custom exception instead of the current approach." *[Provide code or description]*

ChatGPT might respond by pointing out instances of, say, if(condition) { throw new Exception("Inventory issue:..."); } and suggest creating a specific exception class for it. It can even draft the code for the custom exception class and how to implement it.

Additionally, when implementing a new custom exception, you can test how it will be caught. Using ChatGPT's voice interface, you could simulate a conversation like: *"I created a custom exception DataIntegrityException. How can I ensure any uncaught DataIntegrityException results in an email to admins?"* ChatGPT could then suggest approaches like a centralized error handler trigger or using a try-catch at a higher level that sends email, which might inspire your implementation.

Developer Tip: Always include meaningful messages when throwing custom exceptions. This seems obvious, but it's easy to throw new MyException() with no message, which will result in a generic "common.apex.runtime.impl.ExecutionException" message without context if it ever surfaces unhandled. By providing context (like which record or what operation failed), you save future you (or colleagues) a lot of debugging time. Also, document the scenarios for custom exceptions in comments or documentation. For example: "// throws InventoryShortageException if product inventory is insufficient".

System Exceptions, such as those triggered by breaching Salesforce governor limits, cannot be caught by try-catch blocks in Apex. These exceptions immediately halt execution and roll back any changes, making them fundamentally different from custom or standard exceptions you can handle in code. To avoid running into system exceptions, developers should proactively monitor resource usage by leveraging the

Limits class, which provides real-time insight into consumption of limits like CPU time, SOQL queries, and DML statements. By checking these limits before performing potentially risky operations, you can take defensive actions—such as breaking up logic or deferring work—to prevent exceeding thresholds and triggering uncatchable exceptions.

Custom exceptions empower you to make error handling a first-class part of your application design instead of a reactive afterthought. By naming and categorizing failure modes, you make the code's intent clearer and error processing more flexible. Now that we've covered catching exceptions and defining our own, a crucial question remains: once an exception is caught, where does it go? How do we make sure the issue is recorded and noticed? That's where logging and monitoring come into play.

LOGGING EXCEPTIONS STRATEGICALLY IN APEX

Catching an exception is important, but if you just catch it and do nothing visible, the problem might as well have disappeared into a black hole. **Logging** is the practice of recording runtime information, especially errors, so that developers and system administrators can diagnose issues later. In Salesforce, we have debug logs that capture system debug statements and a lot of contextual info, but relying solely on debug logs isn't always enough—especially in production where you might not have logs running for every user or action. In this section, we explore best practices for systematically logging exceptions in Apex: using the Salesforce debug log system effectively, building custom logging frameworks or using existing ones, and ensuring that logs help achieve rapid diagnosis and troubleshooting. We'll also discuss long-term tracking of issues and how logs tie into monitoring and alerting in an enterprise.

Why Logging Matters: Consider a scenario where an Apex batch job fails at 2 AM. If it's unhandled, Salesforce will send an automated error email to the user who scheduled the job, which might contain a stack trace. But if it's handled (caught) and not logged anywhere, you might not know anything went wrong until someone notices data is off the next day. Proper logging ensures that *every exception leaves a trail*. It's like an airplane's black box: when something goes wrong, you want a record of it to investigate the cause. Additionally, logs can reveal patterns (e.g., an error happening 100 times a day that might not be individually reported but adds up to a bigger issue).

Salesforce Debug Logs: The simplest form of logging in Apex is using System.debug() to print information to the debug logs. If a user (or automated process user) has debug logging enabled, these statements will appear in the log. For example, in a catch you might have System.debug('Error processing order: ' + e.getMessage());. In development and testing, this is fine and helps you see issues in real-time. However, in production, you typically do not keep debug logging perpetually on for all users due to log storage limits and performance. You might only turn it on when investigating something. That means solely relying on System.debug for error logging could miss a lot in a live system. **System.debug is not persistent.** Once the log is gone (logs have a limited retention and size), that info is gone. Therefore, for serious error logging, especially in an enterprise context, you should consider more persistent solutions.

Persistent Logging Strategies: There are a few common approaches to persist logs in Salesforce:

Custom Object Logging: Create a custom object, e.g., Error_Log__c, with fields like Message, StackTrace, Severity, Timestamp, and perhaps fields to link to related records or process names. In your catch blocks (particularly in global error handlers or high-level catches), insert a record into this object with the details. This way, every significant error is stored as a record in Salesforce that you can query, report on, and audit over time. You might only do this for certain critical exceptions to avoid flooding the database with minor ones. For example, you might log only exceptions that indicate system faults, not business validation failures that are shown to users (those might be fine not logged or logged at a different level).

Platform Events for Logging: Salesforce platform events provide a way to stream out events. You could define a Platform Event, say Error_Event__e, with fields for error details. In catch blocks, publish an event (which is asynchronous and does not roll back if the main transaction rolls back after the publish has been successfully enqueued). A separate subscriber (maybe a Trigger on the platform event, or an external system via CometD) can listen and record these events. This decouples error logging from the main logic. It's advanced but useful if you want, for example, an external monitoring system to get all errors in real-time. One advantage: platform events can be configured to *not* roll back with the transaction if you call EventBus.publish after the point-of-no-return in your code, although careful – if the whole transaction fails completely, events might not fire. There is also the Database.RaisesPlatformEvents batch interface that guarantees events on batch failures. If you implement that, each failed batch chunk can emit an event with error details, which is a neat way to capture errors from batch jobs.

Third-Party or Open-Source Logging frameworks: As mentioned in the DevOps Launchpad article, why build from scratch when you can use community solutions? Tools like Nebula Logger are open-source libraries specifically designed for Salesforce logging. Nebula Logger provides a framework to log at different levels (INFO, DEBUG, ERROR) and writes logs to custom objects under the hood, with features like logging complex objects, tagging, etc. It's built to handle large volumes and to be used in managed packages or enterprise orgs. By using such a framework, you get a standardized way to log without reinventing wheels. Nebula Logger, for example, supports centralized logging with the ability to capture similar issues together and link logs to specific contexts (like a record or a process). If you're working in an enterprise with many Apex classes, adopting a logging framework early can pay off in consistency. Instead of dozens of different debug statements, you have a unified approach.

Regardless of method, the key elements to log are: **what happened (exception type and message), where (which class/method, maybe line number)**, and **when.** Optionally, **context** (e.g., record Ids or user info) and a **stack trace** if available (Exception.getStackTraceString() in Apex can give a stack trace string). Logging the stack trace is useful for deep debugging. Be mindful not to log sensitive data (like personally identifiable information or secrets). Error messages sometimes contain record Ids or field values; that's usually fine, but don't log entire records with sensitive fields.

Example of Custom Object Logging in Code: Suppose we have a catch at a high level of a trigger handler:

```
catch (Exception e) {

  // We've decided this is a critical error point that we should persist

  Error_Log__c log = new Error_Log__c();

  log.Error_Message__c = e.getMessage();

  log.Exception_Type__c = e.getTypeName();

    log.Stack_Trace__c = e.getStackTraceString().substring(0, 10000); // Stack trace (limit to 10000 chars if needed)

  log.Location__c = 'OrderTriggerHandler.beforeUpdate'; // some identifier of where error occurred

  log.Severity__c = 'High';

  insert log;

  // Optionally rethrow or handle accordingly

  throw e;

}
```

This would record the error for later. Note: In a trigger context, inserting a log record might fail if the main transaction is being rolled back. If we throw e after inserting log, that insert will also roll back (since we rethrew). One trick is not to rethrow after logging, if logging it is more important than aborting. Or you could perform logging in a @future method or Queueable so it runs after the main transaction (only if you decide to swallow the exception to allow main transaction to continue—dangerous unless you truly handled the issue). More commonly, in a trigger, you wouldn't catch and continue; you'd catch, log, and still throw to halt the save (to not commit potentially bad data). In such cases, the log insert will roll back, which is a conundrum. A pattern here: use a platform event or email in addition to trying the insert, or even use the addError() on a record to surface it to the user and then a platform event to log. Apex doesn't allow a separate, independent transaction within the same execution context except via events or async.

This challenge is why a **global error handler** approach is sometimes used: e.g., an AuraHandledException can carry a message to user, and simultaneously you could fire off a platform event or have a separate logging mechanism outside the transaction. In scheduled or batch contexts, you have more freedom after catching an exception (since you're already handling it and can do a DML to log without necessarily rolling back everything).

Developer Tip: To avoid cluttering your code with logging logic, consider using utility methods. For example, a static method LogUtil.logException(Exception e, String location) that internally handles creating the log record or calling the logging framework. Then your catch blocks are one-liners like LogUtil.logException(e, 'OrderService.finalizeOrder');. This centralizes changes—if you later switch to a platform event, you change LogUtil implementation, not every catch.

Monitoring and Alerts: Logging is the first step, but monitoring is what you do with those logs. In an enterprise scenario, you might have admins or an operations team who periodically review the error logs. Better yet, set up alerts for certain kinds of errors. For example, if using a custom object, you could create a scheduled job or even a flow that checks for any new Error_Log__c records of severity "High" and send out an email or post to Chatter. Some teams integrate with external monitoring: for example, using Salesforce's API to pull error log records into Splunk or DataDog for analysis and alerting. The DevOps Launchpad article refers to "dynamic error notification" as the next step after logging. This means notifying the right people at the right time. It suggests using platform events, email, etc., based on severity. For instance, a critical integration failure might send an immediate email to the on-call engineer, whereas a minor nightly batch error might just log for next-day review. Be careful to not alert too often and cause alarm fatigue – calibrate your severity levels.

Using ChatGPT to Analyze Logs: Once you have logs, you can even use ChatGPT to help analyze them. Suppose you export a set of logs (maybe as CSV) and want to find common patterns. ChatGPT's data analysis abilities could be useful. You might copy a portion of a log file and ask, *"What are the most common exceptions in this log and what do they mean?"* ChatGPT could categorize exceptions by type and explain, for example,

"NullPointerException occurs 5 times typically indicating a code trying to use a null object" and so on. This is especially helpful if you have a lot of log data to sift through. If the log data is sensitive, ensure you remove identifying details or use a secure approach (perhaps using the ChatGPT "Advanced Data Analysis" mode which processes data without sharing it externally, and still, with caution).

For a more interactive approach, you can even use **voice input** to query the logs summary. For example, *"Hey ChatGPT, I saw several DmlExceptions in today's logs, what might be causing them?"* ChatGPT, not having direct access to your system, will give general reasons (like validation rules, required field missing, etc.) but that can spur your investigation.

Exception Handling: Logging works best in concert with good exception handling. For instance, if you have custom exceptions, you can incorporate custom context into log entries. If a PaymentFailedException has a field for orderId, log that specifically. Also, decisions like where to catch exceptions affect logging: if you catch too low and log, you might log something but then the higher process doesn't know. Alternatively, you might choose to let exceptions bubble to a single point and log once centrally. That often leads to cleaner logs (one log entry per error occurrence). Redundant logging in multiple catch blocks can make debugging harder ("which layer logged this?"). Therefore, design a strategy: maybe only log at the top of a transaction or in global handlers.

Tester Tip: When testing error scenarios, verify that logging occurs as expected. If you have a test that forces an exception (say by providing bad data), you can follow up by querying the Error_Log__c (if not rolled back) to ensure an entry was created. In test methods, you might need to surround with Test.startTest()/Test.stopTest() if an async job is logging. Also, consider testing that the content of the log is correct (e.g., the error message was recorded, the severity is set properly). Catching errors is good, but if nobody knows they happened, it doesn't help operationally. A tester ensuring "exceptions are not only handled but also auditable" adds a lot of value.

Real-World Example – Nebula Logger: To illustrate a community solution, Nebula Logger (by Jonp, open-source) provides a ready-made logging mechanism. It uses a custom object (Log__c) under the hood and utility Apex to create logs. You might use it like:

```
Logger.error('Order processing failed')
    .setException(e)
    .setContext('OrderId', order.Id)
    .log();
```

This one-liner would create a log record with level ERROR, attach the exception (so message and stack trace stored), and record a context key-value (OrderId). The.log() method handles the DML insert either immediately or via queueable. The benefit is consistency and less boilerplate in your code. When you deploy such a framework, developers just need to remember to call it in catch blocks. Over time, you accumulate an error database you can analyze. The open-source aspect means many Salesforce professionals have vetted it, and it's efficient and bulk-safe.

Product Owner Tip: Logging isn't just a technical nicety; it's an insurance policy for your project. As a product owner, encourage the team to surface metrics from logs: for instance, how often are certain errors happening per release. This can inform you if a particular feature is flaky or needs improvement. It also helps in prioritizing bug fixes – if logs show an error happening 50 times a day even without user complaints, it's a silent issue worth addressing before it becomes a loud one. Using ChatGPT, you could even ask for a summary to present in a meeting: *"Summarize the key error trends from last quarter's logs in non-technical terms,"* which can help communicate to stakeholders where stability has improved or where attention is needed.

In summary, logging is about creating a safety net that catches you after you catch the exception. It ensures no error goes unseen. By employing structured logging, whether via custom code or frameworks, and by implementing processes to monitor those logs, both enterprises and SMBs can significantly reduce mean time to resolution for issues. With logs in hand and maybe a dash of ChatGPT to parse them, you're far better equipped to maintain system health.

Now that we have mechanisms to *capture* and observe exceptions, let's shift focus to contexts where exceptions behave a bit differently: asynchronous processes. Batch jobs, queueables, scheduled Apex, and future methods present unique challenges and opportunities for error handling, which we'll explore in the next section.

HANDLING APEX EXCEPTIONS IN ASYNCHRONOUS PROCESSES

Asynchronous Apex (such as **batch Apex**, **queueable jobs**, **future methods**, and **scheduled Apex**) runs in the background without an immediate user context. These processes are crucial for handling large data volumes and long-running tasks. However, their error handling needs are a bit different from synchronous request-response code. In asynchronous jobs, you typically don't have a user watching the screen to see an error message, and the jobs might continue processing even after encountering some errors. In this section, we'll cover how to manage exceptions in these asynchronous scenarios, ensuring reliability, proper notifications, and minimal disruption.

Characteristics of Asynchronous Errors:

No immediate user feedback: If a batch job fails on a record, you can't show a popup to a user. Instead, you might log it and perhaps notify an admin or simply continue.

Partial failure handling: Batch Apex processes records in chunks (batches). If one chunk fails with an unhandled exception, that chunk is aborted, but other chunks will continue to process in separate transactions. This means your job can partially succeed. You need to decide how to handle records in the failed chunk – perhaps retry them or mark them as failed.

Automatic retry and notifications: Salesforce will *not automatically retry* a batch chunk that fails due to an exception (it just moves on to the next chunk, except for certain exceptions like Governor Limit ones that stop the whole job). However, if a whole batch job aborts (like an exception escapes in the finish method or a limit exception in execute that isn't caught), the batch is marked as failed. For future and queueable, an unhandled exception results in an error email to the initiating user (similar to how unhandled exceptions in future methods send an email to the user who called the future).

Different context, different strategy: The approach you take to handle an error in a batch might be "log and continue", whereas in a trigger (synchronous), it might be "stop everything and show user error". One blog put it nicely: *"the approach you take to handle an error in a batch Apex job can be worlds apart from how you'd manage it in a trigger"*. In triggers or synchronous Apex, immediate feedback and rollback is usually appropriate; in batch, you aim to complete as much as possible and handle failures gracefully offline.

Batch Apex – Fine-grained error control:

Batch Apex provides some built-in tools for handling exceptions:

Try-catch in execute: Within the execute method of a batch, you can wrap your logic in try-catch. If an exception occurs for a particular record or set of records, you catch it, maybe log which record failed, and crucially, you can decide not to rethrow it. This way, your batch will not abort the chunk; it will simply handle that record's error and move on to the next record in the chunk. For example:

```
global void execute(Database.BatchableContext bc, List<Account>
scope) {

  for(Account acct : scope) {

    try {

      processAccount(acct);

    } catch(Exception e) {

        // log the error for this account, maybe mark account with an
error flag

      ErrorLogger.log(acct.Id, e);

      // don't rethrow, so loop continues to next account

    }
```

```
    }
}
```

This ensures one bad account doesn't abort processing of the rest in the scope. However, if a certain exception is irrecoverable (like a governor limit issue), catching it might not be possible (governor limit exceptions like CPU time can't be caught). But those are rarer if code is well-written.

Database.Stateful for aggregates: If you need to gather information across batch chunks (say collect all failed record Ids in an instance variable), implement Database.Stateful in your batch class. Then you can accumulate, for instance, a list of failed IDs. In the finish method, you could do something with that list (like send a summary email or create a report log). This helps report partial failures at the end.

onBatchError events (newer feature): Salesforce introduced a Database.RaisesPlatformEvents interface for batch classes. If you implement this and your batch encounters an unhandled exception in a chunk, Salesforce will publish a platform event containing error information and the records that failed. This is super handy to centralize error handling. For example, you could have a separate listener (maybe another Apex trigger on that event or an external system listening) which automatically retries those failed records or logs them. It's like a safety net for batch failures.

Batch finish method: Use the finish method to handle post-batch actions. If your batch needs to notify about any errors that were caught and accumulated, finish is the place. For instance, you might send a single email: "Batch X finished. 950 succeeded, 50 failed. See error log." If the batch was processing records marked with some "Needs Processing" flag, you might want to mark those that failed for reprocessing later.

The philosophy for batch is *resilience*: don't let one error kill the whole job if you can handle it. As an analogy from the DevOps Launchpad content, *"it's crucial that a single error doesn't bring the whole process to a standstill... handle partial successes"*. In practice, this often means designing the batch to continue despite errors, and dealing with them after.

Queueable and Future – Fire-and-forget with a catch:

Queueable and future methods run once in their own context. They aren't chunked like batch, but they often involve an operation over many records or an external call. For these:

You can include try-catch in the execute() of a queueable or the body of a future method. If something fails, you could perhaps attempt a recovery or schedule another attempt. One technique: If a queueable fails to process some records, you could catch and then enqueue a *new* queueable job for those failed records (maybe after a delay or with exponential backoff logic to avoid immediate re-failure). This is a manual retry mechanism.

If you do not catch an exception in a queueable or future, Salesforce will send an email to the user who initiated the async call (e.g., the user who invoked the future or enqueued the job). This email includes the exception message and stack trace. That is helpful but not always sufficient; if no one monitors that inbox or if it's a generic user, the team might miss it. Therefore, it can be better to catch exceptions and handle them by logging or notifications that your team controls (like sending to a specific Slack channel or creating a Case record).

Scheduled Apex is similar: if a scheduled Apex job throws an unhandled exception, the scheduling user gets an email. If your org schedules jobs under a generic integration user, ensure that email is monitored. Many orgs use a dedicated email or forward those to a support system.

One pattern: use a custom setting or custom metadata to store an email or user to notify for failures, and in your catch for queueable/future, send an email to that. That way, you're not reliant on the default platform email which goes to the running user (which might be a no-reply or unmanned mailbox).

Example: Retrying a Failed Queueable:

Suppose you have a queueable that makes a callout to an external API to sync data. If the API is down (throws a CalloutException or returns an error code), you might want to retry later.

```
public class SyncDataJob implements Queueable, Database.Allows-
Callouts {

  public void execute(QueueableContext ctx) {

    try {

      ExternalService.syncData(); // makes callout

    } catch(CalloutException e) {

      // external service down or timeout

      System.debug('Sync failed: ' + e.getMessage());

       // schedule a retry in 30 minutes (using System.enqueueJob again
with a delay if possible)

      // or use a Platform Event or Scheduled job to retry, depending on
requirements

      RetryHandler.scheduleRetry('SyncDataJob', e);

    } catch(Exception e) {

      // log unexpected exception

      System.debug('Unexpected error in SyncDataJob: ' + e);

      // we might let this bubble to get an email, or handle accordingly

    }

  }

}
```

In this pseudo-code, RetryHandler.scheduleRetry is a made-up utility that could perhaps schedule another execution. There isn't a built-in enqueueJob with delay (except scheduling a job via Cron), but one could use a combination of schedule apex or platform event as a timer. The idea is we caught the callout failure and chose to retry later instead of giving up. This is a form of **retry logic** that enhances reliability – a transient error doesn't permanently stop processing.

User Notification in Async:

Even though asynchronous processes lack a user interface, we often need to inform someone when things go wrong, especially if manual intervention is needed. Who that someone is can vary:

If a batch processing orders fails on some orders, maybe the operations team needs a report of those failed orders to handle manually.

If a nightly integration fails, maybe an IT support or on-call engineer needs an alert.

You have options: email, Chatter post, logging (for later viewing), or even a custom object record that surfaces on a dashboard. In a smaller org, an email to admin might suffice. In an enterprise org, you might integrate with an ITSM (IT Service Management) tool – for instance, create a Case or send to ServiceNow via API when a critical scheduled job fails.

Lightning Platform Events in Async:

Lightning platform events can also be used by queueables or future. For example, if a queueable catches an exception and you want to inform a UI in real-time, you could publish a platform event that a Lightning component is subscribed to (maybe via empAPI in LWC) – then the UI could display a notification. This is more complex but shows the flexibility: asynchronous doesn't have to mean silent.

However, a more common usage is just to log and move on, as there's often no user waiting. In fact, one advantage of asynchronous jobs is you can take your time to deal with errors: maybe try alternative strategies without making a user wait.

Architect Tip: Design your asynchronous processes with a failure management workflow in mind. Ask, for each async job: "If this fails for some records, what happens to those records? Do they vanish into the ether, or do we have a plan B?" A robust design might include tagging failed records (e.g., an Error_Flag__c field set to true) so they can be retried later manually or by another job. Or use a "dead-letter" queue concept: records that could not be processed get put into a custom object or queue for further analysis. This prevents data from being lost when a background process encounters an issue. ChatGPT can help brainstorm such strategies if you describe your process: *"My batch job does X, how can I handle records that fail processing?"* It might suggest the above patterns (stateful lists, error flags, etc.) which you can adapt.

Testing Asynchronous Error Handling:

Writing test classes for async processes involves some tricks. To test a batch's error handling, you might create input data that you know will trigger an exception in execute (for instance, maybe insert a record that violates a condition so your batch code will throw). Run the batch in a test context and use Test.stopTest() to execute it. Then verify outcomes: query the log records or check that the failed record was marked as expected. If you implement Database.Stateful to collect failures, after Test.stopTest(), instantiate your batch from the return of Database.executeBatch and check the stateful member (unfortunately in tests you can't get the instance back directly after executeBatch call easily). Another way is to have the batch insert a summary record in finish listing failures, then verify that in test.

For queueables, use Test.enqueueJob and similarly catch exceptions or check results post-Test.stopTest(). You can simulate callout failures using HTTP callout mocks to force exceptions or error codes so you can test your retry logic path.

The goal is to ensure that even in test, when an exception is supposed to happen, your code catches it and the intended logging or recovery takes place. If your queueable is supposed to enqueue a retry job after failure, you can assert that a new Queueable job was indeed enqueued (perhaps by checking some static flag or outcome, since you can't see the job queue directly in a test but you could track via a static variable or an insert).

ChatGPT can generate example unit tests for you. For instance, you might prompt: *"Write an Apex test method to verify that MyBatchJob logs an error when processing a bad record."* It could produce a framework of a test, which you adapt.

Example: Batch Partial Failures and Recovery:

Imagine a batch job processing 1000 records. 990 processed fine, 10 failed due to some data problem. With good design:

Those 10 are caught in try-catch within execute, perhaps added to a static List failedIds.

In finish(), the batch checks failedIds and maybe enqueues a follow-up queueable to process them differently or sends an email listing them.

The system doesn't lose those 10 records; they are flagged for attention. Meanwhile, 990 records are completed, so the batch delivered most of its work successfully.

One can say the batch *gracefully degrades* its functionality when errors happen – it does as much as possible. This concept of graceful degradation is key in asynchronous processes: if part of the work can't be done, do what can be done and handle the rest separately.

Integration with ChatGPT's Operator (Hypothetical): If you have access to the ChatGPT Operator (the autonomous browsing agent), you might even use it for monitoring asynchronous processes. For example, you could instruct Operator to periodically check a Salesforce dashboard or an error log page for any new errors and then compile a report or attempt certain fixes. This is speculative but shows how AI might assist DevOps: a kind of automated "sentry" watching for failures. Operator could navigate your org (with proper authentication) to retrieve logs, though in practice building that might be complex and raise security considerations. Still, it's an interesting idea for the future of handling exceptions – AI agents that proactively catch and resolve certain errors (like automatically retrying a job after a transient failure, based on predefined rules).

Tester Tip: Test asynchronous processes under load and failure conditions. For example, test a batch with large data to see if you hit any limits (like CPU) and ensure your code either handles it or at least fails gracefully. If you discover via testing or logs that a batch hit a governor limit (like too many SOQL queries), realize that such LimitExceptions *cannot be caught* in Apex. The solution is to refactor code to avoid hitting the limit (bulkify queries, etc.). Sometimes testers will notice an Apex job quietly not processing everything; that could be because it died on a Limit without proper logging. Bringing that up leads to performance fixes rather than just error handling.

By adeptly handling exceptions in asynchronous Apex, we ensure background processes are resilient and maintain data integrity. We allow for partial success, implement retries for transient issues, and notify the right people when needed. In an enterprise, this means nightly integrations don't require constant babysitting—if something goes wrong at midnight, the system either self-recovers or at least logs and alerts so the team can react first thing in the morning. Even an SMB benefits by not having silent failures; a small business might not have a 24/7 admin, but with good logging and perhaps an email alert, they can address issues in a timely manner and keep operations running.

Now that we've covered asynchronous error management, let's extend our discussion to Apex integrations—scenarios where Salesforce interacts with external systems, which often require careful error handling strategies like retries and graceful degradation on failures.

EXCEPTION HANDLING IN APEX INTEGRATIONS

Salesforce Apex frequently serves as the bridge between Salesforce and other systems through APIs and web services. Whether you're making callouts to external services (REST/SOAP) or exposing Apex as a web service (HTTP endpoints via Apex REST/SOAP classes), robust error handling in these integrations is critical. External systems can be unreliable: they might time out, return errors, or be down entirely. Moreover, when things go wrong, it's not just a Salesforce issue—customers or downstream systems may be impacted. In this section, we'll look at how to handle exceptions in Apex integrations, including callouts and API interactions, focusing on graceful degradation, retries, and clear communication of errors.

Challenges in Integrations:

Network and External Failures: Unlike internal Apex operations, callouts depend on network connectivity and external application health. Common exceptions include CalloutException (e.g., timeout, DNS failure) and IOException (if something low-level goes wrong in the HTTP request). The external service might also respond with an error status (like HTTP 500 or 400), which doesn't throw an Apex exception by itself but is still an error that your code needs to handle (you might treat a non-200 HTTP status as an "exception" for business logic).

Partial Transactions: Often, an integration involves multiple steps. Perhaps you create a record in Salesforce and then need to send its data to an external billing system. If the external call fails after the record is created, you have a partial transaction that needs reconciliation.

User expectations: If a user triggers an action that involves a callout (say clicking "Sync to ERP"), they expect either success or a clear message if it failed. You need to manage that user experience despite the external uncertainty.

Security and Idempotency: Retrying integration calls is often necessary for transient failures, but you must consider if the operation can be safely repeated (idempotent). For example, sending the same data twice to a payment gateway might double-charge—so maybe you shouldn't blindly retry those without safeguards.

Handling Callout Exceptions: Always wrap callouts in try-catch. The types of exceptions you can catch include:

CalloutException: covers timeouts (if the service doesn't respond within the set time limit) and other callout issues.

System.HttpResponse[Exception]: actually, the HTTP callouts in Apex don't throw on HTTP error status; you get a HttpResponse object with a status code. So for logical errors (like a 404 or 500), you handle those by checking response.getStatusCode() rather than catching an exception. However, if the endpoint URL is bad or something, Apex can throw.

JSON or XML parsing exceptions: If you callout and then parse the response, be ready to catch JsonException or XmlException if the payload isn't what you expected.

Example:

```
Http http = new Http();

HttpRequest req = new HttpRequest();

req.setEndpoint(remoteUrl);
```

```
req.setMethod('POST');
req.setBody(jsonBody);
HttpResponse res;
try {
  res = http.send(req);
} catch (CalloutException e) {
  // likely a timeout or inability to reach host
    throw new ExternalServiceException('Billing service unreachable', e);
}
if(res.getStatusCode() >= 500) {
  // server error on other side
    throw new ExternalServiceException('Billing service error: ' + res.getStatus());
} else if(res.getStatusCode() >= 400) {
  // client error, maybe our request was bad
  // log details for debugging
  System.debug('Billing service returned 4xx: ' + res.getBody());
    throw new ExternalServiceException('Billing request was rejected. Error ' + res.getStatusCode());
}
// else 200-399 range, consider it success (could be 202 accepted etc.)
```

In this snippet, we created a custom ExternalServiceException to encapsulate any issues with the billing service. We treat callout exceptions and HTTP error statuses similarly by throwing this custom exception. The calling code can catch ExternalServiceException to know something went wrong with the external system, distinct from, say, a Salesforce DML exception.

Retry and Fallback Patterns: Integrations often benefit from retries. If an external service times out or gives a transient error (like 503 Service Unavailable), a simple retry after a short delay might succeed. However, Apex has limitations: you can't pause a running transaction to wait and retry (no sleep/delay in Apex), and you have governor limits on how many callouts you can do in one execution. A pattern for retry might involve:

If a callout fails, schedule a future method or queueable (perhaps with an exponential backoff counter) to try again. We saw an example of this in the asynchronous section with RetryHandler.scheduleRetry. That could be implemented with a custom object "Integration_Job__c" that a scheduled Apex picks up and retries, or simply using System.schedule to run the same call 5 minutes later.

Alternatively, if using platform events, on a callout failure you could publish an event "BillingCalloutFailedEvent" which triggers another piece of logic that waits or retries.

Use a circuit breaker concept: if the external system has been failing repeatedly, maybe stop trying aggressively and mark it as down (to avoid overloading it or spamming errors). This can be done via a custom setting/flag that turns off integration temporarily after X failures, and perhaps notifies ops.

Fallbacks: Sometimes, if an integration is down, you might have an alternate path. For example, if the live pricing API is down, maybe use a cached price or default price so at least the process continues in a degraded mode. Or if an address verification service fails, you can let an order go through unverified and flag it for later review. Fallback strategies are very case-specific but always aim to balance the risk of proceeding without external confirmation vs the benefit of continuing service. Graceful degradation might be: log the failure, inform the user that "External service is unavailable, the action will be completed later" (if you have an async retry) or "... please try again later" if not.

Product Owner Tip: Decide which integrations are mission-critical vs which can wait. This influences error handling. For mission-critical ones (e.g., payment processing), you must handle failure very explicitly (likely no silent fallback except maybe trying a backup gateway). For less critical (like syncing a marketing preference to an external system), you might queue it for later and even not alert the user at all, just ensure it goes eventually. Communicate these differences to your dev team so they implement appropriate responses. Using ChatGPT, you could even draft user-facing wording for error messages: *"How should we phrase an error message when the payment service is down?"* and refine from there.

Consuming External Errors: When Apex is on the receiving end (for example, an Apex REST service that external systems call), you want to handle exceptions and return proper error responses. If an unhandled exception occurs in Apex REST, the external caller will get a generic 500 with maybe an error message (which might expose Apex internals if not caught!). To avoid that:

Use try-catch in your Apex REST methods and return a structured error message or status. You can set RestContext.response.statusCode = 400; and include a message in the response body JSON or XML.

For example, if an external calls your service to create a record and your Apex catches a validation (like required field missing), you might catch DmlException and then do:

RestResponse res = RestContext.response;

res.statusCode = 400;

res.responseBody = Blob.valueOf('{ "error": "Validation failed", "details": "'+ e.getMessage() + '"}');

return null; // since we've set response manually

This way, the external system gets a clean message. Avoid letting raw exceptions propagate out of your web service.

Secure Error Information: Be cautious not to expose sensitive info in error messages across integrations. For example, if a callout returns an authentication error, don't log or expose the full auth token or credentials in any error. Similarly, your Apex REST service shouldn't return a full Apex stack trace to a caller (that could reveal internal schema names or logic). Instead, return a simple error code or message. Internally log the detailed error for debugging.

Graceful API Error Handling: Consider a scenario: Salesforce calls an external shipping API to get rates. If the API is down:

The user trying to get a shipping quote waits a bit, then our callout times out. We catch CalloutException.

We decide to have a fallback: maybe use last known rates or a flat rate. We inform the user: "Real-time shipping rates are currently unavailable, a default rate will be used." The process continues so the user isn't blocked.

Meanwhile, we log this incident (so ops knows the shipping API might be down).

We might also trigger an asynchronous job to keep retrying to get the actual rate in the background. Perhaps after the order is placed, an integration job picks up and tries to update the order with the actual shipping cost once the service is back.

If that happens, maybe notify the user later or adjust billing accordingly.

Such a flow ensures business continuity. The user can place an order (with maybe an approximate shipping cost) rather than being stopped. It requires business agreement that using a default is acceptable when the API fails—these are the design discussions that need to happen.

Using ChatGPT's voice feature, a product owner or architect could even simulate how to handle such a conversation: *"ChatGPT, if our shipping provider API goes down, what are best practices to handle shipping cost calculation?"* The answer might include exactly these strategies (caching, defaults, notification), giving you talking points for stakeholder meetings.

Integration Specific Tools and Settings:

Timeout settings: By default, callouts have a 10-second timeout. If an external API often takes longer, you can increase up to 120 seconds with req.setTimeout(120000). But balancing user wait vs timeout is key. If a user is waiting, 2 minutes might be too long – maybe do the callout asynchronously instead in that case.

Circuit Breaker via Custom Setting: Possibly have a custom setting like "ExternalServiceAvailable = true/false". If you get repeated failures, set it to false (manually or automatically after X tries). Your code can check this setting and if false, skip callouts and immediately throw a user-friendly exception "Service currently unavailable, try later." This prevents making useless callouts when you know the service is down, and gives immediate feedback. It's like tripping a circuit breaker to stop calls until you reset it.

Governor Limit – Callout Limits: You can only have 100 callouts per Apex transaction. If your integration logic loops over records making individual callouts, you risk hitting that. Better to bulk calls if possible (some APIs allow batch requests) or move callouts to async (because in batch each execute gets its own limit of 100 callouts). If a LimitException for too many callouts occurs, you can't catch it , so design to avoid it.

Testing Integrations: Use HttpCalloutMock in tests to simulate various responses: success, error, timeout. For timeout, you can't directly simulate a real timeout but you can simulate an exception by having your mock throw a CalloutException when Http.send is called. Ensure your test covers the branch where the callout fails and that your fallback or error logging logic runs. Ask ChatGPT to generate a mock implementation if unsure. For example: *"How can I simulate a callout timeout in an Apex test?"*, ChatGPT might suggest implementing HttpCalloutMock that returns no response or throws.

Operator usage: If troubleshooting an integration issue, ChatGPT's Operator could potentially fetch the external API documentation or status page if needed (for example, if you get an unfamiliar error code from an API, Operator could search the API's knowledge base for that code). This saves developer time cross-referencing error codes.

Communicating Errors to Users and Systems: We touched on user messages. Equally, if an external system calls Salesforce and something fails, communicating back an error code is polite. For example, if an external system calls your Apex REST to create an invoice in Salesforce, but the data violates a validation rule, catch the DmlException and return a 400 with a message like {"error":"INVALID_FIELD_VALUE", "message":"Invoice Date cannot be in the future."}. External developers will thank you for clear, predictable error responses. It's similar to how the Salesforce REST API returns errors in a structured array with errorCode and message. If you mimic those patterns, integration becomes smoother.

In summary, handling exceptions in integrations is about being prepared for the unpredictability of external dependencies:

Anticipate failures by coding defensively (try-catch and checking response statuses).

Retry sensibly to overcome transient issues but avoid infinite loops or harmful duplicates.

Fail gracefully by falling back or at least not crashing the whole transaction unless absolutely necessary.

Inform clearly both the user (if they initiated the action) and admins/support (through logs/alerts) about the issue.

Maintain data consistency by rolling back changes or marking incomplete data for cleanup if the integration couldn't finish. For instance, if you created a record then a callout failed, maybe delete or deactivate that record and notify someone, or mark it "pending external sync".

Architect Tip: Document your integration error handling flows. For each integration point, write down: "If service X is down or returns error Y, our system will do Z (retry, fallback, etc.)." This not only guides development but is crucial for operations. If an incident occurs, having this reference lets everyone know what the expected behavior was. You can even use ChatGPT to draft a failure mode effects analysis: *"List possible failure scenarios for our Salesforce-Payment Gateway integration and suggest handling for each."* This can serve as a starting checklist to ensure no scenario is overlooked.

With integration patterns addressed, we've covered a large spectrum of Apex exception handling: from internal code and best practices, through asynchronous processing, to cross-system communication. The final piece of the puzzle is to step back and look at proactive strategies – how to avoid exceptions in the first place. That's what we'll explore next, rounding out this chapter with techniques for error prevention and early detection.

PROACTIVE STRATEGIES AND ERROR PREVENTION TECHNIQUES

The best exception is the one that never occurs. While it's vital to handle exceptions gracefully when they happen, it's even better to design your system to minimize the occurrence of errors. Proactive error prevention involves anticipating potential issues and coding defensively to avoid runtime exceptions. It also means leveraging Salesforce's declarative features (like validation rules) and thorough testing to catch problems before they hit production. In this final section, we'll discuss techniques such as input validation, defensive programming, guarding against common pitfalls, and how to embed these practices into development and testing cycles.

Input Validation and Sanitization: Many exceptions in Apex, such as NullPointerException or QueryException, can be prevented by checking inputs and assumptions. For example:

Null Checks: Before you use an object or variable that might be null, check it. Rather than assuming a query returned results, verify !results.isEmpty() before accessing results[0]. If you pass a parameter to a method, assert that it's not null (throw a meaningful exception or return a status if it is). This way, you turn a potential NullPointerException into a controlled flow – possibly throwing a custom exception like InvalidParameterException with a message, which you can catch at a higher level or at least will show clearly in logs.

Bounds Checking: For collections, check indices and sizes. A List<String> indexing error (ListException) can be avoided by verifying the index is < list.size(). Similarly, if expecting at least one record from a SOQL query, perhaps check the query result count; if 0, handle that scenario (maybe throw a custom exception or take alternate action) rather than letting code proceed and blow up later when accessing non-existent elements.

Value Validation: If a method expects a positive number (e.g., a percentage or quantity), validate the argument at the start and throw an IllegalArgumentException (or custom exception) if it's out of range. This is a fail-fast principle – you catch incorrect usage early, which is easier to debug. It's better to throw a controlled exception saying "Quantity must be non-negative" than to let some deeper calculation go wrong and throw a divide-by-zero or some other error.

Use of Schema.Describe for metadata constraints: For instance, if your code accepts an sObject field name from user input (like dynamic SOQL or data mapping), use describe calls to verify the field exists on the target object. That prevents a InvalidFieldException at runtime when running a dynamic query with an invalid field.

Defensive Programming: Defensive programming is about writing code that assumes something might go wrong and handling it gracefully. Some practices:

Use try-catch around risky operations even if you think they "shouldn't fail". For example, if parsing an integer from a string (maybe from user input via API), use try-catch for TypeException or use Integer.valueOf carefully. If you catch it, you can throw a custom exception that's more user-friendly or handle the default.

Avoid hard-coding IDs or assumptions that might break in different orgs or contexts. For instance, code that assumes record types or profile names can cause errors if those aren't present. Use Custom Metadata/Settings or describe info to make the code adaptable.

Guard against SOQL limits preemptively: If you're writing a method that *could* be called in bulk, assume it will be. Instead of doing a SOQL per record in a loop (which would throw LimitException after 100 records), bulkify from the start (query outside the loop, use maps, etc.). This prevents hitting the governor limit exception at all. Similarly, use Limits.getQueries() and related methods if you must dynamically adapt behavior as you approach a limit. For example, if you absolutely must loop with callouts, you could check if(Limits.getCallouts() == Limits.getLimitCallouts() - 1) { // we're about to hit the last allowed callout } – though such designs should be refactored to avoid getting close to the limits in the first place.

Utilize Apex design patterns that inherently reduce errors: e.g., the Bulk State Transition pattern (processing records in chunks with careful state management) can avoid hitting limits. Or the fflib Apex Commons framework encourages separation of concerns which can help ensure, say, all database operations are done in a single unit (reducing chance of scattered DML and related exceptions).

Declarative Enforcement (Clicks not Code): Salesforce provides declarative ways to prevent bad data:

Validation Rules: If certain combinations of data are invalid for processing, a validation rule can prevent the record from being saved in that state. This avoids Apex ever having to handle it. But be cautious: if Apex itself is creating records and triggers a validation rule, that will raise a DmlException. So your Apex either needs to anticipate that (and perhaps fix the data or handle the exception) or you temporarily bypass the validation in Apex if appropriate (not usually recommended except maybe in back-end integration where you intentionally set field values that violate a UI rule).

Flows or Triggers for data corrections: A "before insert" trigger could, for example, default a missing field instead of letting a null cause an error later. Or it could split overly large tasks (like if a user tries to do an operation on 1000 records via a custom UI, maybe the trigger or logic splits it to avoid hitting limits).

Permission checks: Ensure users have the right permissions to avoid exceptions like NoAccessException when running without field access. Check user permissions at the start of operations (via Schema.sObjectType.MyObject.fields.MyField.isAccessible() etc.), and throw a meaningful message if not (or handle accordingly), rather than letting a security exception occur. Properly handled, this never surfaces as an unhandled exception.

Use of Test Classes as Preventive Tools: Thorough testing is one of the best proactive strategies. If your test suite covers edge cases and simulates error conditions, you catch them before deployment. For example:

Write tests that intentionally pass bad data to methods to ensure the method handles it (and perhaps returns an error or throws a specific exception). If the method currently just blows up with NullPointerException, your test will fail, signaling you need to add handling.

If you fix a bug that caused an exception, add a test for it to prevent regression. Use test data that hits governor limits in a controlled environment. For instance, create 101 records and try to insert them in a context without the proper bulk handling, see that it fails, and then adjust code accordingly. Once adjusted, the test should pass (meaning the code either now bulkifies or catches the limit situation gracefully).

Leverage tools like Static Code Analysis (PMD, SonarQube) in the build process. They can catch patterns likely to cause issues: SOQL in loops, empty catch blocks, etc., which are essentially possible error hotspots. Treat those findings seriously and address them before they become runtime problems.

Pattern: Fail Fast vs. Fail Safe: Fail-fast (immediate failure on bad input) often helps expose issues early, whereas fail-safe (try to continue with defaults or corrections) can make the system more robust but potentially mask issues. Use a combination:

In development and testing, favor fail-fast: throw exceptions when something is off. It's better to know and fix the cause.

In production, certain non-critical features can be fail-safe: e.g., if an optional external service is down, log it and continue with a warning instead of failing the whole operation (as we discussed).

Provide toggles (maybe a debug mode) where fail-fast can be enabled to troubleshoot. For instance, if something weird is happening, an admin could turn on "Strict Mode" via a Custom Setting, making the code throw exceptions instead of silently defaulting values, just to see what's going wrong. This is advanced, but can be useful.

Avoiding Common Apex Pitfalls:

Aggregate Query Results Limit: If a SOQL could return over 50k records, design with that in mind (either batch processing or filter further). Otherwise, a runtime exception will occur (QueryException: exceeded 50001 rows). Better, use Batch Apex or query with LIMIT and partial processing in loops. Knowing these limits (50k query, 10k DML in one go, etc.) and coding within them prevents exceptions like LimitException or QueryException.

Email and Callout Limits: If sending emails in code, max 10 per transaction (SingleEmailMessage). If you need more, you'll get a LimitException. Avoid by aggregating or using email-to-case or other mechanisms for bulk notifications.

SOQL on unindexed fields in huge data might not throw an exception but can be super slow or hit timeouts. Use selective queries (consider adding custom indexes, using query filters that hit indexes) to avoid hitting CPU time limits.

Many of these fall under performance, but performance issues often manifest as exceptions (CPU time limit exceeded is a common one that is effectively an uncatchable exception).

Culture of Proactive Error Handling: Encourage developers to think "what could go wrong here?" at each step. One way is code reviews focusing on error scenarios:

When reviewing a pull request, ask "What if this method receives an empty list? What if the external ID doesn't exist? Did we handle that?"

Create a checklist for developers and testers: e.g., "Any time you write a SOQL in code, consider scenario of 0 records, >1 record, or too many records. Any time you do a division, consider divisor 0. Any time you cast, consider wrong type. Any external call, consider it failing."

Pair programming with ChatGPT: A developer can literally ask, *"ChatGPT, review my method X for any potential exceptions or edge cases I might have missed."* The AI might point out, for example, "If the map is empty, map.get(key) will return null leading to NullPointerException on line Y" – a helpful nudge.

Preventive Monitoring: If you have logging and monitoring, use that data not just to react, but to prevent. For instance, logs might show frequent "NULL_FIELD_VALUE" errors on a particular process. That suggests perhaps you can add a check or a default to stop those from happening. Or if a certain integration times out once a week around midnight, maybe schedule it differently or see if you can optimize the call to finish quicker.

Also, Salesforce has features like **Apex Exception Emails** (which send to developers when unhandled exceptions occur in Apex code running via certain contexts like Visualforce or scheduled). If you ever get one, treat it as a bug to fix – even if you "handle" it by adding a catch to suppress, consider how to eliminate the root cause if possible.

Continuous Learning and Improvement: No matter how proactive you are, new types of errors will pop up (especially as features are added or usage patterns change). Make it a habit to periodically review what exceptions occurred (via logs or cases) and update your best practices. Maybe you notice many times your code threw QueryException for non-selective query. Next iteration, proactively add indexes or change queries and mark that as a new standard for large object queries.

Tester Tip: Incorporate negative testing in QA cycles. Beyond the happy path, deliberately feed bad data, simulate outages, etc. Use tools like Postman to call your Apex APIs with wrong inputs to see how they respond. Or in UI tests, try leaving required fields blank to see if Apex or the platform gracefully tells the user. If the system fails ungracefully in any of those, that's a bug to address. Testers can also use ChatGPT to brainstorm edge cases: *"What edge cases should I test for a feature that transfers money between accounts?"* (It might answer: negative amounts, same account transfer, extremely large amounts, etc.) This can be applied to any feature to catch exceptions before code hits production.

Summing up prevention: It's a mindset of being one step ahead of errors:

Use Salesforce features (validation, permissions) to stop bad data.

Write code that checks conditions rather than assuming.

Design for limits and bulk from the get-go.

Test thoroughly and use results to harden the code.

Learn from past incidents to fortify future code.

By implementing these proactive measures, an organization can dramatically reduce the number of exceptions that occur in production, which in turn means fewer user-facing issues and less firefighting. It's often said that 80% of errors come from 20% of the causes – if you identify and address those causes early, you eliminate a majority of exceptions. For those that still slip through, you have all the robust handling strategies from earlier sections to ensure they don't wreak havoc.

Architect Tip: Integrate error prevention into your coding standards and CI/CD pipeline. For example, have automated checks that fail the build if System.assert(false) appears (indicative of a deliberate failure without message) or if any catch block is empty. Also, include guidelines like "all service methods must validate inputs" in your development handbook. ChatGPT can even assist in generating a section of your internal wiki or documentation for "Defensive Coding Guidelines in Apex." Ensuring everyone follows them makes the whole codebase more resilient.

As we conclude this section and the chapter, we've journeyed from reacting to errors to preventing them. Enterprise or SMB, these principles scale: a small org might have simpler logic but still benefits from checking nulls and using validation rules; a large enterprise with complex integrations absolutely needs these practices to keep the system robust.

With proactive strategies in place, you are not just putting out fires – you're installing smoke alarms and fireproofing the building. The combination of strong exception handling and preventative design results in an Apex codebase that is stable, predictable, and easier to maintain.

SUMMARY

In this chapter, we explored how to handle exceptions and errors in Apex comprehensively. We began with the fundamentals: understanding exceptions in Apex and the importance of structured try-catch-finally blocks. We emphasized catching specific exceptions (like DmlException or CalloutException) instead of using broad strokes, and avoiding the mistake of swallowing exceptions without action. Through examples, we saw how ChatGPT can aid in refactoring poorly structured error handling into clearer, more maintainable patterns.

We then considered custom exceptions, learning how creating our own exception classes (such as PaymentFailedException or BusinessLogicException) can make error handling more expressive and tuned to our business needs. This allows different error causes to be handled differently and makes the code's intent more obvious. We provided tips for architects on fitting custom exceptions into overall application architecture and saw how ChatGPT can help brainstorm or implement these classes.

Logging and monitoring formed the next major topic: capturing exceptions in a persistent way so that no error goes unnoticed. We discussed strategies from simple debug logs to sophisticated frameworks like Nebula Logger that centralize error information. We highlighted that logging is invaluable for troubleshooting and that combining it with alerting (emails, platform events, etc.) ensures the right people know about issues. ChatGPT's assistance in analyzing log patterns and even generating visualizations (like charts of error frequency) was noted as a modern way to leverage AI for insights.

Shifting to asynchronous processes, we covered the special considerations for batch jobs, future methods, queueables, and scheduled Apex. The key is designing for partial failures – allowing a batch to continue processing even if some records fail, and capturing those failures for later recovery. We illustrated try-catch within batch execute, the use of stateful batches to aggregate errors, and the importance of notifying or compensating when asynchronous operations encounter problems. This ensures background operations are robust and enterprise-friendly, not requiring constant manual supervision.

For Apex integrations, we addressed the challenges of callouts and external services. We looked at how to handle callout exceptions and HTTP error responses gracefully by wrapping them in custom exceptions or user-friendly messages. We talked about implementing retries for transient issues and fallback strategies when external systems are down, so the Salesforce app can degrade gracefully instead of outright failing. We also pointed out the need to not expose sensitive internal errors to external callers, by catching exceptions and returning clean error messages or codes.

Finally, we closed with proactive approaches to avoid exceptions: validating inputs, coding defensively, using Salesforce's declarative features to maintain data integrity, and testing thoroughly to catch issues early. This proactive mindset, combined with continuous improvement, helps minimize the occurrence of runtime exceptions in the first place. We emphasized how adopting these techniques leads to better system stability and maintainability.

Throughout the chapter, we provided callout tips for various roles – from developers gaining tactical advice (like not to swallow exceptions) to architects setting up enterprise standards (like logging frameworks and exception hierarchies), testers ensuring every error path is tried, and product owners seeing the value in treating error handling as part of the user experience (thus allocating time for it in requirements and testing). We also interwove ways ChatGPT can be used: as a code assistant, a brainstorming partner for error handling strategies, a quick reference for best practices (even via voice queries), and a tool for generating test scenarios or documentation.

In summary, **exception handling and robust error management in Apex** is a multi-faceted discipline. When done right, it means Salesforce applications that can withstand the unexpected – from a missing piece of data to a third-party outage – without derailing business processes. It results in clearer code (because the "what ifs" are explicitly handled), happier users (who see informative messages instead of cryptic errors), and empowered support teams (who have logs and tools to diagnose issues quickly). By leveraging both the platform features and intelligent assistants like ChatGPT, teams can implement these best practices more efficiently and with greater confidence.

As you move forward to the next chapter, you'll carry with you this solid foundation in error management. It will serve you well, because whether the topic shifts to security, performance, or any other aspect of Apex development, the principles of anticipating problems, handling them gracefully, and continually improving are universally applicable. With reliable error handling in place, you can now focus on building the next features, knowing that your application is resilient underneath – a true hallmark of enterprise-grade Salesforce solutions.

7

APEX SECURITY BEST PRACTICES

Security plays a crucial role in developing robust and trustworthy Salesforce applications. In an enterprise environment, where customer data and business processes rely on Apex code, a single vulnerability can undermine the entire organization's trust infrastructure. Even in smaller businesses, insecure Apex code can lead to compromised data, regulatory fines, and loss of customer confidence. This chapter addresses why treating security as a first-class concern in Apex development is non-negotiable for both enterprises and SMBs.

Inadequate security practices in Apex can have serious consequences. Data breaches exposing sensitive customer or employee information may lead to compliance violations (for example, violating GDPR data protection rules or HIPAA regulations on health data) and significant reputational damage. A flaw as simple as a missing permission check or an injectable query string can open the door to attackers, resulting in financial loss and legal penalties. By understanding common pitfalls and implementing best practices, Salesforce architects and developers can prevent these outcomes and protect their organizations.

Readers of this chapter will gain practical guidance on enforcing Apex security standards, protecting sensitive data, implementing secure coding techniques, and leveraging Salesforce's built-in security mechanisms effectively. We will explore how to properly enforce user permissions in code, guard against injection attacks and cross-site scripting, use sharing rules to control record visibility, secure external integrations, encrypt data to meet compliance requirements, and audit your Apex code for vulnerabilities. Each section provides in-depth explanations and real-world examples, ensuring you can apply the advice directly to your Salesforce projects.

Apex-specific security aligns closely with broader Salesforce enterprise security strategies. While Salesforce provides robust declarative security settings (profiles, roles, permission sets, etc.), Apex code runs with powerful privileges that can bypass those settings if not written carefully. That's why architects, developers, testers, and product owners all have a stake in Apex security. Architects must design systems that meet enterprise security standards; developers need to follow secure coding patterns; testers must validate that security controls work correctly; product owners and sponsors should understand the trade-offs and importance of investing in security measures. The content here addresses concerns for each of these roles, from technical implementation details to process and governance considerations.

Throughout the chapter, we also highlight how ChatGPT can assist you in achieving stronger Apex security. As an AI assistant, ChatGPT can help audit Apex code for vulnerabilities, suggest secure coding improvements, and even help ensure ongoing compliance by interpreting complex regulations into actionable steps. Whether you're asking ChatGPT to review a piece of code for field-level security enforcement or using it to brainstorm test cases for injection attacks, integrating this tool into your development workflow can enhance your team's security vigilance.

We will cover seven key areas of Apex security best practices, each in its own section:

Enforcing CRUD and field-level security: We begin with the fundamentals of ensuring your Apex code respects user permissions on objects and fields.

Preventing SOQL injection and XSS: Next, we examine how to protect

Apex from malicious input via injection vulnerabilities and cross-site scripting exploits.

Sharing and record visibility: We discuss the proper use of Apex sharing rules and visibility controls to enforce record-level access consistently.

Secure callouts and integrations: We outline strategies for safeguarding Apex callouts to external services, including managing authentication tokens and using encryption for data in transit.

Data encryption and secure storage: We cover how to encrypt sensitive data at rest using Salesforce Shield and Apex crypto, and how these practices align with GDPR and HIPAA compliance.

Auditing Apex security: We explore Salesforce tools (like Security Health Check and event monitoring) that help audit your Apex code and catch security gaps or anomalies.

Fostering a security-first culture: Finally, we conclude with methods to build a team culture where security is ingrained in every step of the Apex development lifecycle.

Each section provides detailed explanations, extensive ChatGPT prompt examples (including voice, file, and image-based interactions), and role-specific tips so you can put these concepts into practice effectively. With this overview in mind, let's begin by tackling the foundation of Apex security: ensuring CRUD and field-level security checks are properly enforced in your code, so that no user can access data they shouldn't.

ENFORCING CRUD AND FIELD-LEVEL SECURITY (FLS) IN APEX

Every Salesforce developer must ensure that Apex code respects the platform's declarative security settings. **CRUD** (Create, Read, Update, Delete) and **Field-Level Security (FLS)** controls determine what data a user is allowed to access. When Apex runs in *system context*, it ignores these permissions by default. This means that without explicit checks, your code could inadvertently read or modify data the user shouldn't see. Enforcing CRUD and FLS in Apex is essential for securing Salesforce data effectively: it prevents unauthorized data exposure, upholds compliance with regulations, and maintains the principle of least privilege in your applications.

Why Apex Runs in System Mode and Its Implications

By design, most Apex executes in system mode, which means the code has access to all objects and fields, regardless of the running user's permissions. For example, an Apex trigger on the Account object will run and can query or update any field on Account, even if the user who caused the trigger doesn't have visibility into those fields. Without proper enforcement of CRUD/FLS checks:

Data privacy violations can occur: A user might indirectly retrieve sensitive fields (like a Contact's Social Security Number or an Account's confidential rating) via Apex code that exposes them (perhaps in an email or downloadable report) despite not having permission to view those fields in the UI.

Compliance breaches become possible: Laws like GDPR and HIPAA mandate that personal data access be restricted to authorized personnel. If Apex bypasses profile or permission set restrictions, it could lead to audit findings or fines.

Security review failures: If you're developing a managed package or enterprise solution, Salesforce's security review (or internal audits) will flag missing CRUD/FLS enforcement as a vulnerability. It's a common source of issues that architects and developers must address early.

Architect Tip: Make CRUD and FLS enforcement a design requirement. Include security architects in design reviews to ensure every Apex entry point (trigger, class, batch, etc.) has a strategy for permission checks before it handles data.

Best Practices for Enforcing CRUD and FLS

Salesforce provides several mechanisms to check user permissions in Apex code. Here are some best practices and techniques:

Use Schema describe methods: Apex's Schema namespace can check object and field accessibility. For example, before querying or reading a field, use Schema.SObjectType.Account.fields.AnnualRevenue.isAccessible() to verify the current user can read that field. Similarly, use isCreateable(), isUpdateable(), and isDeletable() on sObject or field describe results to check permissions for insert, update, and delete operations. If a check returns false, your code should gracefully handle it – for instance, skip that field or throw a custom exception indicating insufficient access. This manual checking is the traditional approach and gives fine-grained control.

SOQL with WITH SECURITY_ENFORCED: For queries, Salesforce offers a query clause WITH SECURITY_ENFORCED that automatically enforces object and field-level security in SOQL. If the running user doesn't have access to a field or object referenced in the query, the query will throw a runtime exception. This simplifies enforcement – you can wrap your SELECT statements with this clause instead of writing multiple permission checks. However, be prepared to catch exceptions or avoid querying fields the user isn't permitted to see. For example:

```
// This query will fail if the user lacks read access on AnnualRevenue

List<Account> acctList = [SELECT Id, Name, AnnualRevenue

            FROM Account

            WITH SECURITY_ENFORCED];
```

In practice, WITH SECURITY_ENFORCED is useful in web services or utility methods where an exception can be handled or propagated. It ensures you never accidentally expose restricted fields through query results.

Use Security.stripInaccessible(): This Apex method programmatically removes fields that the user has no access to from sObjects. It can be used after a query or before a DML operation to strip out disallowed fields. For example, after querying accounts, you can strip inaccessible fields from the resulting list:

```
List<Account> rawResults = [SELECT Id, Name, AnnualRevenue, Phone FROM Account];

SObjectAccessDecision decision = Security.stripInaccessible(AccessType.READABLE, rawResults);
```

```
List<Account> sanitizedResults = (List<Account>)decision.getRecords();

// 'sanitizedResults' now has null for any field that the user is not allowed to see
```

In this snippet, if the user cannot see AnnualRevenue or Phone, those fields will be nulled out in sanitizedResults. This method is a powerful way to enforce both CRUD and FLS for bulk data operations. Similarly, you can use AccessType.CREATEABLE or UPDATEABLE with stripInaccessible before inserting or updating records to avoid exceptions and to sanitize sObjects that have been deserialized from an untrusted source.

Leverage user mode for database operations: Recent Salesforce enhancements allow developers to run specific operations in user mode rather than system mode. For example, you can perform DML using the Database.insert(records, AccessLevel. USER_MODE) method or the shorter syntax insert as user. When you run DML in user mode, the platform will respect the user's object and field permissions – if the user isn't allowed to create or update a field, the operation fails with an exception. Similarly, *SOQL queries can be executed in user mode* by using methods like Database.query(query, AccessLevel.USER_MODE) or Database.queryWithBinds. These features minimize the need for manual checks by making the platform enforce security on your behalf. Keep in mind that as of this writing, user-mode database operations may need enabling as a feature and proper error handling to catch permission errors.

Apply "with sharing" appropriately: While sharing settings (record-level access) are a separate layer of security (discussed in the next section), remember that using *with sharing* on your Apex classes does *not* enforce CRUD or FLS. It's a common misconception. You must still use one of the methods above to enforce object and field permissions. Think of "with sharing" as controlling *which records* your code can see, and CRUD/FLS checks as controlling *which fields and objects* your code can operate on.

When combining these techniques, choose the approach that fits your context. For a simple utility class, using WITH SECURITY_ENFORCED in queries and insert as user for DML can be quick and effective. In more complex logic, you might do manual checks or use stripInaccessible to handle multiple records and avoid exceptions. The key is that before performing any CRUD action or exposing any field value, your code consciously verifies user permissions.

Developer Tip: Integrate permission checks into your coding routine. For example, start writing queries by default with WITH SECURITY_ENFORCED during development. Use helper functions or a framework method (some teams create a utility like UserPermissions.canEdit(object, field) to centralize these checks) so that enforcing security becomes a one-line call rather than repetitive boilerplate.

USING CHATGPT TO AUDIT AND IMPROVE SECURITY CHECKS

One practical way to ensure your Apex code is properly enforcing CRUD and FLS is to use ChatGPT as a code-review assistant. You can ask ChatGPT to analyze your code and point out where security checks might be missing. For example, a developer might prompt:

Developer: "I have an Apex controller below. Can you check if it respects field-level security and CRUD permissions? If not, what should be fixed?" *[Attaches or pastes Apex code]*

If you provide your Apex class or trigger (either by pasting the code or uploading the file to ChatGPT), the AI can examine each data operation. ChatGPT will identify lines where the code queries or updates sObjects without permission checks and suggest the appropriate solution – such as adding Schema.describeSObject() checks or using Security.stripInaccessible. Always verify the suggestions, but this can save time by catching oversight that a manual review might miss. *(Remember to sanitize any sensitive code or dummy data before sharing, and consider turning off chat history or using secure settings to maintain privacy when uploading code.)*

ChatGPT can also help generate snippets for you. For instance, if you aren't sure how to use stripInaccessible, you could ask:

Developer: "Show me an example of using Security.stripInaccessible in Apex after a SOQL query on Contact."

ChatGPT might then provide a short code example demonstrating the pattern. Such prompt-driven assistance is like having a knowledgeable pair of eyes to double-check your work. You can even use voice input to describe a scenario – for example, verbally explain what your code does – and ChatGPT's voice-capable mode will discuss where you might need a permission check, which is useful if you're multitasking or away from a keyboard.

Developer Tip: Make it a practice to run new Apex code through a security-oriented Q&A with ChatGPT (or another peer review) before finalizing it. This could be as simple as asking, "Have I enforced all required CRUD/FLS checks in this method?" By doing so, you reinforce a security mindset and catch potential gaps early.

Architect Tip: Encourage your team to use automated static code analysis tools in tandem with ChatGPT. Tools like Salesforce Code Analyzer or PMD can flag missing CRUD/FLS enforcement. ChatGPT can then help interpret those findings and recommend fixes. As an architect, you might create a "security checklist" that developers can paste into ChatGPT along with their code to get a quick compliance report.

By diligently enforcing CRUD and FLS, you lay the groundwork for all other security measures in your Apex code. Once your code is only operating on data that the user is allowed to access, you can move on to the next layer of defense: protecting against malicious inputs such as SOQL injections and cross-site scripting attacks.

SECURING APEX AGAINST SOQL INJECTION AND CROSS-SITE SCRIPTING (XSS)

Apex code often handles user-supplied data, whether it's search keywords, form inputs, or values from an integration. Two of the most common web application vulnerabilities, **SOQL injection** and **cross-site scripting (XSS)**, can also threaten your Salesforce org if Apex is not written securely. This section explains these threats and how to defend against them in your Apex code.

Understanding SOQL Injection in Apex

SOQL injection occurs when an attacker manipulates a piece of user input to alter the intended logic of a dynamic SOQL query. If your Apex code concatenates user input into a query string without validation or escaping, a malicious user could input a fragment like ' OR Name != '' to trick the query into returning records they shouldn't see, or worse, cause unintended operations. For example, consider an Apex snippet that builds a query from a search term:

```
String searchTerm = ApexPages.currentPage().getParameters().get('keyword');

String q = 'SELECT Id, Name FROM Account WHERE Name LIKE \'%' + searchTerm + '%\'';

List<Account> results = Database.query(q);
```

If searchTerm is something innocent (e.g. "Acme"), this works. But if an attacker provides %' OR Name != '% as the keyword, the query becomes: SELECT Id, Name FROM Account WHERE Name LIKE '%%' OR Name != '%', effectively dropping the intended filter and exposing all Account names. In this case, the attacker exploited the string building to inject additional SOQL logic.

Preventing SOQL Injection: The most effective way to prevent SOQL injection is to avoid dynamic queries built from untrusted input. Use static SOQL queries with **bind variables** whenever possible, which automatically handle input safely. For the above example, a safe version would be:

```
String searchTerm = ApexPages.currentPage().getParameters().get('keyword');

List<Account> results = [SELECT Id, Name

             FROM Account

             WHERE Name LIKE :('%' + searchTerm + '%')];
```

By using :<variable> in the query, Apex will bind the searchTerm value securely, treating it as a single value (with special characters properly escaped), no matter what the user enters. Even if an attacker submits a string with ' OR..., it will be handled as a literal search term and not break the query syntax.

If you absolutely must construct a dynamic query (for example, building a flexible filter based on multiple conditions), you should:

Whitelist and validate inputs: Only allow specific values or patterns. For instance, if a user selects a field to filter on from a dropdown, don't accept any arbitrary field name—use a predefined list. Validate that text inputs match expected formats (e.g. no SQL/SOQL keywords or symbols like ' or --).

Escape user strings: Use String.escapeSingleQuotes(userInput) to sanitize single quotes in any user-supplied text that will be appended in a dynamic SOQL string. This method adds the proper escape character (\) to prevent breaking out of the string context. Keep in mind escaping alone is not foolproof if other parts of the query (like field names or operators) are injectable.

Use safe query building methods: Consider higher-level frameworks or the built-in Database.query with bind parameters even for dynamic queries. You can sometimes structure Database.query to include dynamic fields but still use bind variables for values.

Using Database.queryWithBinds in Apex allows you to safely execute dynamic SOQL queries with bind variables, which helps prevent injection vulnerabilities. This method separates query structure from user-supplied values, ensuring that special characters in inputs are properly escaped and treated as data rather than executable code. By leveraging queryWithBinds, developers can construct flexible queries while maintaining robust security against SOQL injection attacks.

From a testing perspective, always include negative tests for SOQL injection. A developer or QA tester should attempt to input strings like test' OR 1=1 in search fields (or use tools that automate such attacks) to verify that the application does not return unexpected data or errors.

Developer Tip: Never trust user input in your queries. By default, use bound variables in SOQL and avoid string concatenation. If using dynamic SOQL is unavoidable, implement rigorous validation. Keep an eye on code scans or linting tools – they can often detect dynamic queries and warn you to review them for injection safety.

Avoiding Cross-Site Scripting (XSS) in Apex Applications

Cross-site scripting (XSS) is a vulnerability where an application includes malicious scripts in the output it generates for users. In the Salesforce context, this usually means user-supplied data (perhaps stored in a record or entered in a form) is rendered in a Visualforce page, Lightning component, or other UI without proper encoding, allowing an attacker's script to execute in the victim's browser.

While Salesforce's modern frameworks have built-in protections (for example, Lightning Web Components automatically escape output by default), it's still possible to introduce XSS in Apex-driven features if you're not careful:

Visualforce and Aura Components: Visualforce pages escape the values of apex:outputText by default (escape attribute is true). XSS issues arise when developers intentionally turn off escaping (e.g., using <apex:outputText value="{!comment}" escape="false"/> to render rich HTML) or use mechanisms like <apex:pageMessage> that might display raw strings from Apex. If those strings contain HTML or script, and you haven't sanitized them, you have an XSS risk. Always sanitize or encode user inputs before displaying. Salesforce provides encoding functions such as HTMLENCODE() and JSENCODE() (for use in Visualforce or formulas) to help.

Lightning Components (Aura/LWC): These frameworks auto-encode output in most cases. However, if you use dangerouslySetInnerHTML in LWC or manually insert DOM elements in Aura (like setting innerHTML in JavaScript), you bypass those protections. In such cases, sanitize any HTML fragments on the server side (Apex) or client side before insertion. Apex doesn't have a full HTML sanitizer library out of the box, so you might restrict input to a known safe subset or use a third-party library if needed.

Emails and Integrations: Apex might generate content for emails or external systems. If user data flows into an email template or PDF without encoding, an XSS could potentially execute when that content is viewed in a browser (for HTML emails or rendered PDFs). Use the merge field security settings (which by default escape HTML) in email templates, and be cautious if inserting any dynamic content in Visualforce email templates with escape disabled.

In practice, completely preventing XSS means ensuring any untrusted data is output-encoded for the appropriate context. For HTML, that means converting < to <, " to ", etc., which the platform does automatically in most cases. For JavaScript context, ensure quotes and special characters are handled or use safe JSON serialization.

Tester Tip: To validate that XSS is handled, testers should include test cases where user input contains script tags or HTML. For example, enter <script>alert('XSS')</script> into a text field (or a less obvious payload), then observe if that script executes anywhere in the app. If it does, that's a bug. Tools like browser dev consoles or automated scanners can help identify where unsafe output might be occurring.

USING CHATGPT TO IDENTIFY AND PREVENT INJECTION/XSS ISSUES

ChatGPT can be a valuable helper in spotting potential injection or XSS vulnerabilities in your Apex code. You might ask ChatGPT to review a piece of code and specifically look for dangerous patterns. For instance:

Developer: "Here is an Apex controller method that builds a query from user input. Can you check if it's vulnerable to SOQL injection and suggest improvements?" *[Pastes code snippet]*

ChatGPT will analyze the code and likely point out any string concatenation in queries. It might respond with something like: "You are directly appending user input into a SOQL query, which is a potential SOQL injection vulnerability. Consider using a bind variable or at least escape the input using String.escapeSingleQuotes." This kind of feedback can validate your approach or catch something you overlooked. You can even supply examples of inputs, and ask ChatGPT what the resulting query would be, to see if it exposes a flaw.

For cross-site scripting, you could present ChatGPT with a Visualforce page or Apex code that handles user-supplied HTML and ask if there's any XSS risk. For example:

Developer: "I show a custom message on a Visualforce page using a string from Apex (code below). Could this be prone to XSS?" *[Provides code]*

ChatGPT might inspect it and warn, for instance, "If customMessage contains malicious HTML and you render it with escape=false, it could execute scripts in the user's browser. Make sure to escape or strip script tags."

Beyond code analysis, ChatGPT can help generate test cases. A tester could ask, "What are some malicious inputs I should try to test this search function for injection vulnerabilities?" The AI could list payloads like a%' OR 'x'='x for SOQL or <script>alert()</script> for XSS. This can augment your test plans with creative cases.

Tester Tip: Consider using ChatGPT's Operator tool to simulate attacks in a safe environment. For example, you could instruct the Operator to navigate to a test page in your org and input a known malicious script string into a comment field, then submit the form. This automated browser interaction (available in ChatGPT's interface as a beta feature) might reveal if any error messages or unexpected behavior occurs, indicating a security weakness. Just ensure this is done in a test context, not production, and follow your organization's guidelines for testing security.

Developer Tip: Use ChatGPT to double-check your mitigation logic. After implementing input validation or output encoding, you can describe your solution to ChatGPT (or show a code diff) and ask if it addresses the vulnerability. The AI can act as a rubber duck, reasoning through whether the fix covers all cases. While it shouldn't be the only validation (you should also do code reviews and use security scanners), it adds an extra layer of confidence.

Preventing injection and XSS in Apex is about being vigilant with how you handle data. With these coding practices and the aid of tools like ChatGPT to spot weaknesses, you can significantly reduce the risk of a successful attack. In the next section, we'll build on this by looking at how record-sharing rules and Apex-managed sharing can further ensure users only access what they're supposed to.

SECURE USE OF APEX SHARING AND VISIBILITY CONTROLS

Salesforce's record-level security (often called "sharing") ensures that even if a user has access to an object (CRUD) and a field (FLS), they can only see specific records that they're allowed to. Apex code, running in system context by default, can bypass sharing rules unless explicitly told to enforce them. This section focuses on how to use Apex sharing keywords and programmatic sharing to maintain proper data visibility.

With Sharing, Without Sharing, and Inherited Sharing

When you define an Apex class, you can specify its sharing behavior:

with sharing – The class respects the current user's record-sharing rules. Queries and DML in the class will only operate on records the user has access to (unless the code overrides this by running in system mode via other means).

without sharing – The class does not enforce the user's sharing rules, meaning it runs in system mode for record access. It can see and manipulate all data, regardless of ownership or sharing rules.

inherited sharing – Available for classes defined without an explicit sharing declaration. It means the class will run with sharing if it was called from a class that is with sharing, and without sharing otherwise. For top-level entry points (like Visualforce controllers or Aura controllers), inherited sharing defaults to with sharing behavior. This is a good default for utility classes or services that might be invoked by both kinds of callers.

By default, if you don't specify a keyword, an Apex class behaves as *without sharing* (except for Apex triggers, which always run without sharing). This default can be dangerous: it might inadvertently expose records. For instance, a global Apex REST service without sharing could retrieve and return data that the user behind the API call shouldn't see. Therefore, it's considered a best practice to explicitly use *with sharing* for classes that handle user requests or sensitive queries, unless you have a deliberate reason to do otherwise.

Developer Tip: As a rule of thumb, start classes as "with sharing" and only use "without sharing" in controlled scenarios where you need administrative access in code. If using "without sharing," document why. This makes code reviews and security assessments easier, since everyone can see which classes deliberately bypass sharing.

Apex Managed Sharing and Manual Share Records

Declaring a class "with sharing" ensures Apex respects existing sharing rules (such as role hierarchy, owner-based sharing, and criteria-based sharing rules) for record access. But what about granting access in Apex? Sometimes, you'll need to programmatically share a record – for example, granting a specific user read access to a record via custom logic (often called **Apex managed sharing**).

Salesforce allows Apex to create and adjust share records:

For standard objects and custom objects with "Grant Access Using Hierarchies" turned on, each object has a share object (e.g., AccountShare or MyObject__Share). Your Apex code (running without sharing or with an elevated privilege) can insert or update records in these share tables to grant or revoke access. For example, to give User X read access to a custom object record, you could insert a MyObject__Share record with that User's Id, the record's Id, and a ShareReason/access level (Read or Edit).

Apex managed sharing rules often accompany a custom reason (for custom objects) that indicates the share was created by Apex logic. Salesforce requires that any Apex share you create be maintained by your code. If the conditions for sharing change, your Apex may need to adjust or remove the share. To aid with this, you can define an *Apex Sharing Recalculation* on the object, which runs a piece of Apex (typically batchable) to recalculate your sharing rules periodically or when triggered manually.

Consider an example: you have a custom object **Project__c** and you want to share each Project with an external collaborator based on a lookup field. You might write Apex that, after insert or update of a Project, creates a Project_Share record giving the collaborator read access. If that collaborator field changes, your code needs to remove the old share and create a new one for the new collaborator. Defining an Apex Sharing Recalculation class for Project__c can systematically evaluate all records and ensure only the correct shares exist, which is especially useful if manual changes or data loads might have gotten the shares out of sync.

Architect Tip: Use configuration where possible for record sharing (criteria-based sharing rules, public groups, roles). Resort to Apex managed sharing only for scenarios that declarative sharing cannot cover. When you do implement Apex sharing, design the logic carefully to avoid "over-sharing" data and consider how it will scale (bulk processing, large data volumes, etc.). Also, ensure that any Apex that manipulates share records runs in a context with sufficient privileges (often that means the Apex class itself might need to be without sharing, even if the rest of the logic is gated by permission checks).

Respecting the Sharing Model in Apex Code

To maintain security, Apex developers should always ask: "Am I respecting the sharing model?" In practice:

Use with sharing on controllers, batch classes, schedulable classes – any entry point where a user action or event initiates the Apex. This ensures all downstream operations default to user-level access. If those classes call helper classes marked without sharing, be very cautious because that will open up record access in those helper calls.

Minimize use of System.runAs in non-test code. In Apex tests, System.runAs is invaluable for simulating different user profiles and sharing scenarios. But in production code, there is no direct equivalent – any attempt to manually enforce or bypass sharing should be done through the class's sharing setting or careful SOQL filtering (e.g., adding WHERE criteria that mimic what sharing would do).

When writing SOQL queries in a with sharing class, you typically don't need to add extra filters for ownership or sharing – Salesforce will automatically apply the user's record access. However, be mindful if you switch to without sharing mid-execution (or call a without sharing method); you might inadvertently query more data than the user should see.

Sharing recalculation: if your application uses Apex-managed sharing heavily, incorporate mechanisms to audit and fix sharing inconsistencies. This might involve logging whenever a share is created by Apex, and periodically verifying those shares are still needed. Salesforce's native sharing recalculation jobs (for custom objects) can be scheduled to help with this, as mentioned.

Product Owner Tip: Clearly define data visibility requirements in user stories. Rather than telling developers "make it so users in Team A can also see Team B's records via code," work with your architect to see if Salesforce's declarative sharing can solve it (for example, a shared role, a criteria-based sharing rule, or a public group). If custom Apex sharing is truly needed, ensure the acceptance criteria include thorough testing of who should or should not see the records. This prevents corner cases where a developer might be tempted to use a quick fix like without sharing, which could have broader impacts.

Developer Tip: When debugging or reviewing Apex, pay attention to the class's sharing context. If you find a SOQL query returning records unexpectedly, check if the code is running without sharing. Tools like Salesforce's Security Scanner or PMD can flag classes without sharing. You can even ask ChatGPT to review your class list and highlight any that don't specify sharing — a hint to double-check them. Additionally, consider writing unit tests that verify sharing enforcement: for example, create a record as one user, run your Apex as a different user with lower access, and assert that the second user's context doesn't retrieve the first user's record.

Using ChatGPT for Sharing and Visibility Guidance

ChatGPT can assist in designing and troubleshooting sharing logic:

Reviewing design: If you describe your sharing model to ChatGPT (e.g., "We have object X that managers should see all records of, and object Y that is private unless flagged"), it can suggest whether standard sharing rules suffice or if Apex is needed. While it may not know your org's specifics, it understands Salesforce's sharing mechanisms well and can propose solutions (like "use a hierarchy relationship for managers" or "use a criteria-based sharing rule for flagged records").

Code examples: You can prompt ChatGPT for examples of Apex managed sharing code. For instance, "How do I programmatically share a record with a user in Apex?" ChatGPT can produce a sample snippet inserting a Share record. Always adapt and test the snippet, but it's a good starting point if you're unfamiliar with the syntax.

Diagramming and explaining: To help explain sharing to non-developers, you might ask ChatGPT to generate an analogy or even create a simple diagram of your sharing model. For example, "Generate an illustrative analogy comparing record sharing in Salesforce to a library lending system." You could also use ChatGPT's image creation tools to visualize a role hierarchy or sharing scenario (be careful to abstract any real data). This can be useful for training and governance meetings.

Troubleshooting: If a particular user isn't seeing data they should, or vice versa, and you suspect an Apex sharing issue, you can describe the scenario to ChatGPT. For example, "User A should see record X because of rule Y, but my Apex class isn't returning it under with sharing. What might be wrong?" ChatGPT might remind you to check if the class or any method is declared without sharing, or if there's a specific Owner or Criteria-based rule that isn't met.

Architect Tip: Leverage ChatGPT as a teaching tool for your team. If developers are new to sharing rules, have them ask ChatGPT specific questions (in a group setting) – it's like having a whiteboard consultant. For instance, ask "What are the implications of using without sharing in a helper method?" and discuss the answer it gives. This can reinforce best practices in an interactive way. Always verify the answers with Salesforce's official documentation or your own experience, but it's a great way to spark discussion and learning.

By using Apex sharing settings correctly and aligning with Salesforce's sharing model, your code will honor the fundamental data visibility constraints of the organization. Next, we'll shift focus from internal data access to external communications – exploring how to secure Apex call-outs and integrations with other systems.

IMPLEMENTING SECURE APEX CALLOUTS AND INTEGRATIONS

Integrating Salesforce with external systems is a common requirement and Apex callouts (HTTP requests or SOAP calls from Apex code) make it possible. However, integrations introduce additional security considerations. You need to protect credentials, ensure data is transmitted securely, and handle inbound and outbound information in compliance with policies like GDPR or HIPAA. In this section, we outline best practices for secure Apex callouts and integrations.

Use Named Credentials and Secure Endpoints

One of the strongest tools for securing outbound callouts is *Named Credentials*. A named credential is a Salesforce configuration that stores the details of an external endpoint (the URL) and the authentication needed (such as OAuth tokens, basic auth, etc.). By using a named credential, you avoid hardcoding sensitive credentials in Apex code. Instead, your Apex callout references the named credential by name in the callout URL (for example, call HttpRequest.setEndpoint('callout:MyService/weather')). Salesforce then injects the proper endpoint URL and handles authentication (like injecting an OAuth bearer token) behind the scenes. This means:

Passwords or tokens aren't visible in your code or version control.

Credentials can be updated or revoked in one place (the named credential configuration) without altering code.

You can set permissions so that only certain profiles can modify the named credential, adding an extra layer of administrative control.

Always prefer *HTTPS endpoints* for callouts – in fact, Salesforce requires HTTPS for callouts to external sites (HTTP is not allowed for external callouts). Ensure the endpoint's certificate is valid; Salesforce will reject callouts to sites with untrusted or expired SSL certificates. You don't usually have to worry about verifying the certificate manually (the platform does that), but you should periodically review that the endpoint is using modern TLS protocols. If integrating with an older system, confirm it supports TLS 1.2+ as required by Salesforce.

Beyond using HTTPS, consider the domain allowlist: Salesforce's callout mechanism requires that you set up either a Remote Site Setting or use a Named Credential for any external host you call. This prevents code from unexpectedly sending data to arbitrary endpoints. Only hosts you explicitly configure can be called, which is a security checkpoint.

Architect Tip: Standardize the use of named credentials for all integrations in your org. This not only increases security but also simplifies deployment (no need to remember adding remote site settings). It also makes security audits easier – you can review all configured integration endpoints in one list, seeing which auth mechanism they use. As an architect, enforce guidelines that no callout code should contain literal URLs or credentials.

Protecting Sensitive Data and Tokens

When Apex integrates with another service, you often deal with API keys, access tokens, or other credentials for that service. Here's how to handle them securely:

Never log sensitive info: Avoid using System.debug to print tokens, secrets, or personally identifiable information to logs. Debug logs are accessible to admins and could be downloaded, so treat them as potentially persistent. If you must log something for troubleshooting, redact or truncate it.

Store tokens securely: If your integration involves refreshing tokens or storing an API key, use a protected custom setting or custom metadata, or let the Named Credential handle it. With the newer *Enhanced Named Credentials*, Salesforce can even manage OAuth refresh for you. If you find yourself writing Apex to store a token (for example, after an OAuth flow, storing a refresh token), ensure that field is protected (only accessible to Apex code, not visible on layouts) or encrypt it using Platform Encryption. Under GDPR and HIPAA, access to integration credentials should be limited and monitored just like access to customer data.

Minimize data in transit: Only send the necessary data in callout requests, and only retrieve necessary data in responses. This reduces exposure of sensitive fields. If you need to send sensitive personal data (say, an SSN or health information) to an external service, double-check that the transmission is encrypted (HTTPS covers data in transit). For extra caution, you could encrypt certain payload fields using Apex Crypto before sending, but then the external system would need to decrypt them, so this is only feasible in specific tightly-controlled integrations.

Validate inbound data: If your Apex callout processes a response (JSON or XML) from a third-party, be mindful of what you do with that data. Don't directly insert or update records with unverified external data that might violate your org's data integrity or security rules. While less common, consider scenarios like an external system sending back an unexpected large payload or script content – if your Apex later displays that content somewhere, it could introduce XSS. Parse and handle responses carefully, and use try-catch to handle any content that doesn't match the expected format.

Developer Tip: When writing callout code, always implement error handling. Catch exceptions like HttpResponse[StatusCode] errors or timeouts. In your error handling, do not expose internal details or captured request info in a way that users can see. For example, if an integration fails and you surface an error message back to a user or admin, say "Integration failed, please contact support" rather than exposing the full endpoint or raw token. This prevents potential information leakage about your integration setup.

Compliance Considerations (GDPR, HIPAA) for Integrations

Integrations often mean data is leaving Salesforce to another system or coming in from one. This can raise compliance questions:

Data minimization: Under GDPR's data minimization principle, only transfer data that is necessary for the purpose. If an external system doesn't need a particular field, don't include it in the callout. For instance, if you're sending patient records to a billing system under HIPAA, you might exclude detailed medical notes if only billing codes and personal details are needed.

Audit trails: Keep an audit of integration activities. Salesforce has an "API Callout" event in Event Monitoring that can log callout requests. Also consider logging (securely) within Salesforce when callouts happen and what key actions were taken (e.g., record X was updated via integration Y). Ensure these logs themselves don't contain sensitive payload data, but maybe reference IDs and timestamps for traceability.

Third-party compliance: Ensure the target system of your integration is compliant as well. If you are sending personal data to an external service, you as the data controller (under GDPR terms) need assurances that the service will handle it with adequate protection. That might mean having Business Associate Agreements in place for HIPAA, or GDPR-compliant data processing agreements with the vendor.

Product Owner Tip: When planning an integration, involve your security and compliance teams early. For example, if you propose to integrate with a new cloud service, confirm that the service is certified (ISO, SOC2, HIPAA-compliant, etc.) as needed. Non-functional requirements like "the integration must use OAuth 2.0 and no credentials in code" or "the data must be encrypted at rest on the target system" should be captured in the user story or acceptance criteria. This ensures the development team has clear security targets to meet, and avoids costly rework if a late security review finds an issue.

Using ChatGPT for Integration Security Assistance

ChatGPT can support developers and architects in ensuring integration security:

Reviewing callout code for issues: You can show ChatGPT an Apex callout class (sans any real credentials) and ask, "Do you see any potential security vulnerabilities in this integration code?" It might catch things like lack of timeout handling, or use of a hardcoded URL, and remind you to use named credentials. For example, a developer can paste a snippet where they build a Bearer token header manually, and ChatGPT might respond that using a Named Credential with OAuth would be safer and easier.

Generating secure code templates: If you're unsure how to implement something like OAuth in Apex, you can ask ChatGPT for a code example. E.g., "Give me an Apex callout example that uses a Named Credential and handles a JSON response securely." The AI might provide a scaffold that you can adapt, including parsing logic inside a try-catch.

Troubleshooting integration errors: Sometimes errors from callouts are cryptic. You could feed ChatGPT the error message or logs (after removing sensitive details) and ask for insight. For instance, "I'm getting a '401 Unauthorized' error in my Apex callout. What could be the cause?" ChatGPT might suggest checking if the token is expired or if the Named Credential's user has proper permissions on the external system.

Voice-assisted brainstorming: Using ChatGPT's voice mode, an architect or developer can verbally walk through an integration design, almost like talking to a rubber duck or colleague. For example, you might explain, "We need to send Salesforce data to an external inventory system. I plan to use a named credential with OAuth. The data includes customer addresses. Are there security steps I'm missing?" ChatGPT's voice response could iterate through the checklist (encryption, audit, consent) so you can confirm you've covered everything. This is especially handy in meetings or when multitasking.

Developer Tip: After writing an integration, use ChatGPT as a final check. Ask questions like, "Did I securely handle the auth token in this code?" or "Could an unexpected response crash my logic?" While the answers might not always be perfect, they can highlight areas to double-check. Think of it as an extra pair of eyes on your integration from a security standpoint.

Architect Tip: Encourage a culture where integration security is as much a priority as functionality. Tools like ChatGPT can be included in integration development workflows—perhaps as part of a code review checklist, a developer might note "Consulted ChatGPT on error handling and followed suggestions X and Y." This shows proactiveness. Additionally, have architects periodically review the list of all external integrations (maybe exported from Named Credentials) and consider asking ChatGPT to summarize potential risks ("List common risks of integrating Salesforce with payment gateways", for example) to see if any apply to your case.

Securing your Apex callouts and integrations ensures that extending Salesforce beyond its borders doesn't become the weakest link in your security chain. Next, we'll look at how Salesforce helps you audit and monitor these security measures, particularly focusing on tools for reviewing Apex security in your org.

AUDITING APEX SECURITY WITH SALESFORCE TOOLS

Even with all the best practices implemented, it's important to continuously monitor and audit your Salesforce org for security issues. Salesforce provides various tools to help you review settings, track usage, and catch anomalies. This section will discuss using those tools to audit Apex-related security aspects, ensuring nothing slips through unnoticed.

Salesforce Security Health Check and Config Audits

The *Security Health Check* is a feature in Setup that evaluates your org's security settings (password policies, session settings, etc.) against a baseline and gives you a score. While Health Check is more about org configuration than Apex code, it can highlight weaknesses (for example, too lenient password rules or missing IP restrictions) that could indirectly affect Apex security (e.g., making it easier for an attacker to log in and then exploit an Apex flaw). After all, code is only as secure as the environment around it. An administrator or security analyst should review Health Check periodically. As a developer, be aware that if your org has a low score due to certain settings, your otherwise secure Apex might still be running in a less secure environment.

Beyond Health Check, you should audit *Installed Packages and Apex Classes*:

Ensure any unmanaged or third-party Apex in the org is reviewed. Sometimes security issues come from code you didn't write. Use the *Security Review* guidelines (like those for AppExchange apps) on any external code.

Tester Tip: Treat security auditing as part of the testing process. For instance, after all features are built, a tester (or security team member) can run a mini "audit checklist": run the Health Check, ensure all critical and high risks are addressed or accepted; use a code scanner to analyze Apex for common vulnerabilities; verify that debug logging is turned off in production (or at least not verbose for sensitive operations). Testers can also use Salesforce's free scanners (like PMD-based rules) to catch things like SOQL injection or hardcoded credentials in the code. If you're a tester without coding expertise, tools and Health Check provide a user-friendly way to spot issues.

Event Monitoring and Logs

Salesforce's *Event Monitoring* (part of Shield, but some events are available to all orgs as log files) provides detailed logs of user and system activity. For Apex security auditing, some relevant event types include:

API or Apex Execution events: These can track when large data exports happen via API or when an Apex method consumes unusual resources. While originally meant for performance, they can hint at misuse – for example, a report of an Apex call that retrieved thousands of records could be normal, or it could be a sign of someone running an unexpected data extraction.

Apex Exception events: If enabled, these events log details of unhandled exceptions in Apex. Frequent exceptions might indicate bugs that could be exploited or simply poor error handling that leaks info.

Login and Access events: Who is logging in, from where, and are there repeated login failures? While not Apex-specific, a compromised user account could lead to Apex misuse. Also, *Transaction Security Policies* can be set up (with clicks) to detect certain events (like many records viewed in a short time) and respond (like block the user or alert an admin).

As an auditor, you might download event log files for a period and analyze them for anomalies. For example, using a tool (Salesforce provides an Event Log File browser or you can use Splunk/New Relic connectors), you could spot that a particular user invoked an Apex REST API method 500 times in an hour – is that expected? Or that an Apex batch job is running much longer than usual (maybe someone changed its logic or it's processing more data, which could hint at abuse or just growth).

Product Owner Tip: Schedule regular security reviews of the system. For instance, quarterly, have a "security day" where the team runs Health Check, reviews critical Apex classes, and looks at event logs for unusual patterns. As a product owner, you don't need to dive into code, but you can ensure that time is allocated for these tasks and that any findings are tracked. Treat security tasks as user stories with priority – e.g., "Address findings from Health Check (set stronger session timeout, etc.)" or "Refactor Apex code flagged by scanner for not enforcing FLS." This demonstrates to stakeholders (and auditors) that you have a proactive security maintenance process.

Static Code Analysis and Vulnerability Scanning

Earlier sections mentioned using tools to scan Apex code (for CRUD/FLS, SOQL injection, etc.). Salesforce provides a *Code Analyzer* (SFDX scanner) that incorporates PMD rules and can detect many security issues in Apex. Running this as part of your CI pipeline or even manually before a release is a great audit measure. It might surface something like "Method XYZ performs a query without using security enforced." Each finding should be reviewed:

- False positives can be documented (maybe you know the context makes it safe).
- Real issues should be fixed or mitigated.

Additionally, if you're an ISV or building something for AppExchange, a security review by Salesforce is mandatory – they use tools like Checkmarx. But even for in-house enterprise orgs, consider a one-time or periodic professional code review or using a scanner like SonarQube with Apex rules.

Tester Tip: Security testing often falls between dev and QA. As a tester, you can use tools like ZAP or Burp Suite to do dynamic testing (for example, intercepting an Apex REST call and fuzzing the input to see if it can break). While those tools are more commonly used for web apps, they can call your Salesforce APIs too. If you find any suspicious behavior, log it as you would a functional bug. Also, when new Apex endpoints are introduced (say, a new @RestResource), write tests not only for expected inputs but also deliberately malformed ones to see how the system reacts (do you get a sensible error vs. a crash or data exposure?).

ChatGPT for Log Analysis and Security Insights

AI can assist in parsing through the wealth of information that auditing tools provide:

Interpreting Health Check results: If Health Check flags, say, "Minimum Password Length = 8 (High Risk, should be 12)", you can ask ChatGPT to explain the risk in layman terms. It might respond, "Short passwords are easier to guess or brute force. Increasing to 12 significantly improves security." You can use that explanation when justifying a change to stakeholders.

Summarizing log anomalies: If you have an event log file (CSV or JSON) of, for example, API calls, you could paste a small excerpt into ChatGPT and ask, "Do you notice any unusual patterns?" It might identify that one API user is making far more requests than others. (Be cautious not to paste any sensitive identifiers; anonymize data if possible when using external tools like ChatGPT.)

Security Q&A: ChatGPT can answer Salesforce security questions that come up during audits. For example, "Can I track which user accessed a particular field's data?" It might inform you about Field Audit Trail or Event Monitoring capabilities. Or "How do I detect if someone tried to SOQL inject my Apex?" It could suggest reviewing logs for certain error patterns or using custom validations.

Developer Tip: Incorporate security checks into your development pipeline. You could even script ChatGPT (via API) as a part of code review – for instance, automatically feed new Apex code to a prompt like "List any security weaknesses in this code." While a human should always be in the loop, this could highlight areas in a merge request to pay attention to. At minimum, developers can manually use ChatGPT as a rubber duck: explain the design of a feature and ask, "Do you see any security pitfalls?" Sometimes articulating it (even to an AI) exposes something you hadn't considered.

Product Owner Tip: Use the outputs from these tools and AI summaries to brief non-technical stakeholders. For example, if your CISO asks, "How do we know our Salesforce is secure?", you can provide a report: Health Check score, number of code scan issues found and fixed, event monitoring summaries (e.g., "no abnormal access patterns in last quarter"), etc. ChatGPT can even help format this into a neat summary or presentation. Demonstrating this due diligence not only builds confidence, but it also prepares you for compliance audits where you might need to show evidence of security monitoring.

By regularly auditing with Salesforce's tools and incorporating intelligent analysis, you close the feedback loop on Apex security. You're not just trusting that you wrote secure code – you're verifying it continuously and catching any regressions or new threats. In the final section, we'll discuss how to cultivate a security-first culture in your Apex development team, so that these practices become second nature.

DEVELOPING A SECURITY-FIRST CULTURE IN APEX TEAMS

Technology alone is not enough – the people and processes using that technology are equally important. Cultivating a security-first culture means that everyone involved in building Salesforce solutions (architects, developers, testers, product owners, project sponsors, and beyond) prioritizes security in their daily work. In such a culture, security isn't an afterthought or a box to check at the end; it's baked into planning, development, and deployment.

Promoting Security Awareness and Education

Start by ensuring the team understands *why* security matters. Share examples of Salesforce security incidents (anonymized) or hypothetical scenarios: e.g., "What if an attacker stole a session cookie and accessed our org?" When people grasp the impact – from legal compliance fines to reputational damage – they're more likely to internalize the importance of following best practices.

Regular training is key. New Salesforce developers should go through secure coding training (Salesforce offers Trailhead modules on security and the OWASP Top 10 for web apps). Conduct periodic "lunch and learn" sessions on topics like preventing SOQL injection or using CRUD/FLS checks properly. These can be interactive – for example, walk through a vulnerable piece of code and let the team discuss how to fix it. ChatGPT can assist here: you might ask it to generate a few quiz questions or scenarios ("John writes an Apex trigger without sharing. What are the risks?") and use those to spark discussion.

Encourage a blame-free environment when it comes to reporting security issues. If a developer spots a potential vulnerability in a colleague's code during review, it should be praised as a positive action (preventing a problem) rather than seen as an accusation. Managers and leads play a role in setting this tone. One practice is to include security questions in code reviews and stand-ups ("Are there any security implications we need to consider for this new feature?") so that it's always on the radar.

Architect Tip: Lead by example in design. When crafting solution architecture, explicitly address security in design documents. For instance, include a section for "Security Considerations" for every feature: list how you are handling authentication, access control, encryption, etc., for that component. By normalizing this, you signal to the team that security is an integral design concern. Architects should also stay current on Salesforce security advisories and share relevant news with the team (e.g., if a new platform feature for security is introduced or if there's a known vulnerability in a certain approach).

Integrating Security into the Development Lifecycle

A security-first culture treats security as part of "definition of done" for user stories. This means:

Requirements phase: If a user story involves sensitive data or a new Apex service, ask "What are the security requirements?" Perhaps the story needs to include acceptance criteria like "must enforce profile permissions" or "must log access attempts."

Development phase: Developers use checklists that include security items. Some teams create a simple list (e.g., "CRUD/FLS checked? Sharing enforced? Inputs validated? Output encoded?") that every developer goes through before marking a feature complete. It's like a preflight checklist for pilots.

Testing phase: QA includes security test cases (as discussed in the previous section). Even if you don't have a dedicated security tester, functional testers can verify things like "user without permission X cannot perform action Y" – which is a security test.

Deployment phase: Incorporate static code analysis (as part of CI) and maybe even automated penetration testing for critical apps. Make deployment contingent on addressing high-severity issues.

Encourage developers to not just fix security issues, but to understand how to prevent them next time. If a peer review finds that a SOQL query was vulnerable to injection and gets fixed, take a moment in a team meeting to talk about how to avoid such mistakes going forward. Over time, the team will internalize these lessons and they become habits.

Developer Tip: Embrace security as a quality metric for your code, just like functionality or performance. Challenge yourself to write code that passes security checks on the first go. One practical habit is to run the security scanner (or even just do a manual scan of your code for things like "= :UserInfo.getUserId()" or "without sharing") before committing your code. Another habit: when you learn something new (say, a new safe coding pattern), share it with the team on your chat channel or at daily stand-up. This peer-sharing reinforces collective knowledge.

Continuous Improvement and Threat Modeling

A security-first culture is not static; the team should continuously improve their security practices. One way to do this is through periodic **threat modeling** exercises. For example, when starting a new project or major feature, gather the team and brainstorm: "In this feature, what could go wrong? Where could a bad actor potentially exploit something?" You can draw a simple diagram of the feature's architecture and then systematically think like an attacker. Maybe an attacker could manipulate an Apex REST endpoint's input or try to escalate privileges via an insecure Lightning component. By identifying these early, you can design defenses upfront.

Make these sessions inclusive and creative – not just developers, but also product owners and testers can provide perspectives. A tester might say, "What if someone tries to break the UI and input HTML where only text is expected?" which is essentially thinking about XSS. Write down the potential threats and decide on mitigations for each. This practice not only results in safer design, it also further ingrains a security mindset in everyone. People start to naturally think about "How could someone abuse this?" whenever they work on something.

Another aspect is to celebrate security wins. If the team goes through a release with zero security findings or successfully passes an external audit, acknowledge that achievement. Just as we celebrate meeting functionality deadlines or performance improvements, do the same for security milestones.

Product Owner Tip: Incorporate security into your acceptance criteria and definition of done. When planning sprints, allocate effort for security-related tasks (like updating dependencies, running scans, fixing vulnerabilities) and treat them with equal importance to user features. If velocity or deadlines are tight, resist the temptation to de-prioritize security tasks – skipping them might save a day now but could cost dearly later if an incident occurs. By consistently scheduling and completing these tasks, you signal to the team that security is an intrinsic part of the product's success and not optional.

Manager Tip: Support your team with time and resources for security. This could mean budget for a security training course, time for a developer to attend a Salesforce security webinar, or purchase of a better code scanning tool. In performance reviews, consider highlighting individuals who contribute to security improvements, not just feature delivery. Also, ensure there's a clear process for team members to report security concerns (and maybe even an internal reward or recognition for finding and fixing significant vulnerabilities). When new hires join, instill security from day one by including secure development guidelines in onboarding.

Leveraging ChatGPT to Strengthen Team Security Culture

ChatGPT can be a handy aid in fostering a security-first mindset:

Team learning sessions: Use ChatGPT during meetings as an interactive participant. For example, in a threat modeling meeting, after the team brainstorms, you might ask ChatGPT: "Can you think of any additional security risks in this architecture that we didn't mention?" Its answer might introduce a new angle (perhaps something about denial-of-service or data leakage) that the team can discuss. This can validate your thoroughness or reveal gaps.

Creating training materials: Need a quick refresher document on Apex security best practices for the team? Ask ChatGPT to draft one. For instance, "Summarize the top 5 Apex security best practices all developers should follow." You can then review and tailor the answer, and share it as a checklist or poster in your dev area.

Role-playing exercises: To build empathy and vigilance, you could have ChatGPT role-play as an attacker in a simulation. For example: "Pretend you are a malicious user trying to exploit our Salesforce org. What are some things you might try?" The responses can be eye-opening and serve as a basis for an internal penetration test. This can be a fun but educational game for the team – see if anyone can actually do what ChatGPT suggested (in a safe, sandbox environment) and then fix those weaknesses.

Policy and compliance Q&A: If someone on the team asks, "Do we really need to do X? Is it that important for GDPR?", you could have ChatGPT quickly explain the relevant regulation or consequence. It's like having a compliance advisor on call. Of course, verify with official sources for actual decisions, but it's a quick way to reinforce why certain practices (like encryption or audit logging) are non-negotiable.

Architect Tip: Continuously gauge your team's security maturity. There are frameworks (like BSIMM or OWASP SAMM) for measuring security culture. You can simplify this by occasionally asking ChatGPT, "What are signs of a strong security culture in a software team?" and use the answers as a checklist. Do we do those things? If not, which can we adopt next quarter? Also, consider having ChatGPT help craft a "security vision statement" for your team – a short, inspiring message about keeping customer data safe – and reiterate that vision in planning meetings or retrospectives to keep the mission front and center.

A security-first culture means everyone takes ownership of security. By educating the team, integrating security into every step, encouraging open dialogue about risks, and even using tools like ChatGPT to keep ideas flowing, your Salesforce development team will become not just code experts but security champions. This cultural shift ensures that all the practices we discussed in this chapter – from enforcing permissions to encryption and auditing – are sustained over the long run, no matter how technology or team members change.

SUMMARY

In this chapter, we covered a comprehensive set of best practices to secure Apex in Salesforce applications. We began by ensuring that Apex respects user permissions through diligent enforcement of CRUD operations and field-level security. By using techniques like schema describe checks, WITH SECURITY_ENFORCED in SOQL, and Security.stripInaccessible, you learned how to prevent unauthorized data access at the object and field level.

We then examined how to defend against common application vulnerabilities such as SOQL injection and cross-site scripting. Using bind variables, input validation, and proper output encoding, Apex developers can neutralize malicious inputs and protect both data and users. Next, we reinforced the importance of record-level security by leveraging Apex sharing rules. Using *with sharing* classes and Apex managed sharing ensures that users only see data intended for them, aligning programmatic access with your org's sharing model.

The chapter also addressed secure callouts and integrations, highlighting the need to protect data when it leaves Salesforce. Named Credentials, secure handling of authentication tokens, and strict endpoint governance help maintain trust when connecting to external services. We discussed how data encryption (both with Salesforce Shield Platform Encryption and Apex Crypto) and careful data storage practices fulfill regulatory requirements like GDPR and HIPAA, while keeping sensitive information safe from prying eyes.

To ensure these measures remain effective, we explored Salesforce's auditing tools. Security Health Check, event logs, and code scanners allow you to continuously monitor your Apex and org configuration for weaknesses, catching issues early and proving compliance through logs and reports. Finally, we emphasized building a security-first culture among your team. When architects, developers, testers, and product owners all view security as part of their job, the best practices we've covered become second nature. Regular training, threat modeling, and open communication about security make a tremendous difference in maintaining a robust posture.

By applying these practices, you create a layered defense: from proper code patterns in Apex, to encrypted data, to vigilant monitoring, all supported by a team mindset that values security. This holistic approach will significantly reduce the risk of data breaches, vulnerabilities, and compliance violations in your Salesforce org.

As you move forward to the next chapter, you'll carry this security mindset into the realm of integrations. With Apex security fundamentals in hand, you are ready to tackle Chapter 8, which explains Apex Integration Patterns – ensuring that as you connect Salesforce with other systems, you do so in a way that is not only effective and scalable, but also secure by design.

8

APEX INTEGRATION PATTERNS

Integrations are the connective tissue of enterprise software, enabling Salesforce to work in concert with other systems to deliver comprehensive business solutions. In a Salesforce context, integration means extending the platform's capabilities beyond core CRM features to incorporate data and functionality from external applications such as ERP systems, databases, or cloud services. A well-designed integration allows Salesforce to seamlessly share information with tools like AWS, MuleSoft, SAP, or Workday, ensuring users have a unified experience despite data residing in multiple places. This chapter introduces key integration patterns in Apex and explains why mastering these patterns is essential for building robust, scalable, and maintainable Salesforce applications.

Understanding integration patterns is vital for architects and developers aiming to build systems that can grow with the business. An integration that works for a small user base or limited data might falter under enterprise-scale loads if not designed properly. By following best practices for REST and SOAP callouts, handling responses gracefully, and managing asynchronous processes, teams can prevent common pitfalls such as API limits, timeouts, or data mismatches. Solid integration design not only improves reliability but also reduces maintenance costs – a point that project sponsors and product owners appreciate when considering the total cost of ownership. In essence, strong integration patterns help Salesforce applications remain stable and performant as they connect with external ecosystems.

Throughout this chapter, we'll explore the most important integration mechanisms available to Salesforce developers. You will learn how to perform RESTful callouts from Apex and parse JSON results, how to integrate with legacy SOAP web services, and how to leverage asynchronous Apex features for long-running or high-volume integrations. We'll dive into the use of Platform Events for event-driven architectures, ensuring real-time communication where appropriate. A significant portion of our discussion will also cover data mapping and transformation – the techniques needed to translate data between Salesforce and external schemas – as well as error handling strategies to make integrations fault-tolerant. Finally, we'll cover monitoring and logging practices so that once integrations are deployed, you have visibility into their behavior and can quickly troubleshoot issues.

Integration considerations span technical and business domains, so this chapter highlights perspectives for various stakeholders. **Architects** will gain insight into designing integration flows that align with enterprise architecture and governance. **Developers** will find practical tips and Apex examples for implementing callouts, handling payloads, and writing test classes for integrations. **Testers** will see how to simulate external systems and validate integration logic thoroughly before production. **Product owners** will understand the implications of integration choices on requirements, user experience, and business continuity. For example, a product owner at an SMB might see immediate value in a simple Salesforce-to-Workday integration for employee data, while an enterprise architect will be thinking about how that pattern scales when hundreds of thousands of employee records are involved.

A unique aspect of this book series is the inclusion of ChatGPT as a development partner. Integrations often involve complex configurations and troubleshooting, and this is where ChatGPT's assistance can shine. We'll demonstrate **extensive ChatGPT prompt examples** – showing how you can use natural language prompts to generate Apex callout code, parse error messages, or even document integration processes.

ChatGPT can help design integration logic (for instance, suggesting how to handle a timeout error from an AWS API) and assist in troubleshooting by analyzing stack traces or logs you provide. Remember that ChatGPT supports multiple interaction modes: for example, you might **use voice input** to describe an integration problem and get a quick explanation, or **upload a JSON file** from an API response to have ChatGPT generate a corresponding Apex deserialization class. These non-text interactions (voice, file upload, images, and the upcoming **Operator** tool for running code) can enrich your workflow – just be sure to enable any necessary privacy settings when sharing integration data with the AI.

In the following sections, we break down Apex integration patterns into seven key topics. First, we'll start with the fundamentals of making **REST callouts** from Apex and the best practices that accompany them. From there, we move into **SOAP-based integrations** for systems that use XML web services. Subsequent sections explore **asynchronous integration patterns**, the power of **Platform Events** for event-driven design, and techniques for **data mapping and transformation** between systems. We'll then address how to implement robust **error handling and recovery** mechanisms in integration code, before concluding with strategies for **monitoring and logging** integration activities to maintain visibility. By the end of this chapter, you will have a comprehensive toolkit of Apex integration best practices, with real-world examples involving technologies like AWS, MuleSoft, SAP, and Workday to ground these concepts. Let's dive in to see how Apex and ChatGPT together can help you build integrations that are both powerful and resilient, beginning with one of the most common approaches: integrating via RESTful APIs.

REST-BASED INTEGRATION BEST PRACTICES

Modern web services predominantly use REST (Representational State Transfer) APIs, making REST callouts a cornerstone of Salesforce integrations. In this section, we detail best practices for using REST APIs within Apex – covering correct use of HTTP methods, handling JSON payloads, and managing endpoints securely and efficiently. By following these practices, Salesforce developers can ensure their Apex callouts to services like AWS or MuleSoft are reliable and easy to maintain. We also provide ChatGPT prompt examples that can help validate your REST integration logic, troubleshoot issues, and even assist in writing boiler-plate code for callouts.

Using the Right HTTP Methods and Structure: REST APIs commonly use HTTP methods (GET, POST, PUT, DELETE, etc.) to perform operations. In Apex, you'll use HttpRequest and Http classes to perform these callouts. It's important to use the correct method for the operation at hand (e.g., GET for retrieving data, POST for creating, PATCH for updates, DELETE for deletions) and to construct the request appropriately. Ensure that you set the request endpoint URL exactly as required by the API and include any required headers (for example, content type application/json, authentication tokens, etc.). In Apex, a callout might look like:

```
HttpRequest req = new HttpRequest();

req.setEndpoint('https://api.example.com/v1/resource');

req.setMethod('POST');

req.setHeader('Content-Type', 'application/json');

req.setBody(jsonPayloadString);

Http http = new Http();

HttpResponse res = http.send(req);
```

Notice that we specify the HTTP method and content type, and attach the JSON body for a POST request. Always check the API documentation of the external service for required headers and supported methods. For instance, if integrating with an AWS service through API Gateway, the API might expect an API key in a header or a specific format of URL. If you accidentally use GET where POST is expected (or vice versa), the external service could respond with an error or unexpected behavior.

Managing Endpoints and Authentication Securely: Hard-coding endpoints or credentials in Apex is strongly discouraged. Instead, leverage Salesforce's Named Credentials feature for callouts. Named Credentials allow you to define the endpoint URL and authentication (like OAuth tokens or basic auth) in Salesforce settings, and then simply reference the named credential in your Apex code. This separation means if an endpoint URL changes, you can update it in one place (the Named Credential) without modifying code. Likewise, Named Credentials securely store authentication details and handle automatic token refresh for OAuth, removing the need for developers to embed sensitive info in code. In practice, your Apex endpoint might be "callout:MyNamedCred/some/path" instead of the full URL, and Salesforce will insert the base URL and auth header at runtime. This improves maintainability and security of integrations immensely.

Architect Tip: Consider establishing a Named Credential for each external service (e.g., one for AWS APIs, one for a MuleSoft endpoint proxying SAP, etc.). This not only simplifies your Apex code but also allows architects or admins to manage endpoints and credentials across sandbox refreshes and production without modifying the codebase. It's a best practice to use Named Credentials for all external callouts to avoid storing passwords or tokens in Apex or in custom settings.

When using Named Credentials with OAuth, you typically set up an Auth. Provider in Salesforce and configure the named credential to use it. This way, Salesforce can perform the OAuth handshake and securely maintain a refresh token. From a developer's perspective, the Apex callout doesn't need to manually attach the OAuth token – it's done for you. If the external API uses an API key or static token, Named Credentials can store that as well. This abstraction means your code remains unchanged even if the authentication method updates (say the service moves from API keys to OAuth 2.0): you update the Named Credential configuration, not the Apex classes.

Handling JSON Payloads Effectively: JSON is the lingua franca of REST APIs. Apex provides classes like JSONParser and static methods in the JSON class to serialize and deserialize JSON strings. There are two main approaches to handle JSON responses: parse dynamically or deserialize into a typed Apex class.

Dynamic parsing: You can parse the JSON using JSON.deserializeUntyped into generic Apex types (maps and lists). For example:

Map<String, Object> result = (Map<String, Object>) JSON.deserializeUntyped(res.getBody());

List<Object> records = (List<Object>) result.get('records');

This approach is quick for ad-hoc processing of a response when you only need a couple of fields. However, it can be error-prone if you mistype keys or if the JSON structure changes.

Strongly-typed classes: A more robust approach is to define an Apex class structure that mirrors the JSON and use JSON.deserialize(jsonString, MyClass.class). You can write these classes manually or use tools to generate them. Salesforce offers a tool called **JSON2Apex** (available in the Developer Workbench or third-party sites) which can generate Apex classes from a sample JSON payload. In our context, ChatGPT can help with this too – you could provide a JSON sample and ask ChatGPT to create an Apex class. This yields code that's easier to work with, as you can access data via properties (e.g. myResponseObj.records[0].Name) instead of casting maps.

Using strongly-typed classes makes your integration code more readable and less brittle, especially for complex APIs. For example, if integrating with Workday's REST API (which returns employee data in JSON), you might generate an Employee Apex class with fields like firstName, lastName, title, etc., matching the JSON structure. Then your code simply does Employee emp = (Employee) JSON.deserialize(res.getBody(), Employee.class); and you have a populated object.

Robust Request Formation and Response Parsing: Beyond basic JSON, consider edge cases in forming requests and parsing responses. Always validate or encode any dynamic parts of your endpoint URLs. For example, if you pass an account name in a query parameter to a REST API, ensure its URL-encoded to avoid issues with spaces or special characters. On the response side, check the HTTP status code (res.getStatusCode()) and handle errors appropriately. A common mistake is assuming a 200 OK response – but you should design for 400-series client errors or 500-series server errors from the external system. If res.getStatusCode() is not in the 200–299 range, log the response or take corrective action. For instance, a 401 Unauthorized might mean an expired token – if using Named Credentials with OAuth,

Salesforce will handle token refresh for the next call, but you might need to catch the exception, log it, and show a friendly error to the user or admin.

Speaking of exceptions, note that http.send() will throw a CalloutException if the request fails at a low level (e.g., DNS issue or timeout). Wrap callouts in try-catch blocks and implement retries if appropriate. Salesforce imposes a governor limit of 100 callouts per Apex transaction , and a **cumulative** timeout limit of 120 seconds for all callouts in a transaction. These limits mean that in a single trigger context, for example, you can't call an external service more than 100 times or spend more than 120 seconds waiting on responses. Good practice is to keep callouts few and fast – if you anticipate needing many or slow callouts, use asynchronous patterns. If an AWS Lambda call might take longer than a few seconds, you wouldn't want to do it in a synchronous context; instead you might use a continuation or offload to a Queueable Apex (as discussed later) to avoid hitting the timeout limit.

Developer Tip: Test your REST integration code with various scenarios: typical success responses, error responses (like 404 or 500 status codes), and even unexpected JSON. You can write unit tests using the **HttpCalloutMock** interface to simulate HTTP responses. This lets you verify that your parsing logic handles different payloads correctly. For example, you could have a test where the mock returns a 500 error with an error message in the body – your Apex code should catch the exception or check status and perhaps record an error log. By using HttpCalloutMock in tests, you not only fulfill Salesforce's requirement for callout tests, but also get to validate your error-handling paths.

Common Authentication Scenarios: Many REST APIs require authentication beyond just a static token. Two common approaches are OAuth 2.0 and API keys. If the service is secured by OAuth 2.0 (like many enterprise APIs or something like Google/Microsoft services), the ideal setup in Salesforce is to use an Auth Provider and Named Credential (as mentioned) so that Salesforce handles the token. You initiate the OAuth flow once (e.g., have an admin authenticate via the Named Credential configuration), and then Apex callouts will automatically include the token. You don't have to write code to do the OAuth dance or store tokens – simplifying your integration.

For API keys or custom auth, you might store credentials in Named Credentials (if it's just a single key or basic auth). If that's not possible, consider storing them in a protected Custom Setting or Custom Metadata (accessible to Apex but not visible to users). Never log sensitive info or expose it in error messages. If connecting to a MuleSoft API proxy, for example, you might need to include an API key header; you can set that header in the Named Credential so your Apex code remains clean.

Endpoint Handling and DNS Considerations: Salesforce requires that any external endpoint be added to Remote Site Settings or be a Named Credential. Using Named Credentials covers this requirement automatically. But beyond that, be mindful of endpoints – use consistent base URLs (avoid unnecessary redirects which waste time), and if the endpoint is an AWS service, consider using region-specific endpoints to reduce latency. For example, calling https://dynamodb.us-east-1.amazonaws.com is better from a US-based org than a generic endpoint that might route your request. Also, keep an eye on the response size: by default, the maximum size of callout response that Apex can handle is 6 MB for synchronous calls (and 12 MB for async). If you're dealing with large JSON payloads (perhaps an AWS service returning a big dataset), you may need to page or filter the data to stay within limits.

ChatGPT Prompt Examples – REST Callouts: Let's illustrate how ChatGPT can assist you in REST integrations. Below are some example interactions:

User Prompt: "I have an Apex class doing a REST callout to our SAP system via MuleSoft, but it's failing with a status code 400. Here's the JSON I'm sending and the response. Can you help me figure out what's wrong?"

ChatGPT: *Analyzes the JSON request and response.* "The 400 Bad Request suggests the JSON might be malformed or missing required fields. I notice your JSON payload is missing the customerId field that the API documentation lists as required. Also, there's a trailing comma in the JSON which makes it invalid. You should fix the JSON format and include all required fields. For example: { "customerId": "12345", "orderTotal": 100,... }. Additionally, ensure the Content-Type header is set to application/json.

If using ChatGPT's file upload, you could provide the API documentation or JSON schema for me to verify your payload against it." (The assistant would go on to guide the user through correcting the payload.)

In this prompt, the user could also **upload the JSON file** via ChatGPT's interface instead of copying it into the prompt, which would let ChatGPT parse it without formatting issues. Always remember to scrub any sensitive data before sharing, or use ChatGPT's data privacy features when dealing with real data.

Another example:

User Prompt: "Generate an Apex class named WeatherServiceClient that calls a public Weather API (REST GET request to https://api.weather.example/v1/current?city=CityName) and parses the temperature from the JSON response."

ChatGPT: "Sure! Here's an example Apex class:

```
public class WeatherServiceClient {
  public class WeatherResponse {
    public Current current;
    public class Current { public Decimal temperature; }
  }
  public static Decimal getTemperature(String city) {
    HttpRequest req = new HttpRequest();
      String endpoint = 'https://api.weather.example/v1/current?city=' +
EncodingUtil.urlEncode(city, 'UTF-8');
    req.setEndpoint(endpoint);
    req.setMethod('GET');
    Http http = new Http();
    HttpResponse res = http.send(req);
    if (res.getStatusCode() == 200) {
```

```
            WeatherResponse data = (WeatherResponse) JSON.deserialize(res.getBody(), WeatherResponse.class);

            return data.current.temperature;

        } else {

            // Handle non-200 status

            System.debug('Error from Weather API: ' + res.getStatusCode() + ' ' + res.getStatus());

            return null;

        }

    }

}
```

This class URL-encodes the city name, sends a GET request, and uses a nested Apex class structure to parse the JSON. The JSON is expected to have a structure like {"current": {"temperature": 23.5,...},...}. The code returns the temperature or null if there's an error, logging the error for debugging."

This example shows ChatGPT generating a quick integration class. You could refine the prompt to include specifics (like adding an API key header if required). Notably, ChatGPT created an inner class WeatherResponse to map the JSON – an approach akin to JSON2Apex.

Tester Tip: Use **Postman or a similar API client** to test the external API outside of Salesforce first. This helps you know what responses to expect. Then in Apex tests, use those expected responses in your HttpCalloutMock. You can even use ChatGPT to generate a mock class for you. For example, prompt ChatGPT: "Provide an Apex implementation of HttpCalloutMock that returns a 200 response with a given JSON body for testing." This can produce a template you adapt for various tests. Always test both the successful integration path and failure cases (like a 503 Service Unavailable). Testing failure cases ensures your integration code doesn't break when the external system behaves unexpectedly.

As you implement REST integrations, remember that **efficiency and security** are as important as correctness. Use bulk patterns (process lists of records in one call if the API supports it, rather than one call per record) to stay within callout limits. Also, keep long-running integrations off the main thread – which leads us to asynchronous patterns. Before that, however, many enterprises still rely on SOAP web services. In the next section, we'll switch focus from JSON and REST to XML and SOAP, exploring how Apex handles SOAP-based integrations and what best practices to follow there.

IMPLEMENTING SOAP API INTEGRATIONS

While REST has become the dominant style for web services, SOAP (Simple Object Access Protocol) remains prevalent in many enterprise environments. Systems like older SAP modules or Workday's legacy APIs often expose SOAP endpoints with XML payloads. Salesforce's Apex has robust support for SOAP callouts through WSDL2Apex, which generates Apex classes from a WSDL (Web Services Description Language) file. In this section, we cover best practices for integrating with SOAP-based web services: from generating and using the Apex stubs, to managing sessions and parsing XML responses. We'll also discuss how to handle SOAP-specific challenges such as large messages and error faults. As before, we'll incorporate ChatGPT prompts – for example, to help verify a SOAP integration or to simulate a SOAP response for testing.

Generating Apex from WSDL (WSDL2Apex): The primary way to consume a SOAP service in Apex is by importing its WSDL. Salesforce's platform can convert a WSDL into a set of Apex classes (often referred to as a "stub" or proxy). These Apex classes have methods corresponding to the SOAP operations. For instance, if Workday exposes a SOAP service EmployeeService with an operation GetEmployee, after you import the WSDL, you might get an Apex class like EmployeeServicePort with a static method getEmployee() to call. Under the hood, these generated methods use the WebServiceCallout.invoke method to perform the callout, constructing the SOAP envelope and parsing the response for you.

Best practice is to use WSDL2Apex whenever possible because it saves you from manually crafting XML. To import a WSDL: go to Salesforce Setup, Apex Classes, click "Generate from WSDL", and follow the prompts. If the WSDL is large or has many complex types, be aware of Apex class size limits – the generated code might be huge. Salesforce has some limitations on WSDL import: it might not support certain XML schema constructs (like xsd:choice, or multiple inheritance in complex types). If you encounter an import error, you may need to edit the WSDL to remove unsupported parts or use an external tool to simplify it. There are community tools (for example, FuseIT's WSFT or the open-source WSDL2Apex GitHub project) that can handle more complex WSDLs by generating Apex code with workarounds for Salesforce's limitations.

Developer Tip: Keep track of the WSDL version and source. If the external system's WSDL changes (for example, a new field is added to a Workday API), you'll need to regenerate the Apex classes and update your code. This can be a significant maintenance effort. If the API changes frequently, consider if a middleware (like MuleSoft) can mediate, so Salesforce integration remains stable while the middleware adapts to changes. Also, name your generated classes meaningfully. Salesforce will default to names based on the WSDL, but you might end up with classes like EmployeeServicePort and EmployeeServicePortType. You can refactor these (after generation) for clarity, but be cautious – if you regenerate, your changes might be overwritten. Many developers wrap the generated code in a facade class that they call from application logic, keeping the generated code as-is.

Using the Generated SOAP Stub Classes: Once you have the Apex classes from the WSDL, using them is straightforward. Typically, the generated code includes a class with methods for each SOAP operation. For example:

// Suppose the WSDL had a service for Order processing

OrderServiceSoap orderClient = new OrderServiceSoap();

OrderServiceSoap.OrderRequest req = new OrderServiceSoap.OrderRequest();

req.orderNumber = '12345';

```
// set other request fields...

OrderServiceSoap.OrderResponse resp;

try {

  resp = orderClient.placeOrder(req);

} catch (CalloutException e) {

  // Handle callout exception (timeout, etc.)

}

if (resp != null) {

  // process resp, e.g., resp.status or resp.confirmationNumber

}
```

Behind the scenes, when placeOrder(req) is called, the Apex runtime constructs the SOAP XML with an envelope, body, the OrderRequest serialized in XML, and sends it to the endpoint defined in the WSDL (the generated class contains the endpoint, which you can override if needed). The response is parsed into an OrderResponse object.

One best practice here is to always check if the response is not null and whether it contains any fault information. In SOAP, errors might come back as a SOAP Fault structure. The WSDL2Apex classes often represent faults as exceptions or as part of the returned object (sometimes the stub code will throw an exception if a fault is received). Consult the generated class documentation comments – if the WSDL defined fault messages, the Apex stub might have a custom exception class you should catch. For example, a WSDL might define a fault InvalidEmployeeIdFault which could correspond to an Apex exception class in the generated code; you'd catch it specifically.

Session Management in SOAP Calls: Some SOAP services (like Salesforce's own SOAP API or older SAP services) require a login call that returns a session token, which must be included in subsequent calls (often as a SOAP header). With WSDL2Apex, if the WSDL includes the login operation and defines the session header, the generated Apex should support it. Typically, you would call the login method, get a session ID, then set that session ID in the header member of the stub class before calling other methods. For example:

MyServiceSoap myClient = new MyServiceSoap();

// Assume the WSDL had a Login operation that returns a session token

MyServiceSoap.LoginResult lr = myClient.login('username','password');

myClient.SessionHeader = new MyServiceSoap.SessionHeader_element();

myClient.SessionHeader.sessionId = lr.sessionId;

After this, calling other methods on myClient will include the SessionHeader in the SOAP envelope for authentication. This pattern is common with SOAP integrations. However, be mindful that Apex has no built-in persistence between transactions; if you need to reuse a session token across transactions (e.g., between different runs of a Scheduled Apex), you'd have to store it (perhaps in a Custom Setting or cache) – and manage its expiration. Often, it's easier to just call login each time if that's allowed, or better, use a more stateless integration approach (many modern SOAP services provide an OAuth token or API key alternative to session-based auth).

If the SOAP service uses WS-Security (SOAP headers with encryption, nonces, etc.), Salesforce's WSDL2Apex has limited support. You might have to manually modify the SOAP envelope by editing the generated code (not ideal). In such cases, again, using middleware to handle the SOAP complexity and exposing a simpler REST or token-based interface to Salesforce could save time.

Handling Message Sizes: SOAP XML can get verbose. Large SOAP messages might hit Salesforce's heap limits or size limits. Governor limits note a 6 MB limit on callout response for sync Apex , which applies to the raw SOAP XML. If you anticipate very large payloads (say, retrieving thousands of records from SAP in one call), you may need to use pagination in the SOAP API (if available), or switch to an integration approach that can handle bulk data (Salesforce's Bulk API or an ETL tool, for instance). Additionally, the generated Apex classes themselves count against org code size. If a WSDL import produces, say, 300kB of

Apex code, that's substantial. If only a few operations are needed, see if you can trim the WSDL to just those parts before importing.

Another SOAP-specific consideration is **attachments** (SOAP with MTOM). Salesforce's WSDL2Apex does not support MTOM (binary attachments in SOAP). If an integration requires sending or receiving files via SOAP attachments, you'll likely need to handle that with HTTP callouts manually or again lean on a middleware that can do it.

Error Handling in SOAP Integrations: As mentioned, SOAP faults may manifest as exceptions. For a smooth integration, catch exceptions like CalloutException for transport errors (timeouts, no HTTP response, etc.) and any custom exceptions from the stub. Log the e.getMessage() as well as perhaps e.getStackTraceString() to diagnose issues. The stack trace might include the underlying XML fault string. For example, if calling a SOAP service for SAP returns a fault "Invalid Order ID", that might appear in the exception message. It's useful to propagate a friendly error message up to the user or an admin notification: e.g., "Order sync failed: Invalid Order ID (as reported by SAP)." However, avoid exposing raw XML to end users – parse it or use known fault structures for messages.

Testing SOAP Callouts: Testing can be done with the WebServiceMock interface in Apex. You create a class that implements WebServiceMock and override its doInvoke method to simulate a response. This is a bit more complex than HttpCalloutMock because you have to construct a fake SOAP response XML that matches what the stub expects. One trick is to capture a real SOAP response (using a tool like SOAP UI or by enabling debug logs and seeing the raw response) and then have your mock return that. ChatGPT can assist here: for example, you can prompt "I need to create a WebServiceMock response for the SOAP method X, here is the XML that should be returned, help me write the Apex code to do that." ChatGPT could then produce a doInvoke implementation that returns the XML. Alternatively, if the generated classes have helper methods or you can call the generated parser directly, use that. Some developers find it easier to bypass WSDL2Apex in tests and just simulate the behavior by calling the class methods with known inputs and outputs (since the class methods are static, you might not even need to callout – you could manipulate the response object directly if the call is abstracted enough).

ChatGPT Prompt Examples – Verifying SOAP Integration: Suppose you have imported a WSDL for a MuleSoft-managed SOAP API for an inventory system. You want to ensure your Apex code is correctly setting the SOAP headers. You might ask ChatGPT something like:

User Prompt: "I generated an Apex class from a WSDL for Inventory-Service. I'm calling InventoryServiceSoap.getProductDetails(product-Code) but I suspect I need to set an authentication header in the SOAP. The WSDL mentions a SOAP header AuthToken. How do I set this in Apex?"

ChatGPT: "If the WSDL included the header, your Apex stub should have a member variable for it, typically an inner class representing the header and a member on the service class. For example, if the header is <AuthToken> with a value, the generated class might have something like public AuthToken_element AuthToken in the service class. You would set it before calling the method:

InventoryServiceSoap service = new InventoryServiceSoap();

service.AuthToken = new InventoryServiceSoap.AuthToken_element();

service.AuthToken.token = 'your-auth-token-value';

ProductDetailsResponse response = service.getProductDe-tails(productCode);

Make sure you obtained or stored the token beforehand (perhaps via a login call or from a Named Credential if using one). If the generated class does not have an explicit header member, it could be that the WSDL wasn't imported with the header (some WSDLs separate the header definitions). In that case, you might need to manually add the header by editing the Apex class – which is advanced. But likely, the above will work. Ensure the token string is correct and not expired."

Assistant continues with more guidance...

Another scenario where ChatGPT helps: debugging a SOAP fault.

User Prompt: "I get a CalloutException when calling the SOAP method UpdateEmployee. The error message in the debug log shows a SOAP fault: <faultstring>Invalid Department ID</faultstring>. How should I handle this error in Apex?"

ChatGPT: "This indicates the external service rejected the call due to an invalid department ID. In Apex, you should catch the exception around the callout. For example:

```
try {
  svc.updateEmployee(emp);
} catch (CalloutException e) {
  if (e.getMessage().contains('Invalid Department ID')) {
    // Handle this specific error, maybe mark the record or notify
    System.debug('Employee update failed due to invalid department.');
  } else {
    System.debug('Unexpected error: ' + e.getMessage());
  }
}
```

Ideally, you'd prevent this by validating department IDs before calling, but in any case, handle it gracefully. You might log an error record or send a Platform Event (which we'll discuss in a later section) to notify an integration monitoring process. Also, consider having ChatGPT or another tool parse the WSDL fault section to see if there's a custom exception class you should be catching instead of a generic CalloutException."

This example shows how ChatGPT can parse an error message and suggest a handling strategy. You might even feed the entire exception message or a snippet of debug log to ChatGPT via file upload (ensuring no sensitive info) and ask for analysis. ChatGPT's large context understanding can help spot if the error is due to missing header, wrong endpoint, or data issues.

Tester Tip: If you have the option, test the SOAP integration outside Salesforce first. Use a tool like SoapUI or Postman (which can do SOAP) with the WSDL to send requests. This helps you identify issues with the request payload or credentials without the added complexity of Apex. Once you get a request working in SoapUI (for example, you successfully call SAP's SOAP service to create an order), take that request XML and compare it to what Apex is sending. Salesforce debug logs, when set to finest for callouts, will show the outbound XML (though sometimes truncated if very large). You can also capture traffic with a tool like WireMock if allowed. Ensuring parity between a known-good request and your Apex request is key to resolving integration issues. If differences are found, adjust the Apex (or the WSDL import). ChatGPT can be handy in comparing two XML payloads or WSDL files if you feed them in and ask for differences.

Special Guidance for Developers and Testers: When validating SOAP integrations, remember that the heavy lifting is done by generated code. As a developer, focus on correctly populating the request objects and interpreting the response objects. Don't be tempted to modify the autogenerated classes unless absolutely necessary – they can be complex. Instead, write wrapper methods in a service class that call the stub and handle exceptions, translating them into application-specific messages or actions. This separation means in your org's logic, you call your wrapper, which internally calls the stub. This is a cleaner design and helps testers as well: testers can focus on the wrapper's behavior (given input X, did we ultimately update the Salesforce records correctly after the SOAP call?). Testers should use mock callouts to simulate both success and fault responses from the SOAP API. It's often useful to have sample XML files of SOAP responses (perhaps provided by the API documentation or captured from a test system) to use in your mocks. Ensure your test covers that the integration code properly processes the SOAP response – e.g., if the response should update a Salesforce record, assert that the record is updated in the test after the mock callout.

Having tackled SOAP, which often involves synchronous request/reply patterns, we recognize not all integrations happen in real-time. In many cases, especially when dealing with large data volumes or long-running operations, an asynchronous approach is needed. Salesforce offers asynchronous Apex features (future methods, queueables, batches, schedules) to manage such scenarios. Next, we will explore Asynchronous

Integration Patterns – how to design integrations that don't block the user or that run in the background, thereby improving reliability and user experience for complex processes.

ASYNCHRONOUS INTEGRATION PATTERNS

Real-world integrations sometimes need to happen behind the scenes or on a schedule, rather than immediately in response to a user action. Salesforce's Apex provides several asynchronous mechanisms – Future methods, Queueable Apex, Batch Apex, and Scheduled Apex – which are essential for integration patterns that involve long-running tasks, large data transfers, or decoupling from the transaction that triggers the integration. In this section, we introduce these asynchronous patterns and offer best practices for using them in integration scenarios. We'll look at when to choose each approach, how to handle job chaining and retries, and how to design for error recovery in asynchronous flows. ChatGPT prompts will demonstrate how you might get advice on optimizing an asynchronous process or have it analyze asynchronous code for potential issues. We'll also provide targeted recommendations for architects and developers on selecting the right asynchronous pattern for specific integration needs.

Why Asynchronous? The need for asynchronous integration arises for a few common reasons:

Long Running Calls: Some external calls might take longer than Salesforce's synchronous limit (which, as noted, is 120 seconds max for callouts). If Salesforce needs to wait for an external process (say a complex financial calculation on an external server, or an external batch job to complete), it can't hold a transaction open indefinitely. Asynchronous Apex can allow Salesforce to initiate a process and check results later, or simply not block the user while waiting.

Bulk Operations: If you need to integrate a large volume of data – e.g., sync 10,000 records with SAP – doing this in a single transaction might be impossible or unwise due to governor limits. Batch Apex can process large data in chunks (batches of up to 2000 for queries, 200 for DML or callouts), enabling large-scale integrations that respect limits.

Decoupling from User Trigger: Perhaps when a record is created in Salesforce, you need to send it to an external system, but you don't want the user's save operation to fail just because the external system is down. An asynchronous "fire-and-forget" pattern can be used: commit the Salesforce transaction, and then in a future or queueable context, attempt the callout. This way, the user isn't kept waiting or given an error due to external issues.

Scheduled or Periodic Sync: Some integrations run on a schedule – e.g., nightly sync of currency exchange rates from an external service, or hourly fetch of new orders from an e-commerce system into Salesforce. Scheduled Apex can kick off such processes.

Let's outline the main asynchronous Apex tools and their integration use cases:

Future methods (@future): Easiest way to run async – mark a static method with @future. It will run later (usually quickly, but no timing guarantee) in its own context. It's often used for callouts from triggers (since triggers can't do callouts after a DML without async). *Use case:* After inserting a Contact, call a future method to send the contact data to an external mailing list system. Future methods are simple but have limitations: they can't return values, you can't chain them (one future can't directly call another), and you're limited in number per transaction (currently 50 future calls per transaction). Salesforce architects nowadays often prefer Queueable over future for new development.

Queueable Apex: Think of it as an improved future method. You implement the Queueable interface, which has a single execute() method similar to a execute() in a Batch context. You enqueue a job via System.enqueueJob(new YourQueueable(params));. Queueable jobs can be chained: at the end of one queueable's execute, you can enqueue another. They also allow more complex data (you can have instance state in the queueable object beyond primitive types). For integration, Queueables are great because you can, for example, break a process into steps: first queueable gets an auth token from an API, then chains a second queueable to use that token to pull data, then chains a third to process the data in Salesforce. Each step runs asynchronously in sequence. You're limited to 50 enqueues per transaction (similar to future), and only one level of chaining at a time (though you can chain again in the next one, effectively making a sequence). Queueable is ideal for medium-weight processes and is widely used in integration patterns.

Batch Apex: The heavyweight champion for large jobs. Batch Apex lets you query a large number of records and process them in chunks of 1 to 200, each chunk running in a separate transaction. It's commonly used for data cleansing or mass updates, but also for integrations where you need to sync or send a lot of data. For example, you might have a Batch job that runs nightly to send all updated Salesforce records to an SAP system. In each batch execution (say 100 records each), you perform a callout to send those 100 in one go (if the external API supports batch updates) or loop within execute to send one by one (being mindful of callout limits). Batch Apex allows for 100 callouts per batch execution, so overall you could call out many times, partitioned across batches. You can also implement retry logic: e.g., if one batch chunk fails due to a callout exception, it only rolls back that chunk, and you could catch exceptions and record which records failed, to retry in another batch or send to a queue.

Scheduled Apex: This isn't a separate execution mode, but rather a way to run code on a schedule (which itself often invokes a batch or queueable). You implement the Schedulable interface. Use case: schedule a Batch to run every night at 2 AM. For integration, Scheduled Apex can pull data regularly from external systems (like import new leads from a partner system every morning). It's also useful for delayed retries – e.g., if an integration fails, schedule a retry in 15 minutes via a one-off scheduled job.

Continuations (for long callouts): A bit of a special case: if you have a Visualforce or Lightning component that needs to call out without blocking the request, Apex Continuation can be used. It allows an Apex controller to make an async callout and resume later when the callout completes. Continuations are not widely used unless you have a specific UI use case, but it's worth knowing as an option for long-running callouts that still need to return data to a user session (perhaps outside the main scope of this chapter, which is more about backend processing).

Best Practices for Asynchronous Integrations:

Bulkify and Batch if Possible: If an external API supports processing multiple records in one call, take advantage of that in your Apex integration. For example, AWS might expose a bulk endpoint or MuleSoft could provide a batch API to upsert multiple records to SAP. Instead of 100 separate callouts, one batch callout with a bulk payload can drastically reduce overhead and avoid hitting callout limits. Batch Apex can collect records and call out in bulk. Even in a single queueable, you might accumulate a list of objects and send in one JSON payload. This reduces network chatter and is often faster.

Governor Limit Management: Each asynchronous execution has its own governor limits (with some differences, e.g., 60sec CPU for async vs 10sec for sync). But you still must manage within those. For instance, a batch's execute method gets fresh limits, so you can afford up to 100 callouts *per batch chunk*, not overall. However, don't abuse that – hitting 100 callouts in each of 50 batches = 5000 callouts, which might overwhelm the external system or network. Often, there are also external limits (maybe the API allows only X calls per hour). A good integration pattern is to incorporate pause or rate limiting if needed (perhaps via a static counter or by spacing out a scheduled job).

Chaining and Order of Operations: Sometimes integrations require multi-step processes. For example, you need to send data to Service A, then send the result of that to Service B. You can implement this via chained queueables or by using the Batch's finish() method to trigger another process after all batches complete. A common pattern: have a Scheduled Apex start a batch. In the batch's finish, enqueue a queueable to process results or send a notification. This way, steps happen in sequence. Ensure each step's data is stored or passed along. Queueables can pass simple state to the next by setting properties on the next queueable when constructing it (or storing intermediate data in a record or Custom Setting if needed).

Retries and Recovery: Asynchronous jobs might fail due to external issues (e.g., a queueable making a callout gets a timeout). You should design a strategy to retry. This could be as simple as catching the exception and enqueuing a new queueable for that specific record or API call after some delay. Or using a combination of Platform Events to log failures that trigger another attempt. If using Batch, you might create a follow-up batch for failures. Avoid infinite retry loops – have a cut-off or escalation (like after 3 tries, log a persistent error and notify humans).

Idempotency: Ensure that if an asynchronous job runs twice (maybe due to a retry or manual re-trigger), it doesn't create duplicate data or perform the action twice incorrectly. This often means using external IDs or tokens. For example, if you are syncing an Opportunity to SAP via a queueable, you might mark the record with a "Sent to SAP" timestamp or an external ID returned from SAP. If a retry happens, your code can check if it's already processed. Idempotency is a concept the external side should ideally support too (e.g., Workday might ignore duplicate calls with the same IDempotency-Key header). But from Salesforce side, do what you can: perhaps store a unique identifier of the request in a custom object log to avoid sending it again.

Selecting the Right Tool: As an architect or developer, decide which asynchronous mechanism fits the scenario:

Use future for very simple, one-step fire-and-forget callouts that don't need complex coordination (though queueable could do the same, queueable is now generally favored due to added flexibility).

Use queueable for multi-step processes, or when you need to pass complex types or chain jobs, or simply to logically separate a unit of work off the main thread.

Use batch when dealing with large data sets or if you need to break work into smaller pieces to handle limits. Batch is also useful when the integration has to process each record individually but you want it orchestrated automatically.

Use scheduled to run things at certain times or to orchestrate periodic jobs.

Architect Tip: Establish guidelines in your Salesforce team about when to use which async pattern. For example, "if any job might process more than 2000 records or run longer than a minute, implement it as batch; if it's primarily an immediate extension of a trigger (like callout after insert), use queueable." These guidelines prevent misuse (e.g., someone trying to do a massive data sync in a single future call, which would likely fail). Additionally, consider using Platform Events or an external queue (like AWS SQS or MuleSoft's queueing mechanism) for extremely large-scale or cross-system asynchronous coordination. For instance, rather than schedule a batch to poll for updates, you might have external system push events.

ChatGPT Prompt Examples – Optimizing Async Logic: Suppose you have a future method that sometimes fails due to too many callouts. You might ask:

User Prompt: "I have a @future method doing callouts in a loop (it sends updates for up to 200 records, one callout each). It's hitting limits. How can I improve it?"

ChatGPT: "It sounds like your future method might exceed the 100 callouts per transaction limit if more than 100 records need updates. Also, sending 200 sequential callouts could be slow. Consider switching to a Batch Apex job. In Batch Apex, you can process records in chunks (say 50 per batch) which gives you separate transactions and limit pools for each chunk. Within each chunk's execute, you can also combine records into a single callout if the external API supports bulk updates (for example, send an array of records in one JSON). If bulk API isn't an option, Batch will at least spread out the callouts across transactions. Another approach is to use Queueable and chain jobs: split the list of records into two sets of 100 and call enqueueJob twice (chain the second from the first) to stay under limits. The most robust solution is Batch Apex for this scenario due to its built-in handling of large record sets. Additionally, ensure your code is bulkified: constructing one HTTP request per record might be unavoidable if the external system only accepts one at a time, but if it can accept bulk, implement that. You could ask the external API team if a bulk endpoint exists."

This response (which ChatGPT might give) encapsulates a best practice: if a future is not enough, move to batch or queueable for scale.

Another example:

User Prompt: "Can you provide a pattern in Apex for chaining queueable jobs? For example, first callout to Service A then use its result to call Service B."

ChatGPT: "Yes. You can achieve this by making your Queueable class chain another Queueable at the end of its execution. Here's a simplified example:

```
public class FirstCalloutJob implements Queueable, Database.AllowCallouts {
  public void execute(QueueableContext context) {
    // Callout to Service A
    HttpResponse res = //... perform callout
    String result = res.getBody();
    // Process result...
    // Chain next job with result
    SecondCalloutJob nextJob = new SecondCalloutJob(result);
    System.enqueueJob(nextJob);
  }
}

public class SecondCalloutJob implements Queueable, Database.AllowCallouts {
  String input;
  public SecondCalloutJob(String resultFromA) {
    this.input = resultFromA;
  }
  public void execute(QueueableContext ctx) {
    // Use input to call Service B
    //... perform callout to Service B
  }
}
```

In this pattern, once FirstCalloutJob finishes, it enqueues SecondCalloutJob. You can chain further if needed, but avoid very deep chains or recursive chains that could potentially loop forever. Each job gets a fresh set of limits, and the state (result) is passed via the constructor. Remember that Queueable jobs execute asynchronously and nearly independently, so consider what happens if the second job fails – you might need error handling to retry or compensate (more on that in the error handling section). Also, note the Database.AllowCallouts marker interface on the Queueable if you're doing callouts inside it."

Such an example from ChatGPT provides a template and best practices comment (like including AllowCallouts interface which is required to allow callouts in Queueable or Batch contexts).

Recommendations for Specific Scenarios

If integrating with SAP or Workday where large data syncs are needed (like syncing all employees nightly), use Batch Apex. One batch can retrieve the employees from Salesforce (or an external query via callout in each execute) and then update/create records. Batch's finish() could call a web service to report completion or chain another batch to handle related data (e.g., after employees, sync their department info).

If integrating with an AWS service that might have intermittent slowness, using Queueable with retries makes sense. For example, a Queueable calls an AWS Lambda through API Gateway. If it times out or fails, catch the exception and use System.enqueueJob(this) to re-enqueue itself (perhaps after updating a retry count property and an if (retryCount < maxRetries) check). Add a small delay mechanism if possible (though Apex doesn't have sleep, you might space retries by using a scheduled job for a later retry if needed).

For MuleSoft orchestrations, you might just call MuleSoft's endpoint and MuleSoft handles further distribution. In that case, a simple future or queueable is enough from Salesforce side, because MuleSoft will do the heavy lifting externally (e.g., Salesforce fires an event or makes one call, MuleSoft then calls multiple downstream systems). Ensure to mark such calls as asynchronous if they don't need to complete before giving control back to the user.

Product Owner Tip: Asynchronous integrations can affect user expectations. If a salesperson saves a record and an external integration is supposed to enrich that record (maybe pulling credit score from an external system), asynchronous design means the enrichment will happen a bit later, not immediately on save. It's important to set the right expectations in the UI, perhaps with a status field or message like "Data will be updated shortly." Product owners should be aware of these implications: asynchronous processes improve reliability and performance, but introduce slight delays. They should define acceptance criteria for those delays (e.g., "the data from System X should be reflected in Salesforce within 5 minutes of the trigger") and ensure there's a way to monitor or notify if the integration fails (so that from a user perspective, they know if the data didn't arrive). Balancing user needs with system constraints is key – sometimes you might start with a synchronous integration for simplicity, then move to async after explaining to stakeholders why it's necessary (governor limits, external system latency, etc.).

With asynchronous patterns covered, you have the tools to build integration processes that run reliably in the background, handle large data volumes, and minimize user impact. Asynchronous Apex is powerful, but sometimes a different approach is needed: an event-driven architecture where systems communicate through events rather than direct callouts. In Salesforce, **Platform Events** serve this role. In the next section, we'll explore how Platform Events can be used for robust, loosely-coupled integrations that allow Salesforce and external systems to stay in sync in real-time via publish/subscribe patterns.

USING PLATFORM EVENTS FOR ROBUST INTEGRATION

Platform Events in Salesforce enable an **event-driven integration** approach, which can make your architecture more resilient and scalable by decoupling the communication between systems. Unlike request-response callouts, events allow a "fire-and-forget" style of integration: Salesforce publishes an event, and one or many subscribers (which could be external systems or other Salesforce orgs) react to it. Similarly, external systems can publish events to Salesforce, which Salesforce can handle through triggers or Flow. In this section, we outline how to strategically use Platform Events for real-time integrations both within Salesforce and with external systems. We'll discuss event design principles, the publish/subscribe model, and best practices for managing event lifecycles, including error handling for events. We'll also touch on the role of ChatGPT in evaluating event-based integration logic (for example, checking that an event payload contains all necessary fields for subscribers). Throughout, we'll emphasize how platform events can improve business agility and responsiveness – a point of interest for developers, testers, and product owners alike.

Platform Events Overview: A Platform Event is a special Salesforce entity that is defined by a schema (fields) and can be published to the Salesforce event bus. Once an event is published, any subscriber who is listening will receive it (near real-time). Subscribers can be within Salesforce (e.g., a Trigger on the event, or a Flow, or even another org via Change Data Capture or event relays) or outside Salesforce (external apps can subscribe via the CometD-based Streaming API or tools like MuleSoft or AWS EventBridge connectors). Platform Events support a **publish-subscribe model:** the publisher and subscriber are decoupled – the publisher doesn't know who or what will consume the event. This is great for scalability; you can add new subscribers without changing the publishing logic.

Salesforce Platform Events are **secure and scalable.** Standard volume events can deliver on the order of thousands of events per hour, and high-volume events (with a special add-on license) can scale to hundreds of thousands of events per hour. They are stored temporarily (24-hour retention for standard events, 72 hours for high-volume) for replay if needed. They are also immutable and transactional: if you publish an event as part of a transaction that rolls back, the event doesn't get published (ensuring consistency).

Use Cases for Platform Events in Integration

Notification of Data Changes: Instead of a Salesforce trigger making a callout to inform an external system of a change, it can publish a platform event. External system(s) subscribed will get the event and act on it. For example, when an Opportunity is marked "Closed Won", publish an event Opportunity_Won__e with details. MuleSoft or an AWS Lambda can subscribe and then take that event to update an ERP system, generate an invoice, etc. This way, Salesforce doesn't directly call those systems; it just emits the event.

Cross-System Workflows: Complex business processes often span multiple systems. Platform events can serve as a glue. Suppose a new hire process involves Workday (HR), Salesforce (IT provisioning), and an email system. A Workday integration could publish a "NewHireCreated" platform event into Salesforce (via middleware or directly if Workday can call an API). Salesforce receives it and triggers provisioning tasks. If another system also cares about new hires, it can subscribe as well. Everyone gets the memo, no direct coupling.

Error or Exception Events: If an integration fails or a certain condition arises, you can publish a platform event to flag it. For instance, if a callout to SAP fails after retries, an Integration_Error__e event could be published. An external monitoring system or an admin notification process could subscribe to that to alert the team. This pattern creates a uniform way to handle errors across different integrations (we'll expand on error handling in the next section).

Streaming Data Feeds: Sometimes you want a continuous feed of data. For example, streaming IoT data or transactions into Salesforce. Instead of making API calls for each record, an external system could publish events rapidly and Salesforce just listens and updates records. This is particularly useful for high-frequency data where a REST API approach might be too choppy or create bottlenecks.

Designing Effective Platform Events

When creating a Platform Event (in Setup under Integrations -> Platform Events), you define fields much like you would for a custom object. Best practices include:

Only include necessary fields that subscribers need. Events have a payload size limit (currently around 1-2 KB for standard-volume events). Keep them lean. For example, if notifying about an Opportunity, you might include Opp Id, Name, Amount, Status, and maybe a URL – not every field of the object.

Include a field for a record identifier (like an external Id or Salesforce Id depending on context) so subscribers can relate the event to their own data. If an external system publishes an event about an Order, maybe include the Salesforce Account Id it pertains to, if known.

Consider a field for source system or context, if multiple sources might publish the same event type. E.g., SourceSystem = "SAP" or "AWS" so subscribers know where it came from if necessary.

If ordering matters between events, design with that in mind. Platform events have a replay and sequence IDs, but cross-application ordering is not guaranteed. If a certain sequence is required (like two events must be processed in order), you might need to incorporate logic (like a sequence number in the event payload and have the subscriber buffer if out-of-order).

Idempotency for events: similar to other integrations, ensure consuming the same event twice (which could happen in some fail/retry scenarios) won't cause harm. Maybe include a unique Event UUID field (though Salesforce events have a GUID internally and replayId). External subscribers can track last seen replayId to avoid duplicates.

Platform Events are **published** by Apex (using EventBus.publish()), by Flows, or even from outside via API (API can create platform event objects through the sObject API). Apex triggers on regular objects can publish events as part of their logic. Notably, publishing an event is also subject to governor limits: it counts as a DML statement (you can publish events up to the DML rows limit – which is 10k events per transaction for high-volume events, and much less for standard events because they count against the 2k DML row limit).

Subscribing to Events Externally: External systems usually subscribe via **CometD** streaming. Salesforce provides a RESTful Streaming API endpoint ("/event/YourEvent__e"). Tools like MuleSoft have connectors that make this easy – you configure it to listen to the platform event. AWS has an integration called the **Amazon EventBridge Salesforce connector**, which can forward platform events to AWS services (so you can, for example, have a rule in EventBridge that whenever Salesforce emits an event, it triggers an AWS Lambda or sends to an SQS queue). This is a powerful way to connect Salesforce with AWS in real-time without polling. Another method is using **EMP Connector**, a Java library from Salesforce, if you want a custom app to listen.

Subscribing Internally: Within Salesforce, you can have a Trigger on the platform event (syntax is after insert only, since events are insert-only). For example, you might have trigger OnOrderEvent on Order_Created__e (after insert) {... } which runs whenever an external system's event comes in, allowing Salesforce to, say, create an Order record in Salesforce. These triggers run asynchronously, separate from other transactions (and don't consume DML limits from the origin transaction). One important note: if a platform event trigger fails (uncaught exception), the event is not replayed or retained – it's effectively lost (unless you have a custom retry). So handle errors in event triggers carefully (you might catch exceptions and, for example, publish another event or create an error log record to capture the failure).

Managing Event Lifecycles and Error Conditions: Unlike a direct callout integration, with events you need to consider what happens if the subscriber is temporarily unavailable. For instance, if MuleSoft is down for an hour, events published in that time could be missed unless MuleSoft reconnects and uses the last replayId to fetch missed events (Salesforce can store events for replay up to 24 hours by default, which can be extended to 72 with certain configurations). Therefore, it's critical to implement replay logic in subscribers or use **durable streaming**. MuleSoft's connector and tools like CometD support specifying a replay option (like -2 for all new events from now, -1 for only new from now, or a specific replayId to resume). Best practice is that external subscribers persist the last successful replayId they processed, so if they crash and restart, they ask Salesforce for events from that ID forward, thus catching any they missed.

Error handling with events is different: if Salesforce publishes an event, it doesn't know if the subscriber handled it or not. So the onus is on the subscriber side to handle failures (and possibly publish its own events or send back an error if needed). For example, if an external system receives an event but fails to process it due to its internal error, Salesforce isn't automatically aware. A robust pattern might be that the external system, upon failure, calls a Salesforce API or publishes a separate event like ExternalProcessFailed__e that Salesforce or a monitoring system picks up. Essentially, design a feedback mechanism if acknowledgement is required. Otherwise, it's fire-and-forget.

Developer Tip: Use the **Trailhead tools or CLI** to test platform events. Salesforce provides a Trailhead module to play with platform events and a CLI command (sfdx force:org:open -p /lightning/o/YourEvent__e/list) to open a streaming monitor. Also, the Workbench has a streaming push topic viewer that can subscribe to events for testing. For a quick internal test, you can write an Apex anonymous script to publish a test event and see if your trigger runs.

ChatGPT Prompt Examples – Platform Events Design: If you're unsure about how to structure an event, you might ask:

User Prompt: "I want to create a platform event for when a Case is escalated. What fields should I include in the event, and how to ensure external systems can reconcile it with their data?"

ChatGPT: "For an Escalation event, include the key identifiers: Case Id (perhaps the 18-char Salesforce Id or a case number if external systems use that), maybe a correlation Id if one exists (like if the case came from an external source originally), Escalation Level or a flag indicating it's escalated, and perhaps the time of escalation. Including the Case's Priority or Severity might be useful if external needs it, but keep the payload lean. External systems will likely use the Case Id or Case Number to fetch more details via API if needed; the event should tell them an escalation occurred but not necessarily carry all case details. To help reconcile, if external has its own ID for this case (say cases are mirrored in an external support system with a different ID), include that too if known in Salesforce (like an ExternalId field on Case). In summary, fields: ExternalCaseId__c (if exists), CaseId (Salesforce), Status, EscalationLevel, maybe Subject or a short description. Also consider using an event field for 'SourceSystem' if both Salesforce and say ServiceNow could raise similar events – in your scenario, Salesforce is likely the sole source for this event, so it's fine."

Such advice from ChatGPT reinforces focusing on identifiers and minimal info in the event.

Another example:

User Prompt: "How can I use ChatGPT to verify that my platform event trigger logic is correct? I have a trigger that creates an Order record when an Order_Placed__e event comes in."

ChatGPT: "You can describe the logic of your trigger to me, and I can check for obvious issues. Also, you could provide a snippet of the trigger code. For instance, ensure you're handling the event batch correctly: platform events can be delivered in batches of up to 2000 events in one trigger context, so your trigger should iterate over Trigger.new. If your logic assumes a single event at a time, that might fail when multiple events come in together. Also, make sure to use only after insert triggers for events (which you are, given the context). If your Order creation logic performs DML on standard objects, it might be wise to use a future/queueable inside the trigger if you expect heavy load, to avoid hitting governor limits in the trigger context (because if 2000 events come in, and you try to insert 2000 Order records, that's fine, but if each event led to some SOQL or callout, you could run into limits – using a queueable for processing can distribute load). I can also help write unit tests for your event trigger: you'll insert an Order_Placed__e record in test context and verify an Order was created. Be sure to set Test.startTest(); EventBus.publish(...); Test.stopTest(); to fire the trigger in tests."

The assistant would continue by possibly looking at code if provided, or suggesting improvements like bulk handling, error catching in the trigger, etc.

The mention that platform event triggers deliver events in batches is an important best practice: always bulk-handle events as you would records.

Tester Tip: Testing end-to-end event flows might require integration testing beyond Salesforce. Within Salesforce, you can test that publishing an event triggers the internal logic (via unit test as described). But to test external subscription, you'd likely use a staging environment where a test external client is subscribed. Tools exist (e.g., MuleSoft's testing or a simple Node.js script using CometD). Verify that when an event is published, the external system indeed receives and processes it. Conversely, test that if the external system publishes an event via API, your Salesforce trigger runs as expected. Use the Event Replay Id to simulate catching up: e.g., disconnect your subscriber, publish some events, reconnect subscriber with last known replayId to ensure it receives the backlog. This tests your recovery from downtime. Also test volume: if you plan on say 500 events/hour, try to simulate that in a full-copy sandbox to see if triggers and processing can handle it. Salesforce can throttle event delivery if subscribers can't keep up – observe platform event usage in the monitoring.

Impact on Business Agility and Real-time Response: For product owners and sponsors, platform events can be a selling point: they enable **real-time integration** without tight coupling. This means new requirements can be implemented by adding new subscribers or publishers without altering the core. For example, if a business wants to plug in a new analytics system to get notified of all Case escalations, they can just subscribe to the event – no changes in the case management code needed. This agility speeds up delivering new integrations and reduces regression risk on existing logic. Additionally, real-time events improve user experience; instead of waiting for a batch job or manual export, systems update each other almost instantly. In a sales context, that could mean Salesforce knows about an inventory update from SAP within seconds, allowing a salesperson to give accurate info to a customer.

However, it's worth noting to stakeholders that implementing event-driven architecture requires careful planning (monitoring and error handling as discussed). Once events are flying around, you need to ensure nothing falls through the cracks silently.

With platform events, we've added a powerful tool to our integration toolkit that complements REST, SOAP, and asynchronous Apex. As we design these integration solutions, one constant challenge is ensuring that data is transformed and mapped correctly between systems – a topic we'll dive into next. In the following section, we will focus on **Data Mapping and Transformation** in Apex integrations, showing how to maintain data integrity and consistency when moving data between Salesforce and external applications.

DATA MAPPING AND TRANSFORMATION IN APEX INTEGRATIONS

When integrating Salesforce with other systems, one of the most important tasks is ensuring that data is translated correctly between the systems. Different applications often have different data models – field names, data types, formats, and conventions can vary widely. For example, Salesforce might represent a state or province as "California" while an external ERP stores it as "CA", or currency values might need conversion. This section discusses best practices for handling data mappings and transformations in Apex during integration processes. We'll cover techniques for parsing and formatting data, normalizing values, and maintaining data integrity between systems. Whether you are populating Salesforce records from an AWS-hosted database or sending Salesforce data to an SAP module, careful mapping ensures consistency. We'll also provide ChatGPT prompt examples on how to assist with mapping logic (like generating mapping code or verifying transformations). Additionally, we'll include advice for developers, testers, and architects on validating that mappings are correct throughout development and deployment.

Understanding Source and Target Schemas: The first step in any integration mapping exercise is to clearly understand both the Salesforce data model and the external system's data model. This often means having a data dictionary or field mapping document. For instance, consider integrating Workday (an HR system) with Salesforce for user provisioning. Workday might have a field "Worker ID" that should map to Salesforce "Employee_ID__c", and a complex structure for names (first, middle, last) that must map to a Salesforce Contact name fields. As a best practice, create a mapping table (even if just documentation or a spreadsheet) listing each field in Salesforce, where its data comes from externally (or goes to), what transformations are needed (e.g., "External Country Code" -> "Full Country Name in Salesforce").

In some projects, especially at enterprise scale, architects will use **canonical data models** or middleware mapping logic. MuleSoft's Anypoint, for example, might perform transformations using DataWeave scripts, delivering already-mapped data to Salesforce. However, in Apex-centric integrations, you might handle it within Apex. Even if using middleware, understanding the mapping is crucial for testing and perhaps for Apex to do final touches or validations.

Techniques for Data Transformation in Apex

Apex Code (Manual Transformations): The simplest approach is writing code to convert values. For example, if integrating with SAP which uses numeric codes for order status, you might have a mapping in Apex:

```
Map<String, String> statusMap = new Map<String,String>{ '0' => 'New', '1' => 'Processing', '2' => 'Complete' };

order.Status__c = statusMap.get(externalStatus);
```

Or for formats: parsing a date string from external (like "2025-05-10T18:30:00Z") into a Salesforce DateTime field. Apex's Datetime.parse() can handle ISO8601 formats often, but you might need custom parsing for others. Always ensure you account for time zones and use Datetime.newInstanceGMT or adjust to GMT as needed because Salesforce stores DateTimes in UTC internally.

Custom Metadata/Custom Settings for Mappings: If you have many mappings or ones that may change, consider storing them in a Custom Metadata Type or List Custom Setting. For example, a mapping of country codes to country names can be a Custom Metadata with fields "External_Code" and "Salesforce_Value". Your Apex then queries this metadata (which is cached efficiently) to translate codes. This avoids hardcoding values in Apex and allows administrators to update mappings without deploying code. It's also easier to test various mapping scenarios by adjusting metadata in a sandbox.

Normalization and Validation: Integration often requires normalizing data. For instance, external data might be free-form text with inconsistent casing or extra whitespace. Apex provides string methods to clean that (trim(), toUpperCase(), etc.). For addresses, you might want to capitalize names or ensure zip codes conform to a pattern. Plan these normalizations to match what Salesforce expects (and what users expect in Salesforce UI). Also, validate data before upserting to Salesforce. If an external source might send an Account with no name, you might want to default it or reject it before attempting DML, to avoid partial failures.

Complex Transformations (XML/JSON within fields): Sometimes data might come as a blob or a nested structure that doesn't directly map to Salesforce fields. For example, an external system might provide a single field that contains a JSON array of items. In Apex, you could use JSON.deserialize to break that into records or related objects. Or if a field comes in as one string like "Lastname, Firstname", you might split it. The key is to encapsulate this logic in well-named methods or classes (like a DataTransformer class) so it's reusable and testable. For complicated logic, write unit tests that feed in sample inputs and verify outputs.

Ensuring Data Integrity and Consistency

Data integrity means the data remains accurate and consistent during integration. Some tips:

Use external IDs in Salesforce to avoid duplicate records when syncing. For instance, if importing Accounts from an external DB, designate a field as External ID in Salesforce and use upsert by that ID. This way, if the same record comes twice, it updates rather than creating duplicate.

Handle rounding or precision issues: different systems might have different numeric precision (e.g., Salesforce currency fields might allow 2 decimal places, but an external could have 3). Decide on a rounding strategy to avoid minor discrepancies. For example, always round to 2 decimals before saving to Salesforce if needed, and be aware of potential accumulated differences.

If splitting a single external record into multiple Salesforce records, ensure referential integrity. For instance, an external Order might include customer and order items together. In Salesforce you may need to create an Account (customer) and multiple Order/OrderLine records. You'd handle this by first creating the Account (perhaps caching the external ID to Salesforce ID in a map), then linking the order and lines. Use transaction control (all in one execute if possible) or store intermediate mapping so subsequent transactions know the linkage.

Conversely, if combining multiple Salesforce records into one external structure, ensure you gather all required pieces before sending. For example, sending an invoice might need Account info + related invoice lines; make sure your Apex queries gather all that and format accordingly. You might create a composite JSON with nested structures to send, which Apex can do by constructing maps of maps or using JSON.serialize on a custom Apex class structure that mirrors what the external expects.

Common Transformations Examples

Date/Time and Time Zone: If Salesforce is in UTC and external in local time, convert appropriately. Perhaps store all dates in Salesforce as UTC and have external send UTC, or have a known offset.

Encoding: External data might have characters that need escaping in XML/JSON. Apex's JSON methods handle escaping automatically, but if constructing XML, use String.escapeXml for inserting dynamic data into an XML template.

Units Conversion: Maybe Salesforce stores lengths in inches and external in centimeters. The integration needs to multiply/divide and maybe round. Keep a consistent approach and clearly document these conversions to avoid confusion. It might be wise to store both values if needed to avoid repeated conversions (but only if necessary).

Lookup or Reference Data: Suppose Salesforce has a picklist for "Product Category" but external gives a numeric code. You might need a mapping table as mentioned. Or you might call back to the external system for additional info (though calling back mid-integration flow can complicate things; better to have all info in the initial payload or a predefined mapping).

Developer Tip: Externalize mapping logic from the main integration code. Instead of peppering your callout code with if/else for various value conversions, create a utility method or class. For example, a CountryMapper.toSalesforceCountry(externalCountryCode) that returns the proper string. This separation makes your code cleaner and easier to adjust mappings later. Moreover, leverage Salesforce features like formula fields or validation rules when possible – not all transformation must be in Apex. If all incoming dates must be in future, you can have a validation rule on the object to ensure that, catching any mapping issues where perhaps an external date wasn't converted to UTC correctly. Use every tool available: for instance, a **Translation Workbench** if dealing with multi-language data, though that's more for UI translations than integration.

ChatGPT Prompt Examples – Assisting with Data Mapping:

User Prompt: "I have to map Salesforce Opportunity fields to an SAP API fields. Could you help create a mapping outline or Apex pseudocode for this? For example, SFDC Opportunity.Name -> SAP 'OpportunityTitle', SFDC Opportunity.Amount -> SAP 'TotalValue' (which expects an integer in cents)."

ChatGPT: "Certainly! First, here's an outline of the field mappings based on what you described:

Opportunity.Name → OpportunityTitle (String, no transformation except maybe truncation if SAP has length limit).

Opportunity.Amount → TotalValue (SAP expects an integer representing cents, so take Salesforce Amount (decimal) and multiply by 100).

Opportunity.CloseDate → CloseDate (formatting to SAP's expected date format, e.g., YYYYMMDD).

Opportunity.StageName → StageCode (assuming SAP expects a code; you'd need a mapping of stage to code).

Opportunity.AccountId → CustID (map the Salesforce Account to an SAP customer ID, likely via a lookup table).

Now in Apex pseudocode:

SapOpportunity sapOp = new SapOpportunity();

sapOp.OpportunityTitle = opp.Name;

sapOp.TotalValue = opp.Amount != null ? (Integer) (opp.Amount * 100) : null;

sapOp.CloseDate = opp.CloseDate != null ? opp.CloseDate.format('yyyy-MMdd') : null;

sapOp.StageCode = stageMap.get(opp.StageName); // stageMap defined elsewhere sapOp.CustID = accountMap.get(opp.AccountId); // accountMap perhaps provided from a prior step that looked up SAP IDs

Make sure to handle nulls to avoid sending default zeros if that's not desired. Also, check SAP field length limits – if OpportunityTitle has a 50 char limit in SAP and SF Names can be 120, you might truncate or split.

This pseudocode assumes you have stageMap and accountMap. stageMap could be a constant map of Salesforce Stage to SAP stage code. accountMap might be built by querying a custom field on Account that stores the SAP ID or by an external lookup.

In summary, doing the transformations: Amount to cents (int) and date formatting are key here. ChatGPT can help refine the code once you have specifics like stage mappings."

The assistant might then refine each mapping with any info you provide about format, or ask for clarification if needed.

Another prompt:

User Prompt: "I get data from an external service in JSON: {"fullname": "Doe, John", "annual_revenue": "1,234,567.89", "active": "yes"}. I need to map this to a Salesforce Contact (FirstName, LastName) and an Account Revenue (Number) and Active (Checkbox). What transformations are needed and how to do them in Apex?"

ChatGPT: "From the JSON example:

fullname: It's a single string with Lastname, Firstname. You need to split it. Likely split on comma. In Apex:

```
String fullname = rec.get('fullname');

String[] parts = fullname != null ? fullname.split(',') : new List<String>();

String lastName = parts.size() > 0 ? parts[0].trim() : null;

String firstName = parts.size() > 1 ? parts[1].trim() : null;
```

This will give you first and last names to map to Contact.FirstName, Contact.LastName. (Be cautious of names without a comma or extra parts like middle names).

annual_revenue: It has commas and a decimal point as a string. You'd need to remove formatting to convert to a number. One approach:

```
String revStr = rec.get('annual_revenue');
if(revStr != null) {
  revStr = revStr.replaceAll(',', ''); // remove commas
  Decimal revenue = Decimal.valueOf(revStr);
  acct.AnnualRevenue = revenue;
}
```

This will turn "1,234,567.89" into Decimal 1234567.89. Decimal.valueOf handles the string to decimal conversion once commas are removed.

active: It's "yes" vs perhaps "no". For a Salesforce Checkbox (boolean), you can map "yes" to true, "no" (or anything else) to false. For example:

```
String activeStr = rec.get('active');
contact.Active__c = (activeStr != null && activeStr.toLowerCase() == 'yes');
```

This sets Active__c to true if the string is "yes" (case-insensitive match), false otherwise (including if it's "no" or null).

Additional considerations: If the external JSON might use different words like "Y"/"N" or "true"/"false", handle those in the logic (maybe check for "y", "yes", "true" as true values).

These transformations ensure the data fits Salesforce field types: splitting name, numeric parsing, and boolean conversion."

The assistant may go on to suggest creating a small Apex wrapper to do this conversion for each JSON record, possibly using JSON.deserializeUntyped as implied by rec.get usage (i.e., treating deserialized JSON as a Map).

This example from ChatGPT highlights string manipulation and type conversion, common tasks in transformation.

Tester Tip: To validate mappings, create test cases for each mapping rule. If you have a mapping document with dozens of fields, consider writing a comprehensive Apex test where you construct an external input (maybe as a JSON or XML string), run it through your mapping code (which could be invoked via a test service method), and then assert that the resulting Salesforce objects have expected values. Essentially simulate an integration transaction in a test. Also, test boundary conditions: if external sends an unexpected value (e.g., "yes" in different language, or an unknown status code), ensure your code handles it (maybe defaults or errors out clearly). In integration testing (perhaps using a full sandbox connected to a test instance of the external system), verify a sample of records end-to-end: data created in external -> flows through integration -> ends up in Salesforce correctly mapped. Also verify reverse if applicable (Salesforce to external).

It's wise for testers to also verify that data hasn't lost integrity: e.g., if a high precision number in external got truncated in Salesforce, or character encoding issues (like accented characters not transferring properly). These subtle issues can be caught by carefully reviewing a set of records.

Architect Tip: Consider the overall data flow and whether transformations should be done in Salesforce or elsewhere. If multiple systems need the same data from Salesforce but in different formats, duplicating mapping logic in multiple places can lead to inconsistencies. An architect might introduce a middleware or a common library for transformations. If Apex must do it, ensure that logic is consolidated (like an Apex Service class) and not spread across triggers, batches, etc. Reusability and single source of truth for mapping rules help long-term maintainability. Also, plan for changes: e.g., if a code list changes in external, how quickly can Salesforce adapt? If using Custom Metadata, that's a simple deployment or even a UI edit; if hardcoded, requires code change. Align such decisions with how volatile or client-specific the data mapping is.

By rigorously applying mapping and transformation best practices, you maintain **data consistency across integrated systems**. The result is that users and stakeholders can trust that, say, a customer's information in Salesforce matches what's in SAP or Workday. However, even with perfect mapping, things can go wrong in integrations – networks fail, data can be incomplete, or systems return errors. Thus, it's crucial to build robust error handling and recovery strategies. In the next section, we will focus on **Error Handling and Recovery in Integrations**, outlining how to anticipate failures and design integrations that gracefully handle errors without data loss or user confusion.

ERROR HANDLING AND RECOVERY IN INTEGRATIONS

No integration is 100% error-proof. Systems go down, credentials expire, data conflicts occur – the list of potential issues is long. What distinguishes a robust integration from a fragile one is how gracefully it handles these errors and recovers from them. In this section, we detail critical practices for managing integration errors in Apex: from proactive error detection and graceful degradation of functionality, to implementing retry logic for transient issues. We'll explore common integration error scenarios (like timeouts, API rate limits, and data validation failures) and how to handle them in Apex code. Effective error handling also means communicating issues clearly – to logs, to system administrators, or even to end-users when necessary – without exposing sensitive information. We'll discuss patterns like fallback workflows and the use of Platform Events or custom objects to capture errors. As always, ChatGPT prompt examples will be included, showing how you might get advice on improving error handling or diagnosing an issue. We'll also give specialized guidance for architects and developers on designing integration flows that anticipate external system failures and ensure the Salesforce system remains reliable even when external dependencies falter.

Proactive Error Detection: The first step is to anticipate what could go wrong and detect it when it does. Common categories of integration errors:

Callout Failures: These include timeouts, DNS failures, refused connections. In Apex, these manifest as exceptions (CalloutException). They can also include HTTP error statuses from the remote service (which are not exceptions, but error codes in the response).

Data Errors: The external system may respond but with an error message (e.g., a SOAP Fault or a JSON error object). Or the data you send might violate external business rules (like trying to add a duplicate record).

Salesforce-side Issues: Running into Salesforce governor limits mid-integration (like too many SOQLs in a batch integration), or hitting unhandled exceptions in triggers that run after data insertion from an external source.

Process Logic Errors: Sequence mistakes (calling step 2 before step 1's result is ready), or missing data causing null pointer exceptions in Apex.

External Limits: Perhaps the external API has its own rate limit (only 100 calls per minute allowed) and your integration exceeds that, getting a 429 Too Many Requests.

To detect these, your code should include checks after each critical operation. For example:

After an HTTP callout, if response.getStatusCode() is not in the expected range, treat it as an error path. Maybe examine response.getBody() for an error message.

Use try-catch around callouts and parsing. Catch specific exceptions if possible. E.g., catch JSONExcepton around JSON.deserialize to catch malformed JSON.

If using SOAP stub classes, check for null responses or catch exceptions from the stub.

Validate responses: if you expected a field to be present in the JSON and it's missing, that's a problem (maybe the API changed). Better to detect and log a clear error than let a null cause a NPE later.

Graceful Degradation: When an integration fails, how can you degrade the functionality gracefully? For synchronous user-triggered integrations (like a button click that calls out to get real-time info), graceful degradation might mean showing a friendly error message: "We're unable to retrieve the latest info right now. Please save your work and try again later." Or perhaps defaulting data: if a currency conversion rate API fails, maybe fallback to the last known rate stored in Salesforce, so at least the system continues to operate with possibly slightly stale data. In asynchronous processes, graceful degradation often means marking the record or transaction with a status like "Integration Pending" or "Integration Failed – Will Retry" rather than just leaving it in limbo. For example, if an account sync to an external system fails, you might set a checkbox "Sync_Failed__c = true" on the account, which triggers an alert to an admin or queues the account for retry, but lets the user continue using Salesforce.

One approach is to design a **retry queue** (like a custom object Integration_Job__c or similar) where you log attempts. If something fails, insert a record describing the failure. A scheduled job or a manual process can later pick those up to retry or investigate. This decouples error handling from the immediate flow, which is good especially for asynchronous flows.

Retry Logic: Many integration errors are transient (especially callouts). Implementing a retry mechanism can dramatically improve reliability:

Same transaction retry: Not often possible because if a callout failed in one execution, doing another immediately might not help (and you risk hitting limits or duplicating actions).

Queued retry: More common: catch the exception and schedule a new attempt shortly after. For example, in a Queueable, catch CalloutException and if the retryCount < 3, do System.enqueueJob(new YourQueueable(params, retryCount+1)); possibly after a small delay. You can't delay in code directly, but you could insert a Platform Event with a small delay if needed (Platform Events don't exactly have a delay, but a scheduled job could read a record and trigger after some time).

Exponential backoff: If an external service is rate limiting you, hitting it repeatedly will make things worse. It can help to wait progressively longer: e.g., 1 minute, then 5 minutes, then 30 minutes. In Salesforce, you might achieve this by scheduling Apex in the future (like use System.schedule to schedule a retry job X minutes later). For instance, if a batch fails to callout to Workday due to a maintenance window, schedule a retry for an hour later via System.scheduleBatch in the finish() method with a cron an hour ahead.

Maximum retries and fallback: Decide how many times to retry before giving up. For example, try 3 times then mark permanently failed and notify. Ensure that each retry is logged to avoid infinite loops. Possibly increment a field "RetryCount__c" on a record or pass a state in a job. If after max retries it's still failing, escalate: send an email to admins, create a Task for someone to manually intervene, etc.

Salesforce does not automatically retry failed callouts – it's on us to implement. However, some managed integrations or External Services might have built-in retry, but assume we are custom building here.

Handling Common Integration Errors

Authentication failures (e.g., 401 Unauthorized): If using Named Credentials with OAuth, this shouldn't happen often (SF will refresh token). If it does or if using custom auth, catch it. Often solution is to refresh token (maybe call a refresh token flow) or alert that credentials need update. Could implement logic to attempt one refresh (if refresh token stored) and then retry original call.

404 Not Found: Possibly a reference to a missing resource (e.g., trying to update a record in external that's deleted). Decide if you treat that as a non-critical error (log and move on) or user needs to know. Perhaps create a task for an admin to reconcile the data mismatch.

500 Internal Server Error from external: Usually just log and retry later. It's on the external side, nothing you can fix in Salesforce.

Timeout (likely CalloutException): Increase timeout if it's a frequent issue and if longer waits are acceptable (set req.setTimeout() up to 120k ms). Also implement retry, maybe the external just took too long. Could also break the call into smaller chunks if possible (maybe the query was too big).

Governor Limit exceptions on Salesforce side (like Too many SOQL): This is a bug in our integration code. Handle by reviewing code and optimizing or splitting the job. Try-catch won't catch governor exceptions (they kill the transaction). So best is to avoid them; this is where testing and design come in, not so much runtime handling except preventing them.

Communicating Issues

Logging: At minimum, use System.debug for error details in dev/testing, but in production consider a more persistent logging. You can create a custom object like Integration_Log__c to record errors (and successes if needed). This object could capture fields like Error Type, Message, StackTrace, Related Record Id, Timestamp, etc. Since writing to a custom object is a DML, don't do it excessively in a single transaction (and avoid in bulk triggers unless needed, or use future to log to reduce impact). But having a record of errors is invaluable for support. Alternatively, publish a Platform Event for errors as mentioned. That can trigger a notification mechanism or be caught by an external monitoring tool.

User Notifications: In some cases, you may want to notify end users of issues. For example, if a user clicks "Sync to ERP" and it fails, you should show a Visualforce/Lightning error message. If an async process fails to sync an order, you might post a Chatter @mention to the owner: "Order 123 failed to sync with ERP. Support has been notified." This might be overkill in many cases, so choose user notifications sparingly, for errors that block user tasks.

Admin/Support Notifications: More commonly, notify system admin or integration owner. This could be an automated email from Apex (EmailException can be used to send an email to a configured address, or better, use a platform event and have an external system send email to avoid hardcoding emails in Apex), or a dashboard where an admin can see "Integration Errors in last 24h". Salesforce Shield Event Monitoring (if available) could track API calls and failures too, but a custom approach is usually implemented for app-specific integration issues.

Do not expose sensitive data: When logging or messaging, ensure you're not including things like passwords or personally identifiable info unnecessarily. For instance, if an external system returns an error that includes its SQL query or stack trace, you might want to strip that out of any user-facing message, perhaps only keep it in a secure log.

Developer Tip: Centralize error handling logic. If you have multiple integrations, consider a framework or utility. For example, a method IntegrationUtils.handleCalloutException(Exception e, String context) that logs appropriately and maybe returns a standardized response or throws a custom exception. Use custom exceptions in Apex to differentiate integration errors – e.g., define ExternalSystemException extends Exception and throw that when something specific occurs. It allows catch blocks up the chain to specifically catch integration issues separate from other logic.

You can also use a state object to accumulate errors in a batch process. For example, in Batch Apex, have a member like List<Id> failedRecordIds or List<String> errorMessages that you populate during execute for any failed record, then in finish() you do something with that (like send an email summary: "5 records failed to sync: [list ids]").

ChatGPT Prompt Examples – Improving Error Handling

User Prompt: "My Apex integration code just throws exceptions when callouts fail. How can I improve its error handling? What should I do when an external API returns an error?"

ChatGPT: "You should catch exceptions and handle error HTTP statuses instead of letting them bubble up unhandled. For callouts:

Use try-catch around the Http send. In the catch, log the error (maybe to a custom object or at least debug log) with context like which record or operation failed. You can also implement a retry from catch for certain exceptions (e.g., timeout).

After receiving a response, if res.getStatusCode() indicates error (e.g., 400 or 500 range), handle it. For 400 (client errors), likely there's something to fix in the request or data – log the response body which might contain an error detail. For 500 (server error), it's usually transient – consider a retry later.

Don't expose raw error to end user. Instead, surface a friendly message or mark a status. For example, if this callout was triggered by a user action in UI, use ApexPages.addMessage(new ApexPages.Message(...)) in a Visualforce context or throw a AuraHandledException in Lightning with a user-friendly message.

Implement a fallback. Suppose the external system is down – maybe your code can catch that and instead queue the request (store it for later). Or if read operation fails, maybe use cached data if available.

Also ensure that any database changes in Salesforce that were supposed to happen after a successful integration do not happen on failure. Use a boolean flag or check in your code: only commit changes if external call succeeded. If using a trigger that both updates Salesforce and calls out, consider splitting those – because if external fails, you might need to roll back the SF change (which is automatic if in same transaction and exception uncaught, but if you caught it, you might inadvertently keep partial changes). One approach is to do external call first (if logic allows) then do DML, so if call fails, you simply don't do the DML.

In summary: catch, log, retry appropriately, and keep user and admin informed in a controlled way. You might create a small error handling framework. I could help outline one if needed, for example a custom exception class to wrap external errors with codes, and a helper that takes that and maybe posts a Platform Event for integration errors that an admin dashboard subscribes to."

The assistant gives structured guidance as above, possibly offering to dive deeper into any point.

Another prompt

User Prompt: "I have a batch job that syncs contacts to MailChimp via API. If MailChimp is down, my batch fails and stops. How can I make it so that it continues for other contacts and retries later for the failed ones?"

ChatGPT: "Batch Apex processes records in chunks, and if an unhandled exception occurs in one execute, that batch chunk is rolled back but others continue. To handle MailChimp downtime:

Wrap the callout in try-catch in the execute. If MailChimp's API is down (e.g., you catch a CalloutException or get a 503 status), do not throw after catching; instead handle within. For instance, mark those Contact records in Salesforce with a field "Needs_Resync__c = true" or add their IDs to a static list of failedIds in the batch state.

Allow the batch to continue processing other records. Those that fail are just skipped (maybe they weren't updated because we caught exception and didn't perform whatever success logic).

In the batch's finish() method, check if failedIds list is not empty. If yes, you could schedule a new batch to retry those contacts in, say, 30 minutes (using System.scheduleBatch). Alternatively, you could create Platform Events for each failed contact, which a separate recovery process listens to and retries.

Also consider partial failures: e.g., MailChimp up but returns error for some contacts (maybe invalid email format). Your code should catch that, maybe remove those contacts from further processing or mark them as failed with reason (e.g., store "Invalid Email" in a custom field or log object). That way those don't get retried infinitely.

By doing this, one batch chunk failing doesn't crash the whole job; you handle it and move on. The key is to ensure the error doesn't propagate as an exception beyond execute.

Also, implement an alert: if the whole batch found MailChimp down (e.g., almost all contacts failed), you might want finish() to send an admin email: "MailChimp sync failed for 90% of contacts – possibly service outage." This proactive notice can prompt someone to check MailChimp status.

This approach ensures resilience: other contacts still sync, and failed ones are queued for later when hopefully MailChimp is back online."

The assistant lays out a clear retry strategy within a batch context.

Architect Tip: Design an end-to-end error handling process. This could mean integrating Salesforce error events with a broader monitoring solution. For example, if using MuleSoft, consider sending Salesforce integration errors to MuleSoft's monitoring or Splunk. Ensure that every error has an "owner" – someone or some process that will address it. Unattended error logs help no one. Many enterprises set up daily or weekly error reports. As an architect, you could implement a nightly scheduled job that scans the Integration_Log__c for any new errors and emails a summary to the integration team, ensuring visibility.

Additionally, consider transactional integrity. Salesforce and external systems are usually not in a distributed transaction (no two-phase commit). That means partial successes can happen: Salesforce might commit a change thinking external will, but external fails. One strategy is to design with eventual consistency in mind. For example, if an Opportunity should create an Order in SAP, but SAP is down, the Opportunity is still closed won in Salesforce – you must have a plan to create that Order later (and perhaps mark in Salesforce that order creation is pending). The system should tolerate that state. In some cases, you might even roll back Salesforce transaction on external failure (if the requirement is external and internal must be in sync or not at all). You can achieve that by throwing an exception in a trigger after catching an external error – which aborts the whole transaction, meaning the user's action fails. That's "fail fast" approach. But often user would prefer their action succeed and the integration happens later. Deciding where to apply which approach is an architectural decision based on business impact.

By implementing thoughtful error handling and recovery, your integrations become resilient – they can survive outages and hiccups with minimal manual intervention, and they inform the right people when attention is needed. Once errors are handled, it's equally important to have ongoing **monitoring and logging** in place to observe integration health and performance over time. In the final section, we'll discuss how to monitor and log integration activities effectively, ensuring you can identify and address issues proactively and maintain transparency of data flow across systems.

MONITORING AND LOGGING INTEGRATION ACTIVITIES

After building integrations with robust error handling, the next step is to set up monitoring and logging so you have visibility into those integration processes. Monitoring allows you to detect issues early (perhaps even before end-users notice) and to analyze trends like how often an integration is used or failing. Logging provides an audit trail of integration events and can greatly aid debugging when something goes wrong. In this section, we'll explain the importance of thorough monitoring and logging for Salesforce Apex integrations and provide best practices on how to implement it. We will cover the tools Salesforce offers (such as debug logs, Event Monitoring, and custom logging mechanisms) as well as how you can integrate Salesforce logs with external monitoring systems. We'll also outline strategies like structured logging (logging in a consistent format), setting up alerts for certain conditions, and tracking key performance indicators (KPIs) for integrations (e.g., success rate, average response time). ChatGPT prompt examples will show how to evaluate current logging strategies or generate ideas for improving monitoring. Finally, we'll give role-specific recommendations so product owners, architects, developers, and testers know how to use logged information and monitoring insights to keep integrations healthy.

Why Monitor? Integrations often involve multiple moving parts. Without monitoring, you might only find out about an issue when data is already inconsistent or when a user complains that something is not updated. Monitoring helps answer questions like: Is the integration running as expected? How long do calls take on average? How many records were synced today? Did any calls fail? It also helps in capacity planning – if you see call volume doubling each month, you can proactively adjust for limits or performance.

Salesforce Debug Logs and Monitoring Tools: In Salesforce, every transaction can produce a debug log if a user (or an automated process user) is under debug logging. However, leaving debug logs on for integrations continuously can be impractical (they might generate too much data and you have limited log storage). Instead, targeted logging is better:

Use System.debug for important info, but consider controlling when it's active (perhaps via a Custom Setting "Enable Integration Debug" that you check in code, so you can turn on verbose logging temporarily without a code deploy).

Salesforce's Event Monitoring (part of Shield, or also some events available via API) can track certain events like API calls. For example, you can get an "API Event" log that shows every callout made or every API call to Salesforce. This might help identify integration usage. However, Event Monitoring mostly covers things like logins, API usage by user, etc., not the internal details of your Apex logic.

Custom Object for Logs: As mentioned earlier, creating a custom object like Integration_Log__c is a common approach. You can log entries for important events: e.g., "Sent 50 accounts to SAP", "Received response for

Order 123", or "Error syncing Case 456: [error message]". Decide what level of detail to log – too little and it's not useful, too much and you overwhelm storage. Often, logging only errors and key milestones is sufficient, not every single successful call. Or you might summarize: e.g., "Batch job completed, 245 records synced, 5 errors (see error log entries)". Each log record can have fields for type (info/error), timestamp, maybe external system involved (a picklist: SAP, AWS, etc.), maybe a correlation ID or request ID if you have one (some advanced integrations tag each request/response with an ID for tracing).

Field History and Audit fields: If you integrate changes, enabling Field History Tracking on critical fields can indirectly serve as a log of changes coming from integrations. For example, if an external price updates a Product's Price field in Salesforce daily, field history will show when it changed and from what to what. It won't tell you why (no error info), but it's useful for audit.

Platform Events for Logging: You could publish a custom Platform Event like Integration_Transaction__e on each significant integration action. Those events can then be consumed by an external system (like a Node.js app or Splunk forwarder) to store in a central log system. The advantage is non-intrusive collection: you don't store logs in Salesforce, just stream them out. But an external listener needs to be always on to catch them (or replay within 24h). If you already have a company-wide logging infrastructure, this approach can funnel Salesforce integration logs into it in real-time.

Structured Logging and Correlation IDs: When logging, consistency helps. For instance, always include certain pieces of information such as:

Timestamp (Salesforce will have CreatedDate if using an object, or you include it in log lines).

The integration context (which external system or API endpoint).

Record identifiers (if the log relates to specific records, include their Id or external id).

If you have a flow that spans multiple transactions (like a callout and then a callback later), include an ID that ties them. Some systems use a correlation ID header. You can generate a UUID in Apex (not directly, but you could use Crypto.getRandomUUID() if available, or generate one via Crypto random bytes) and attach it to requests (maybe as a header or part of the payload) and log it. External system should include it in the response or callback, so you can match the request log and response log. This is more advanced but helpful for tracing end-to-end in complex multi-step integrations.

Log levels: not necessary in code, but conceptually mark logs as INFO, WARN, ERROR, etc. You might choose to only persist ERRORs to a custom object and not INFO to reduce clutter, for instance.

Using External Monitoring Services: Many organizations aggregate logs in systems like Splunk, Datadog, or CloudWatch (if using AWS). You can integrate Salesforce logs into those:

Splunk has an app for Salesforce that can ingest Event Log Files (which include things like API calls logs). It might not catch custom debug logs easily, though.

One low-tech method: have Apex send logs via HTTP to a logging service. For example, make a callout to a webhook with the log data. This must be non-critical (you wouldn't want logging itself to fail and break your integration). Perhaps do it asynchronously or batch logs together to send occasionally.

Datadog and others also might connect via the API to pull out Custom Object log entries on a schedule.

If using MuleSoft as middleware, MuleSoft itself can track all messages that go through it, which might suffice for parts of integration that go through MuleSoft. It might not see things wholly within Salesforce (like a trigger that did something).

Performance Monitoring: Logging not only errors but performance metrics can be useful. For instance, log how long an external call took. You can capture time before and after callout using Datetime.now() differences or Limits.getCpuTime() etc. If you notice calls usually take 2 seconds but suddenly now take 10 seconds, it could indicate an issue on external side. Tracking average processing time per batch, or the size of data transferred (maybe response payload sizes) can help tune integration. In absence of a sophisticated APM (Application Performance

Management) tool, even simple debug logs can reveal patterns when analyzed.

Interpretation and Proactive Use of Logs

Dashboards for Integration Metrics: You could build a Salesforce dashboard on top of Integration_Log__c object, showing number of successful syncs vs failures by day, or average time per integration. Product owners and managers like to see such metrics to understand usage and reliability. For example, a dashboard component: "Workday to Salesforce user sync: 1000 records/day, 99% success" gives confidence.

Alerts on thresholds: If certain errors appear in logs, set up alerts. Without external tools, you could schedule an Apex job to run daily and if error count > X or same error repeated Y times, send an email notification. Or use an external tool to trigger alerts (e.g., Splunk queries or Datadog monitors on log data).

Regular review: Testers or developers should regularly review logs, especially after deploying a new integration or change. In the first week after deployment, keep an eye on logs for any pattern of unexpected errors.

Retention: Decide how long to keep logs. Custom object logs will consume storage. Perhaps you archive or purge after 30 days or 90 days, keeping only recent ones available (unless needed for audit compliance). You can automate purging with a scheduled job.

ChatGPT Prompt Examples – Evaluating Logging Strategy

User Prompt: "I have no centralized logging for my integrations. What's a simple way to start logging integration events in Salesforce so I can troubleshoot issues better?"

ChatGPT: "A straightforward approach is to create a Custom Object, say Integration_Log__c, with fields like Name (maybe use an auto-number or a descriptive name), Timestamp, Integration_Type (text or picklist to indicate which integration), Status (Success, Error, Info), Message (long text for details), Related_Record_Id (lookup or text), maybe an Error_Details (long text). Then, in your Apex integration code, wherever you have a catch or even after a successful important step, insert a new Integration_Log__c record.

For example:

```
try {
// callout
Integration_Log__c log = new Integration_Log__c();
log.Integration_Type__c = 'SAP Order Sync';
log.Status__c = 'Info';
log.Message__c = 'Order ' + orderId + ' sent to SAP successfully.';
log.Related_Record_Id__c = orderId;
insert log;
} catch(Exception e) {
Integration_Log__c log = new Integration_Log__c();
log.Integration_Type__c = 'SAP Order Sync';
log.Status__c = 'Error';
log.Message__c = 'Failed to send Order ' + orderId;
log.Error_Details__c = e.getMessage().substring(0, 255);
log.Related_Record_Id__c = orderId;
insert log;
// maybe additional error handling
}
```

This will record an entry for each attempt. You might not want to insert inside a loop of many records (could hit DML limits), but in such cases, you can bulk accumulate and insert in batches or use a future method to insert logs outside of a tight loop.

Once logs are in Salesforce, you can create list views or reports on Integration_Log__c to see recent errors or volumes. Make sure to filter out or limit debug logs in production if performance is a concern; you might make logging optional via a setting as mentioned.

This simple logging is a good start. Over time, you might refine it by adding more structured info or integrating with external logging systems. But having these records will greatly aid troubleshooting because you have persistent evidence of what happened, even after the fact."

Another example:

User Prompt: "How can I monitor if my integration to AWS is running slower than usual? It sometimes times out but I want to catch performance issues early."

ChatGPT: "You can instrument your integration code to record the duration of calls. For instance, record a timestamp before http.send() and after, then log the difference in milliseconds:

Long startTime = Limits.getTimestamp();

HttpResponse res = http.send(req);

Long endTime = Limits.getTimestamp();

Long duration = endTime - startTime; // milliseconds

System.debug('AWS integration call took ' + duration + ' ms');

// Or better, log it to Integration_Log__c as a field e.g., Duration__c

Now, to monitor, you could:

Set up a report on Integration_Log__c to show average Duration per day. If you see it rising over time from, say, 500ms to 5000ms, that's a trend to investigate.

Alternatively, if not using a log object, you could use Salesforce's **Event Log** for API calls. If the AWS call is a callout, Salesforce's event logs might not detail the time the external service took, so better to measure in Apex as above.

You could even create a custom Lightning page or dashboard showing the last X call durations or a gauge of today's average vs yesterday.

For real-time alert, you could add logic: if duration > threshold, maybe publish an Integration_Alert__e event or send an email. But be careful not to spam alerts for one-off slow calls. Perhaps alert if, say, more than 5 calls in an hour exceed 5 seconds.

Since AWS is external, also coordinate with AWS CloudWatch if possible – CloudWatch could have metrics on the API side. Combining both perspectives (Salesforce timing and AWS's own logs) gives full picture.

By actively measuring and logging performance, you turn what is normally invisible (time spent in external call) into observable data. This helps you proactively inform the external team if their service is degrading, or to decide if you need to raise Salesforce timeout limits for that callout."

Tester/Developer Tip: Use your logs during testing! When running integration tests in a sandbox, check that the logs capture what you expect. Simulate errors to see that they log properly. Perhaps create a unit test that forces an error (maybe by using a mock that returns 500) and then query the Integration_Log__c to ensure an error entry was written. This both tests your logging and ensures the integration gracefully handled it (since the transaction didn't crash if log was written).

Product Owner Tip: Encourage the team to surface integration health metrics in a way you can understand. For example, a weekly summary email: "1000 records synced to ERP, 5 failures auto-retried, 0 require manual intervention." This gives confidence. If you suddenly see "100 failures require manual intervention," you know there's a serious issue to prioritize. Having these metrics also helps in planning improvements or communicating system reliability to stakeholders.

Architect Tip: Logging and monitoring should be part of the design, not an afterthought. Define SLAs (Service Level Agreements) or SLOs (Service Level Objectives) for your integrations (e.g., 99% of transactions < 2 sec, or <1% error rate). Then monitor against those. Use tools like Salesforce's built-in **Flow Orchestration logs** if using integrations via Flow, or external APM tools that can instrument at the platform level. Ensure compliance requirements are met too – for instance, if logging personal data, ensure its handled per policy (maybe mask certain fields in logs).

In conclusion, a well-monitored integration is easier to maintain and instills trust across the organization. By implementing structured logging, leveraging Salesforce and external monitoring tools, and proactively analyzing the integration logs, you can quickly pinpoint issues and continuously improve your integration performance and reliability.

SUMMARY

In this chapter, we covered the spectrum of Apex integration patterns: from REST and SOAP callouts to asynchronous jobs and platform events, through to data mapping, error handling, and monitoring. A key takeaway is that robust integrations are not just about writing code to connect systems – they involve thoughtful design to handle data differences, anticipate failures, and provide visibility and control over the data exchange. By applying these best practices and using tools like ChatGPT for guidance and automation, Salesforce architects and teams can ensure their integrations significantly enhance the functionality and user experience of Salesforce-based applications. In the next chapter, we will build on these integration concepts and **Asynchronous Apex Best Practices**, examining in greater detail how to manage complex, large-scale operations in Salesforce without impacting performance – a natural extension of the topics we've explored, focusing further on Apex's asynchronous processing capabilities and patterns for scalability.

9

ASYNCHRONOUS APEX BEST PRACTICES

As Salesforce applications grow in complexity and data volume, **asynchronous Apex** becomes an essential tool in the architect's toolkit. Asynchronous Apex refers to Apex code that runs in the background, on a separate thread or schedule, rather than in the user's immediate transaction. This approach allows Salesforce applications to scale and perform complex tasks efficiently. By offloading long-running or resource-intensive processes from the immediate user interaction, asynchronous Apex helps maintain a responsive user experience. In this chapter, we will explore how running processes "behind the scenes" can significantly improve both system performance and user satisfaction.

Apex offers several mechanisms for asynchronous processing, each suited to different scenarios. Common use cases include processing large data sets (such as updating millions of records or recalculating data in bulk), making external callouts to web services (for example, sending data to an ERP system after a record is saved), and scheduling recurring tasks (like nightly data synchronization or weekly report generation). In each of these situations, asynchronous Apex ensures that users don't have to wait for the operation to complete. A record insert that triggers complex logic can return immediately, while the heavy lifting continues in the background. This design is crucial for enterprise-scale Salesforce apps, as it prevents timeouts and governor limit errors in interactive processes and enables the handling of **large volumes** of work that would be impossible in a single, synchronous transaction.

This chapter guides readers through mastering the four main types of asynchronous Apex: **Batch Apex, future methods, Queueable Apex, and Scheduled Apex.** We will review proper use cases and implementation patterns for each. By the end of the chapter, you will understand how to choose the right asynchronous approach for a given job, how to implement it following best practices, and how to ensure these background processes run reliably and efficiently. In addition to these core techniques, we'll address crucial related topics: managing Salesforce's governor limits within asynchronous jobs, implementing robust error handling and recovery strategies for background processes, and monitoring/auditing asynchronous jobs to maintain visibility into what's happening behind the scenes.

Mastering asynchronous Apex is beneficial to every role involved in a Salesforce project. **Architects** can design more scalable and resilient systems by distributing workload properly and avoiding bottlenecks in user-facing operations. **Developers** gain the ability to meet complex requirements (like processing thousands of records or integrating with external systems) without running into limits or degrading performance, all while keeping code maintainable. **Testers** will find that understanding async processes helps in creating effective test cases and ensuring that background jobs do what they're supposed to do (and handle errors gracefully). **Product owners** also benefit: when asynchronous processes are used appropriately, end-users get faster, smoother interactions and the business can tackle large-scale tasks (such as data cleanup or bulk updates) reliably in the background. In short, asynchronous Apex, used wisely, leads to reliable, scalable solutions in enterprise Salesforce implementations.

Throughout the chapter, we will also highlight how **ChatGPT** can support you in applying these best practices. ChatGPT can serve as a virtual pair-programmer or architectural assistant to refine your asynchronous implementations. For example, if you're unsure about how to structure a Batch Apex job, you can describe your scenario to ChatGPT (perhaps even using voice input for a quick brainstorming session) and get suggestions on the optimal batch design. When you encounter a tricky governor limit error in a Queueable job, you might upload the error log and ask ChatGPT to troubleshoot the issue (ensuring your file's privacy settings are set so that only you and ChatGPT can view it). By leveraging ChatGPT's knowledge, you can quickly iterate on solutions to common pitfalls like optimizing batch sizes or handling exceptions in background jobs. ChatGPT can also help generate code snippets, pseudo-code, or even simple diagrams on request, which can clarify how different asynchronous components should work together.

What to expect in this chapter: We will begin with the fundamentals of **Batch Apex**, exploring patterns and practices for processing large volumes of data efficiently. Next, we'll discuss **future methods**, a straightforward way to execute Apex asynchronously, including when to use them and important considerations. We will then move on to **Queueable Apex**, a more advanced asynchronous technique that offers greater flexibility and control than future methods. After that, the chapter covers **Scheduled Apex** for automating recurring processes on a schedule. With the main techniques covered, we will address how to **manage governor limits** in asynchronous contexts to keep your jobs within Salesforce's execution limits. We will also cover **error handling and recovery strategies** specific to asynchronous execution, ensuring your background jobs are resilient and fail gracefully. Finally, we will look at **monitoring and auditing asynchronous jobs**, so you can keep track of what your background processes are doing and quickly detect any issues. With these topics mastered, you'll be equipped to design and implement asynchronous Apex solutions that are robust, maintainable, and performant, setting the stage for the next chapter's deep dive into advanced Apex performance optimization techniques.

(Now that we've set the stage for asynchronous processing, let's dive into the first key topic: how to leverage Batch Apex patterns and practices for large-scale data operations.)

BATCH APEX PATTERNS AND PRACTICES

Batch Apex is the workhorse of asynchronous processing in Salesforce when dealing with **large datasets or intensive operations**. A Batch

Apex class in Salesforce lets you process potentially millions of records by breaking the job into manageable chunks that execute separately. This pattern is crucial when an operation would exceed normal limits if done in a single transaction – for example, updating every contact's data in an org with millions of contacts, or performing complex calculations on a large set of records. By processing in batches, your Apex code gets a fresh set of governor limits for each chunk of records, allowing far greater total throughput. In this section, we explain the role and advantages of Batch Apex, and then detail best practices for structuring batch jobs, including choosing batch sizes, chaining batches, managing state between batch executions, and handling limits within batches.

At its core, a Batch Apex class implements the Database.Batchable interface, which requires three methods: start, execute, and finish. The **start** method is where you define the scope of data to be processed – typically by returning a QueryLocator or an Iterable that represents all records to work on. Salesforce then divides this record set into chunks (called "batches") and invokes the **execute** method for each batch, passing a subset of records (up to a specified batch size) to be processed in that transaction. Finally, after all batches have run, the **finish** method is called once, which is useful for any post-processing or cleanup (such as sending a summary email or chaining another job). This design means a single batch job actually runs as multiple transactions, each handling a portion of the data. For example, if you have 10,000 records to process and use the default batch size of 200, Salesforce will execute the execute method 50 times (each time with 200 records) to complete the job, followed by one call to finish. From a governor limits perspective, each of those 50 executions is like a fresh context: you get a new set of limits (SOQL queries, DML operations, CPU time, etc.) for each chunk. Batch Apex thus enables processing of volumes and complexities that would be impossible in a single transaction.

Best practices for structuring Batch Apex:

Choosing an optimal batch size: The default batch size is 200 records, but you can specify a different size (up to 2000). Generally, 200 is a good balance for most cases. However, if the processing per record is very lightweight, a larger batch size (say 500 or 1000) might improve throughput by reducing overhead. Conversely, if each record requires heavy processing or callouts to external services, a smaller batch size (like 50 or even 10) can help avoid hitting limits. For instance, if each record processing includes an HTTP callout, you must stay under the callout limit (100 callouts per transaction). In that case, limiting the batch to 50 records ensures at most 50 callouts in one execute, staying well within limits. You might use ChatGPT to help determine a safe batch size: *ChatGPT prompt example:* "I have a Batch Apex job that makes an HTTP callout for each record. Is a batch size of 200 too high, and what size would you recommend?" By asking such a question (via text or even by speaking to ChatGPT using voice input), you can get guidance on adjusting batch size to balance efficiency and governor limits.

Bulkified, efficient code: Within the execute method, treat the batch of records in a bulk manner just as you would in a trigger context. Even though batch processes records in chunks, you still want to perform operations in bulk (e.g. one SOQL query to fetch related data for all records in the batch, not one per record). Write your code to handle a list of records, not one record at a time. This ensures you're using the resources of that transaction efficiently. If you find yourself writing loops inside execute that call queries or DML for each record, consider refactoring – and you can ask ChatGPT to review your batch code for any signs of inefficient queries or DML inside loops. *For example, you might upload your Batch Apex class file to ChatGPT and prompt:* "Please review this batch class for any SOQL-in-loop or other governor limit issues." *(Always remember to keep your code file private when uploading for analysis.)* ChatGPT can highlight portions of your code that could be bulkified or optimized, helping you improve performance and stay within limits.

Managing state between batch executions: By default, each execution of a batch is stateless – static and instance variables are not preserved between one batch execute call and the next. If you need to retain information across batches (for example, aggregating a total count of processed records or collecting errors), implement the Database.Stateful interface in your batch class. This will allow instance variables to persist values throughout the batch job's lifecycle. Use this feature judiciously: storing too much state can increase memory usage (heap size) across executions. For example, an architect tip: *Only use Database.Stateful to collect necessary summary information (like aggregated results or error lists) rather than storing all processed records or large objects in memory.* A good pattern is to accumulate small pieces of info (IDs of failed records, count of successes, etc.) in instance variables and then, in the finish method, handle that information (perhaps by logging it or updating a summary record). For instance, an ErrorCollector list could gather IDs of records that failed during execute calls. In finish, you might insert these into a custom "Batch_Error_Log__c" object for administrators to review. Developer Tip: If you implement Database.Stateful, be mindful of the Apex heap size governor limit (which is 12 MB in asynchronous transactions). Don't let your accumulated state grow too large; if you need to track an extensive amount of data, consider logging to a persistent store within each execute chunk instead of keeping everything in memory.

Chaining batch jobs: Sometimes one batch job isn't enough to accomplish a complex task. You might need to run a series of batch processes in sequence (for example, first batch job recalculates some scores on accounts, and when it's done, a second batch job updates related contacts). Salesforce allows batch chaining – you can call Database.executeBatch from the finish method of a batch to start another batch. This is a powerful pattern for multi-step processes. However, plan batch chains carefully. It's best not to create an endless chain; ensure there is a clear end. Also, remember that each batch job (including those started by another) counts against the concurrent batch job limits. At any given time, Salesforce allows only a certain number of batch jobs to be queued or processing (commonly 5 active batch jobs at once, with up to 100 in a holding queue). Architect Tip: When designing a system that might enqueue multiple batch jobs (chained or in parallel), consider the overall load on the org. Stagger batch jobs if possible (maybe using Scheduled Apex, which we'll cover later) to avoid hitting the concurrency limits. For instance, instead of chaining five heavy batches all to run back-to-back during business hours, an architect might schedule some to run overnight or in sequence with delays, ensuring the org always remains within its processing capacity. If you're unsure about how to schedule or chain without conflict, you could ask ChatGPT: *"What's the best way to chain batch jobs or schedule them to avoid hitting the max concurrent limit?"* for tailored advice.

Error handling in batch jobs: Batch Apex provides some built-in resilience. If one record fails to process (e.g., a DML operation on one record causes an error), it won't stop the entire batch job if you handle it properly. By using partial DML operations (e.g., Database.insert(records, false) which allows partial success) within execute, you can catch Database.SaveResult for each failed record and handle it (log it, or perhaps add to that error list in stateful context). If an unhandled exception is thrown in the execute method, that particular batch execution will be aborted and Salesforce will mark the whole batch job as failed, possibly preventing remaining batches from running. To prevent a minor issue from aborting the entire job, wrap your execute logic in try-catch blocks and handle exceptions gracefully. For example, if processing one chunk of 200 records throws an unexpected null-pointer exception, catch it, log enough details about that chunk for debugging, and decide whether to continue with the next chunks or abort. Often, it's wise to allow the job to continue after logging the error, so other unrelated records still get processed in later chunks. We will dive deeper into error handling strategies later in this book, but keep in mind that for batch jobs, robust error handling is key to achieving "partial success". A well-written batch class might process 99% of records successfully and isolate the 1% that failed for later review, rather than failing everything.

Governor limit management in batch: Even though batches give you higher overall limits by splitting work, each execute still has to obey per-transaction limits (like 100 SOQL queries, 150 DML statements, 50,000 retrieved rows, etc.). Therefore, design each batch execution to work within those limits. If you find that a single batch of 200 records might exceed the limits (perhaps because each record needs multiple queries), you have a few options: refactor the logic to be more efficient (e.g., combine queries, use maps to avoid repeated work), or reduce the batch size as mentioned earlier, so each execution does less work. Developer Tip: Use the Limits class inside your execute method if needed to monitor consumption – for example, Limits.getQueries() can tell you how many queries have been used so far in that execution. This can be useful for debugging or even conditional logic (though typically, hitting limits indicates a need for code changes). If you approach a limit frequently, consider redesigning the job. You can also leverage ChatGPT to scan your batch code for potential governor limit issues:

"Here is my batch execute logic [paste or upload]. Do you see any risk of hitting governor limits, and how can I mitigate it?" ChatGPT might suggest using aggregation techniques or moving certain operations out of the loop.

Planning batch operations strategically: An often overlooked aspect of batch jobs is *when* to run them. Running multiple heavy batches during peak user hours can contend for resources and potentially slow other operations (Salesforce has a "fair usage" algorithm to give each org a slice of processing time, but heavy async loads could still introduce some contention). If a batch job isn't time-sensitive, consider scheduling it during off-peak hours (early morning or late night). For example, a daily data warehouse sync batch might be scheduled at 2 AM when few users are active. It's worth noting here as part of batch best practices: Architects should collaborate with Product Owners to determine appropriate timing for large batch jobs, aligning technical processing with business needs (e.g., have data ready by start of business day, but process it overnight).

Now let's consider how ChatGPT can assist with Batch Apex beyond code review. You can use ChatGPT to **brainstorm patterns**. Suppose you're not sure whether to use one batch job or multiple jobs in sequence. You could ask: *"I need to transform and then load 100,000 records. Should I use a single batch or multiple batches, and how to design them?"* ChatGPT can outline possible approaches (maybe one batch to transform data and a second batch to load it, or even using the finish method to start the second phase). Additionally, you might find it helpful to have ChatGPT generate **documentation** or a **diagram** of your batch process. For instance, you could say: *"ChatGPT, create a simple flowchart of a Batch Apex job that includes start, multiple execute chunks, error logging, and a finish step emailing a summary."* While ChatGPT's textual description can be quite vivid, you might even use an image generation tool connected to ChatGPT to visualize the flow (some advanced ChatGPT integrations can generate diagrams from descriptions). Generating such an image can be a novel way to communicate the batch process to stakeholders who prefer visual aids. If using that approach, ensure you describe the flow clearly and verify the output for accuracy.

Example ChatGPT prompts for Batch Apex

"Explain how to decide the batch size for a Batch Apex job that updates opportunities and sends an email for each – what factors should I consider?"

"Review my Batch Apex code (attached as a file) and suggest any improvements for efficiency or error handling."

"How can I chain Batch Apex jobs? Give me a sample scenario and code snippet using finish method to chain another batch."

"What is the best way to handle a batch job that needs to skip processing records that meet certain criteria? (I only want to process records that haven't been updated in 30 days.)"

By asking these questions, whether by typing or using voice input to quickly query while you're reviewing code, you can leverage ChatGPT to deepen your understanding and refine your batch implementations.

Batch Apex, when used with these patterns and practices, can dramatically extend the capabilities of your Salesforce org. Properly implemented, batch jobs enable the processing of massive data sets and complex logic without running into limits or impacting user experience. They are a key component of building scalable, enterprise-ready solutions. With a solid grasp of Batch Apex under our belt, we can now move to another asynchronous tool: future methods. In the next section, we'll discuss when to use @future methods and how to apply them effectively.

FUTURE METHODS – USAGE AND CONSIDERATIONS

Future methods in Apex provide a simple way to run code asynchronously. By annotating a static method with @future, you instruct Salesforce to execute that method at a later time, independently of the current transaction. The primary use cases for future methods are scenarios where you have a relatively **lightweight task or an external callout** that should not delay the user's action. For example, imagine a trigger that fires when an Account is updated and needs to make a callout to an external billing system. A trigger cannot complete an HTTP callout if the callout is not completed before the transaction ends, and doing a callout within a trigger might hit timing issues or prolong the save operation for the user. By moving the callout to an @future method, the trigger can enqueue the callout to happen asynchronously and then quickly finish the user's transaction. The user sees their Account save complete without waiting for the external system response. Similarly, future methods are handy for any **post-transaction processing** that isn't urgent. This could include tasks like updating related records in a non-urgent way, sending notifications, or performing additional calculations after a record insert/update.

Common scenarios for future methods:

Asynchronous Callouts: As mentioned, one of the most common uses of future methods is to handle callouts to external web services. You mark the method with @future(callout=true) to indicate it will perform a callout, allowing Salesforce to execute it in a context that permits callouts. This is especially useful in trigger contexts, where performing a callout directly could be problematic. The future method can safely call an external API after the database transaction commits.

Resource-Intensive Operations on Small Scale: If you have a task that is heavy in terms of CPU or queries but doesn't involve a large number of records, a future method can offload it. For instance, maybe when a single record is updated, you need to perform an expensive recalculation affecting many related records or do multiple queries to gather info. Doing that in a future method means the synchronous transaction (e.g., the user updating the record) isn't burdened by those extra queries or processing. Note, however, that if you have a truly large number of records to handle or a complex job, Batch Apex or Queueable Apex (discussed in the next section) might be more appropriate. Future methods are best for relatively self-contained tasks that don't warrant the overhead of a batch job.

Fire-and-forget operations: Sometimes you need to kick off an action that doesn't need to give any immediate feedback. Future methods are by nature "fire-and-forget" – once you call them, you cannot easily track their progress or result (there is no built-in callback or easy monitoring beyond checking the Apex Jobs page). For example, a future method might be used to initiate an operation like logging an event to a custom object after a record is saved, where even if it fails, it doesn't impact the user's transaction. The system will try to run it, and if it fails, it's usually logged as an error in the Apex Jobs, but the user isn't directly affected at that moment.

Given these use cases, let's talk about practical guidelines for implementing future methods:

Method signature constraints: A future method must be a static method, return void (they cannot return values because there's no caller to receive a result in the future), and can only accept certain types of parameters. The allowed parameter types are primitives, collections of primitives, or sObjects (and collections of sObjects). You cannot pass complex types like custom Apex objects or objects that aren't serializable. For example, you can pass a List or List to a future method, but you cannot pass a custom Apex class instance. This is because the parameters need to be serialized and stored by Salesforce to run later. Developer Tip: If you need to provide complex data to a future method, consider passing an Id (or list of Ids) and then having the future method re-query or reconstruct the needed data. Alternatively, use Queueable Apex which can handle complex member variables (we'll discuss that soon). It's important to note that future methods run in their own context, so if you pass in record IDs and then update those records in the current transaction, the future method will see the updated values when it runs (since it executes after the original transaction commits).

Using @future(callout=true): By default, future methods cannot make callouts unless this flag is set. Always remember to include callout=true in the @future annotation if the method will invoke an external web service or HTTP call. A common oversight is forgetting this flag, which leads to runtime exceptions when the callout is attempted. If you're unsure, you could ask ChatGPT something like: "*What does @future(callout=true) do, and when do I need it?*" to reinforce your understanding. The answer would clarify that without callout=true, the future method isn't allowed to perform web service requests.

Avoiding overlap and excessive futures: Salesforce imposes a limit of 50 future calls per Salesforce transaction. This means if you're in, say, an Apex trigger handling a bulk load of 100 records, and your code invokes a future method for each record, you could easily hit that limit (100 attempts where only 50 are allowed). Best practice: bulkify your future method calls. Instead of calling a future 100 times for 100 records, modify your future method to accept a list of record identifiers, and call it once (passing the list of 100 IDs). Inside the future method, you can iterate or bulk-process those IDs. This way you use just 1 of the 50 allowed future calls instead of 100 (which would have thrown an error). If you suspect your code might be doing too many future calls, ChatGPT can help review that pattern. *For example: "I have a trigger that calls a future method inside a loop of records. How can I refactor it to avoid hitting limits?"* ChatGPT would likely suggest collecting the records and making a single future call, which aligns with Salesforce best practices. Additionally, be mindful not to call a future method from another future method or from within a batch or queueable context – chaining async methods in that way is generally not allowed or not advisable. Salesforce will throw an error if you try to start a future from a future (it's not permitted to launch a new future in an asynchronously executing context). If you find yourself needing to do that, consider using Queueable jobs (which can chain, as we'll see in the next section).

Error handling in future methods: Unlike batch or queueable, a future method has no built-in mechanism to catch or report errors back to something like a finish method. If an exception is uncaught in a future method, the asynchronous job will fail and be noted in the Apex Jobs with an error status. It's a good practice to handle exceptions inside your future method. Use try-catch blocks around the critical operations (e.g., the external callout or the database updates it performs). If something fails, you might decide to retry (though automatic retry isn't built-in for future calls) or at least log the error. Since you can't directly notify the user in real-time (they've moved on after their original transaction), consider logging to a custom object or sending an email to an admin in the catch block. For example, if a callout in a future fails due to a timeout, catch that exception and perhaps insert a record into a custom "Integration_Error__c" object with details so it can be reviewed and potentially reprocessed later. Remember that futures are "fire-and-forget" – you need a strategy to not lose errors in the void. Tester Tip: When testing future methods, design tests to verify outcomes rather than direct results. Because you can't get a return value, your test might assert that the side effects occurred. For example, if the future method is supposed to set a field on some records, your test (after running the future via Test.stopTest()) should query those records and assert the field is set as expected, or if the future is supposed to create a log record on error, your test can simulate an error and then check that the log record was created.

Thorough testing of future methods: Salesforce has some special considerations for testing asynchronous code. In a test class, when you call a future method, that method's code does not run immediately at that line. You need to wrap the call between Test.startTest() and Test.stopTest() to ensure the future executes during the test. Here's a common pattern in Apex tests:

// setup test data

Test.startTest();

MyClass.triggeredFutureMethod(recordIds); // call the future method

Test.stopTest();

// after Test.stopTest(), the future method has executed

// now query for any changes the future method should have made and assert

The Test.stopTest() forces all queued future methods (and other async jobs like queueables or batch executes) to run to completion. After that, you can verify the results. Ensure your test covers both the normal path and error scenarios if possible. For example, you might create a situation where the future method would hit an exception (perhaps by supplying bad data or using a mocked callout response) to see that your error handling code in the future runs properly. **Developer Tip:** Because future methods are static, they can be a bit harder to stub or inject faults into. Sometimes developers use flags or conditions that only trigger in a test context. For instance, you might have something like if(Test.isRunningTest() && someCondition) throw new CustomException("test"); inside the future method to simulate an error pathway during a test run, which you wouldn't do in production. Use such techniques cautiously, as you want to maintain realistic logic in tests.

When to use future vs. other async options: Salesforce now has multiple ways to run async (Batch, Queueable, Scheduled, Platform Events, etc.). Future methods were the original simple tool. They are still very useful for quick jobs and callouts. However, Salesforce recommends using Queueable Apex for new developments that require more complex behavior, because queueables can do everything a future can and more (like job chaining, more flexible parameter passing). Still, future methods have the advantage of simplicity and a very low overhead to implement. If you need a "one and done" async call, a future might be perfectly fine. Architect Tip: Establish guidelines in your team about when to use future methods vs. queueable. For example, an architect might say: use @future only for trivial operations or callouts where we don't need monitoring or complex chaining, and use Queueable for anything that might need tracking or for larger processes. This ensures consistency in the codebase. If maintaining older code, consider refactoring heavily used future methods into queueable jobs if you need the extra capabilities.

To get more tailored advice or to help decide between approaches, you can leverage ChatGPT's knowledge. Here are a few **ChatGPT prompt examples for future methods**:

"I have a trigger on Case that needs to do a callout to an external system after insert. Should I use a future method or a queueable, and why?" – ChatGPT can discuss the trade-offs, highlighting simplicity vs. advanced features.

"Can you show me an example of a future method that updates a list of contacts' fields? What are the best practices in that code?" – ChatGPT could write a sample code snippet, demonstrating proper usage of collections and error handling.

"I got a governor limit error 'Too many future calls' in my code. Here's the scenario... How can I fix it?" – Presenting the scenario, possibly by describing your code or even sharing a sanitized snippet, can get you an analysis of why you hit the limit and how to restructure your code.

"How do I test a future method that makes an HTTP callout? Are there special considerations?" – ChatGPT can explain the use of HTTPCalloutMock in tests and the Test.startTest/stopTest pattern.

Additionally, consider **using alternative input methods** to interact with ChatGPT about future methods. For example, you might **upload a document** containing your Apex trigger and future method code, and ask ChatGPT to review them together to ensure the trigger-to-future integration is correct (with privacy mode on for your uploaded files). Or, if you prefer talking through a design, you could use voice input to explain your plan ("ChatGPT, I want to use a future method to handle X, is that a good idea?") and hear the explanation in audio if you're using a voice-enabled device. This can sometimes make the learning process more engaging.

In summary, future methods are a straightforward yet powerful feature of Apex for asynchronous processing. They shine in scenarios requiring immediate relief from governor limits in the current transaction or when performing callouts post-transaction. By following best practices—bulkifying calls, handling errors, and testing thoroughly—you can confidently use future methods to improve system responsiveness. Just be aware of their limitations (lack of direct monitoring and chaining) and plan accordingly. *(Next, we will explore Queueable Apex, which builds on the concepts of future methods and offers greater control and flexibility for asynchronous jobs.)*

EFFECTIVE USE OF QUEUEABLE APEX

Salesforce's **Queueable Apex** is an evolution of the future method concept that provides more advanced features for asynchronous processing. A Queueable Apex class is an Apex class that implements the Queueable interface, with a single method execute similar to the execute in Batch but designed for a one-time job. You submit a Queueable job by calling System.enqueueJob(new YourQueueableClass(params)). This enqueues the job for execution, very much like how a future method is enqueued, but with added benefits: you get a **job ID you can monitor**, you can **chain queueable jobs** (starting another job from within a running job), and you can pass **complex types** into a queueable by using member variables (since the whole object is serialized for the job).

Unique strengths of Queueable Apex

Chaining jobs: Perhaps the biggest advantage of Queueable over future methods is the ability to chain. After a queueable job finishes its work, it can enqueue another queueable job. This allows you to sequence asynchronous operations. For example, suppose you have a process that involves multiple steps: first, generate some data, next, call an external API with that data, then process the response and update records. With future methods alone, orchestrating this would be tricky (you might end up having one future call another which isn't allowed, or using a scheduled intermediary). With Queueable, you can implement Step1 in one queueable class and at the end of its execute method, enqueue a Step2 queueable (passing along necessary info via parameters). Step2 can then enqueue Step3, and so on. Each step will run in order, each as a separate transaction. This structured asynchronous execution is incredibly useful for building multi-step workflows that don't need user interaction. It's like a mini process flow but all in Apex on the server side.

Job monitoring and identification: When you enqueue a queueable job using System.enqueueJob, it returns an ID (AsyncApexJob Id). This ID can be used to monitor the job's status in the Apex Jobs in Setup or via SOQL on AsyncApexJob object. For instance, you can query SELECT Status, NumberOfErrors, JobItemsProcessed FROM AsyncApexJob WHERE Id = :jobId. Future methods lack this capability entirely – you have no direct handle on a future once called. With queueable, you at least have an identifier. You might not always need to monitor programmatically, but the mere ability means if something goes wrong, you can find the job and see if it failed or is still queued. Developer Tip: It's often helpful to store that job ID somewhere if the context requires tracking. For example, if a user clicks a button that triggers a queueable job, you might store the job ID on a record so that you could display status to the user (though you'd need a way to poll or check it, which might involve another async or a Lightning component querying status). In many backend processes you may not need this, but for long chains or critical jobs, knowing the job ID and perhaps logging it can be invaluable in debugging. You could even use ChatGPT by providing a job Id and asking, "This Queueable job ID shows Status 'Failed' with error message X, what are common causes of that error?" to get some insights into troubleshooting.

Passing complex data: With queueable, you are not limited to primitive types in method parameters. Your queueable class can have member variables that hold complex data structures. For example, you could create a Queueable class that takes in a list of sObjects or a custom Apex object as a constructor parameter and stores it in a member variable. When the job runs, Salesforce will have serialized that object for you and you can use it directly in the execute. This means if you have, say, a list of Opportunity IDs and a map of calculations, you could set those as fields on the queueable class and they'll be available in execute. Note: There are still some limits – the data has to be serializable, and there's a limit on the total size of the serialized state (which ties into the 12MB async heap limit). But practically, this lets you avoid re-querying or recalculating some info that you already had when you enqueued the job. It's more flexible design-wise.

Greater limits and separate transaction: Like all async Apex, queueable jobs get their own context with the higher governor limits (e.g., 60 seconds CPU, 12MB heap). So if you had trouble doing something in a single trigger, moving it to queueable not only frees the user's wait, but gives you more breathing room for resource consumption. However, remember that launching a queueable is subject to limits too – you can only enqueue a certain number (50) per transaction. Also, you can chain jobs, but Salesforce does have a chain depth limit to prevent infinite loops: you can chain up to a certain depth (commonly documented as a limit of 50 jobs in a single chain in an org, though rarely hit unless something is wrong). Architect Tip: If you foresee needing to chain more than a few jobs (say a sequence of 10 queueables), consider whether Batch Apex might be a better fit, or break up the logic differently. Very deep chains can become hard to maintain and monitor. An architect might use queueable chaining for a sequence of logically dependent steps, but not for processing a massive list by sequentially queueing sub-lists (that's what batch is for).

Now, best practices for implementing queueable classes:

Implementing the Queueable interface properly: A Queueable class looks like:

```
public class MyQueueableJob implements Queueable, Database.Allow-
Callouts {

  private List<Id> recordIds;

  public MyQueueableJob(List<Id> ids) {

    this.recordIds = ids;

  }

  public void execute(QueueableContext context) {

    // your logic here, e.g. query records by recordIds and process

    // maybe chain another job:

    System.enqueueJob(new NextStepQueueable(someData));

  }

}
```

You see above Database.AllowCallouts can be added if you intend to callout in this queueable. Similar to batch, if a queueable needs to perform callouts, you must implement that marker interface. Remember to do so; it's a common oversight that leads to runtime exceptions if you attempt a callout without it. As illustrated, you can pass data (list of Ids in this case) through the constructor. Within execute, you then perform the work. If chaining, make sure to conditionally enqueue next steps if appropriate. One tip: if you chain a job at the end, it will run almost immediately after the first job finishes (depending on system resources, but it goes to the queue right away).

Avoid using too many queueable jobs in parallel from one context: You might be tempted to use queueable in bulk contexts by enqueueing jobs for chunks of records (like how batch automatically chunks for you). For example, in a trigger with 500 records, one might think to enqueue 5 queueable jobs each handling 100 of those records. This can work as long as you don't exceed 50 enqueues and overall it might be fine. But be cautious: firing off many queueables at once can flood the asynchronous queue and potentially throttle your org (Salesforce might delay some if too many are in flight). If the logic truly requires parallel asynchronous processing, Batch Apex is a more controlled way to do it with built-in chunking. Use queueables primarily when you need a *specific sequence or single-unit of work* that doesn't fit well into batch (or when the overhead of batch is not justified).

Error handling in queueable jobs: Unlike batch Apex, queueable doesn't have a finish method for cleanup or a direct partial success concept. It's just one execute. If an error occurs, the job can fail. If you don't catch exceptions, the AsyncApexJob will show a failed status. Therefore, incorporate try-catch in your queueable's execute logic for any operations that could throw exceptions (almost anything: DML, callouts, etc.). If an error is caught, you have options on what to do:

You could log the error to a custom object, similar to the future method approach.

You could attempt a recovery: for example, if a callout failed due to a timeout, you might decide to enqueue a *new instance of the same job* to retry. This is a simple retry mechanism—just be cautious not to create an infinite loop of retries. Perhaps keep a counter of attempts (which you can pass via member variable). E.g., your queueable class might have Integer attempt as a field, and only retry if attempt < 3, incrementing it each time.

Another best practice is to implement exponential backoff when retrying failed queueable jobs. Exponential backoff means that after each failure, you wait for an increasingly longer period before retrying, which helps prevent overwhelming the system with repeated immediate attempts. For example, you might schedule the first retry after one minute, the second after two minutes, then four minutes, and so on, allowing time for transient issues to resolve and reducing contention on system resources.

If the job is part of a chain and it fails, your subsequent jobs won't run because you never enqueued them. To handle this, one pattern is to enqueue a *finalizer* or error-handler queueable as a separate catch. Another approach is to use the new Platform Events or error event logging (though out of scope to implement fully here). We'll discuss more in error handling section, but in short, you should design queueable sequences with the possibility in mind that one step could fail. For critical sequences, perhaps add checks (if a prerequisite record wasn't updated because a prior job failed, handle that scenario in the next job if it still runs). In a direct chain, if one job fails completely (exception not caught), the chain breaks there.

Maintaining readability and manageability: Because queueable chains can become complex, document and structure them clearly. Perhaps different classes for each step (with clear names like ImportDataStep1, ImportDataStep2, etc.). Or if using one queueable class to do multiple things depending on internal state, be very clear with comments. Developer Tip: Leverage ChatGPT to generate documentation for your queueable processes. For instance, after implementing, ask: *"Help me document the flow of these three queueable classes that chain together: first does X, second does Y, third does Z. Provide a concise summary I can put in comments."* This can produce a nice explanation to include in your code or design docs, ensuring anyone reading the code later (including you, months from now) can follow the logic.

When to choose Queueable over Batch: Queueable is generally best when you need to handle asynchronous logic that is not just processing a massive number of records in a uniform way (batch is better for massive volumes), but rather orchestrating a workflow or performing a task on a moderate number of records. For example, if you have 10,000 records to process in bulk, a batch is usually better. But if you have a process where you handle maybe a few hundred records or less, or a mix of tasks (query some records, call an API, then process results), queueable is great. Also, queueable is often used for follow-up processing after a batch. A pattern might be: run a batch to process all data and in its finish, enqueue a queueable to do a final step (like update a dashboard or send notifications) because at that point you might want one consolidated action which is easier as a queueable job with all info at hand (Batch's finish could do it too, but sometimes separating concerns is cleaner).

Let's talk about how ChatGPT can assist with Queueable Apex. Many of the prompt ideas are similar to future methods but focusing on the enhanced aspects:

Design and conversion: If you have existing future method code that has outgrown its usefulness (maybe you now need to chain or handle more data), you can ask: *"How can I convert this future method into a Queueable? Here is the future method code..."* ChatGPT can help outline the new class structure and highlight what changes (e.g., making it an instance with member variables instead of static parameters).

Chaining strategy: "I want to chain queueable jobs. Can you provide an example pattern for chaining two queueables, including how to pass data from one to the next?" The answer could give a mini blueprint, for example by storing the next step data in a custom object or simply passing an ID through the chain.

Debugging help: If you encounter an issue like the second job in a chain didn't run, you might share the code of your first job and ask, *"Why might the next job not be enqueued?"* ChatGPT might spot that you forgot to call System.enqueueJob or perhaps an exception occurred before that line.

Using advanced tools: If you have a complex chain and want to ensure its correct, consider using ChatGPT's reasoning by providing an outline of the steps. Even uploading a diagram or drawing of the desired queueable flow (if you sketched one) could be interpreted by ChatGPT (with its vision capabilities) to verify logic – though a simpler way might be describing the flow in text. If you have access to a visual whiteboard or diagram, you might generate an image to illustrate the chain and ask ChatGPT if it covers all error scenarios.

We should also mention one more method of interacting aside from text: using an **"Operator" integration**. Some development teams integrate ChatGPT into their workflow using tools or bots (often called operators) that can run within development environments or chat ops. For example, an engineering team might have a Slack bot (an "Operator") where you can post a piece of code or question and it replies using ChatGPT under the hood. You could use such an operator to quickly get code reviews or suggestions on your queueable classes directly from your IDE or chat, without having to copy-paste into the ChatGPT web interface. This kind of integration (whether you call it an operator or a plugin) allows you to use ChatGPT in a more seamless way during development.

Always ensure that any such integration is secure and complies with data policies (for instance, do not inadvertently expose sensitive code or data without proper agreements and privacy controls).

Architect Tip: Use Queueable Apex strategically. Decide early in the design which asynchronous mechanism fits each requirement. Architects should guide developers: for workflows that have multiple asynchronous steps or need monitoring, prefer queueables or even a combination of batch + queueable. For simple one-step async tasks, a future might suffice. This governance avoids confusion and ensures the system is built on the right patterns. Additionally, consider the governor limit that only 50 queueable jobs can be added in one transaction (just like futures). If an architectural design might risk exceeding that (e.g., a trigger that could spawn dozens of queueables), then redesign to use batch or different triggers to mitigate that. An architect might also put in place a pattern that any time a queueable is enqueued, it's logged (maybe via debug or a custom object record) with context, so there's an audit trail of what was kicked off – useful in diagnosing issues in complex chains.

Developer Tip: Embrace the flexibility of queueable but remain disciplined. Don't abuse chaining to the point of creating spaghetti asynchronous flows that no one can follow. Also, be careful with global classes: If a queueable needs to be invoked from global context (like from another package or an LWC via imperatively calling, which is rare), it would need to be global, but most of the time public is enough. Keep your Queueable classes focused: do one thing or one step of a process. This makes testing easier too – you can test each queueable in isolation by enqueuing it within a test (again using Test.startTest/stopTest around the enqueue call) and verifying its work. If testing a chain, you might need to simulate the chain by either invoking subsequent classes directly or by actually letting the first enqueue the second (which will run during Test.stopTest as well, since it enqueues immediately). Writing comprehensive tests for a chain can be tricky; you might test steps individually and one integration test for the chaining mechanism.

Testing a chain of Queueable jobs in Salesforce can be challenging because the unit-test runtime doesn't always mirror production scheduling. You can test each Queueable step by enqueuing it between Test.startTest() and Test.stopTest() and asserting the outcomes. You can also assert that step one enqueues step two, but fully executing a longer multi-step chain end-to-end in a single test method is often unreliable or impractical. In production, a running Queueable can enqueue only one child Queueable, and in tests the Test.stopTest() flush behavior often makes chained execution difficult to model cleanly. A pragmatic approach is to write focused unit tests for each step, plus a higher-level test that verifies the chaining decision points (for example, that the next job is enqueued when conditions are met) within the platform's constraints.

With Queueable Apex, you gain fine-grained control over asynchronous execution in Salesforce. By following these best practices – using chaining responsibly, handling errors, and understanding its limits – you can create robust background processes that extend Salesforce beyond the request-response model. Many complex enterprise requirements (like multi-step order processing, offloading computational tasks, or scheduled pipelines of work) can be elegantly handled with queueables. And with ChatGPT's help for optimization and troubleshooting, you can implement these advanced patterns with confidence.

SCHEDULED APEX FOR RECURRING PROCESSES

Not all Apex needs to run in response to a user action; often, you have tasks that must run at specific times or regular intervals. **Scheduled Apex** allows developers to schedule Apex classes to run in the future, either once or on a recurring schedule (daily, weekly, etc.). This is analogous to a cron job in traditional systems. With Scheduled Apex, you can automate maintenance tasks, daily summaries, data syncs, or any process that needs to happen periodically. In this section, we detail best practices for using Scheduled Apex, including how to schedule jobs, patterns for flexible scheduling, monitoring scheduled executions, and handling potential scheduling conflicts or failures.

To use Scheduled Apex, you write an Apex class that implements the Schedulable interface, which requires you to implement one method: execute(SchedulableContext context). This is the entry point that runs when the schedule triggers. You then schedule this class to run at a certain time using either the Salesforce user interface (Setup > Apex Classes > Schedule Apex) or more commonly via code using System.schedule. The System.schedule method takes three arguments: a job name, a CRON expression string, and an instance of your schedulable class. The CRON expression is a string that follows a specific format to define the schedule (fields for Seconds, Minutes, Hours, Day-of-month, Month, Day-of-week, and optional Year). For example, System.schedule('NightlyBackupJob', '0 0 2 * * ? *', new DataBackupJob()); would schedule the DataBackupJob (which implements Schedulable) to run every day at 2:00 AM.

One practical note on cron expressions is worth calling out explicitly. Salesforce now supports defining cron expressions directly through the Setup interface as well as via System.schedule, which makes Scheduled Apex feel much closer to traditional cron-based systems. That power cuts both ways: cron syntax is compact, expressive, and very easy to get subtly wrong. While AI tools like ChatGPT can help explain what a cron expression does or draft a starting point, they are not a substitute for a purpose-built cron validator or visual editor. For anything that matters in production, it is a best practice to generate and verify cron expressions using a dedicated cron tool that lets you see the schedule spelled out in plain language and confirms that it behaves as expected. Treat cron expressions as configuration that deserves the same level of validation as code, because a single misplaced field can turn a nightly job into one that runs every minute, or never runs at all.

Best practices for Scheduled Apex

Use cron expressions wisely: Cron strings in Salesforce can be a bit daunting. Ensure you specify them correctly, as a mistake can result in jobs not running or running at unexpected times. It may help to use tools or ChatGPT to double-check a cron expression. For instance, *"ChatGPT, I need a cron expression for every Monday at 5:30am UTC."* ChatGPT can generate the string (e.g., "0 30 5 ? * MON *" for that schedule). Always double-check time zones — Salesforce CRON by default uses the org's default timezone for execution if scheduled via UI, but via System.schedule it uses the user's timezone context who invokes it (if running in Apex). To avoid confusion, explicitly define the schedule in terms of hours/minutes that make sense in context (and document whether it's UTC or local). If needed, you can calculate offsets for daylight savings if that matters, though CRON in Salesforce doesn't adjust for DST automatically.

Scheduling flexibility: If you need more dynamic scheduling (like user-defined or data-driven schedules), you might store desired schedule info in Custom Metadata or a Custom Object, and have a single schedulable class that reads those and decides what to execute. However, in many cases you'll know ahead of time that e.g. "this job runs every night". Product Owner Tip: Involve product owners or stakeholders when deciding schedule timings. They might have preferences such as "end of month reports should run at 1 AM on the first day of next month" or "don't run heavy jobs during our usage peak at noon". Such inputs ensure the technical scheduling aligns with business operations. It might be useful to generate a small calendar or timeline to visualize the scheduled jobs. You can even ask ChatGPT to help outline a schedule: *"We have 3 jobs: one daily at midnight, one hourly, one every Friday 6pm. Could these conflict? What's the best way to stagger them if needed?"* The answer might highlight if two jobs both run at midnight Friday, for example, and suggest staggering by a few minutes.

Combining Scheduled Apex with Batch/Queueable: Often, the work you want to perform on a schedule is best done via a batch or queueable. Scheduled Apex can act as a launching mechanism. For instance, you might have a class NightlyRecalculationScheduler that implements Schedulable. In its execute method, it simply invokes a Batch Apex (using Database.executeBatch) or enqueues a Queueable to do the actual heavy lifting. This pattern separates the scheduling logic from the processing logic. It's a best practice because it keeps the schedulable code small (just scheduling) and leverages the proper async tool for processing. It also allows reusability – you could run that batch on an ad-hoc basis outside the schedule if needed (like a manual trigger) reusing the same batch class. Developer Tip: When scheduling batch jobs, ensure you monitor for any overlaps. If the previous day's batch hasn't finished and the next run starts, you might have two batches running concurrently. Salesforce will queue them, but logically it might not be what you want (for example, two daily syncs running at once). A defensive approach is to check in the schedulable if the previous job is still running. You can query AsyncApexJob for your batch class's last instance status. Or maintain a custom setting flag "Job in progress". Alternatively, design the batch to gracefully handle being started again (maybe it finds there's nothing new to do if already running). We'll talk more about avoiding conflicts soon.

Handling scheduling conflicts and limits: Salesforce typically allows up to 100 scheduled Apex jobs at one time per org. If you exceed that, scheduling a new job fails. In an enterprise scenario, it's easy to accumulate many scheduled jobs (some might be one-off scheduled runs left behind). Architect Tip: Regularly audit scheduled jobs in the org. Ensure obsolete ones are removed. For example, after deploying changes, you might inadvertently leave an old schedule running – those should be unscheduled if not needed. Tools for auditing include the Setup UI (listing Apex Scheduled Jobs) or querying CronTrigger object via SOQL. Also, be mindful of jobs that might run close together. If two jobs run at the exact same time and both are resource-intensive, they will both attempt to use resources. Salesforce can handle parallel execution, but they share overall CPU limits on the instance. It's wise to offset heavy jobs a bit. For example, instead of two jobs at 00:00, schedule one at 00:00 and another at 00:15. This staggering is a minor adjustment that can reduce contention. If you're not sure about potential contention, you can test in a sandbox by scheduling them and observing. Or simply consult ChatGPT: *"If I schedule two batch jobs to start at the same time nightly, what are the risks?"* The response will likely mention resource contention and possibly hitting the concurrency limit for batch jobs (only 5 can run at once, as noted earlier).

Recovery from missed schedules or failures: If a scheduled job fails (throws an exception and doesn't handle it), that particular execution will be marked failed, but it will still attempt the next scheduled run at the next interval. However, the work intended for that run may not have completed. If the job is critical (say, closing financial records nightly), you need a plan for if it fails. One approach is to build retry logic inside. But since scheduling means time moves on, a better approach might be to have an alert on failure so an admin can manually rerun something. In the execute of schedulable, you can catch exceptions around your main logic and send an email or post to an error logging system if something goes wrong. This at least notifies someone to take action. Some orgs use a second scheduled job as a watchdog – for instance, a weekly job that checks if all daily jobs ran successfully by looking at log records, and if not, raises an alarm. This might be overkill for smaller orgs, but for mission-critical processes it's worth considering. Another pattern: if a job should run and complete something by a certain time, and it fails, perhaps the next run should detect that last run's data is incomplete and attempt to complete it (if it's idempotent). This requires careful design to avoid doing things twice incorrectly, but it's part of resilient design.

Testing Scheduled Apex: Testing the schedule itself (i.e., System.schedule) is possible but usually you don't test that the platform cron works (that's Salesforce's responsibility). Instead, you test the functionality of the schedulable's execute. You can call the execute method of your Schedulable class directly in a test, passing a dummy SchedulableContext (you can create an instance of an SchedulableContext via a stubbing approach, or sometimes just pass null if the code doesn't use it). Alternatively, you can actually schedule the job in a test using System.schedule, but note: in a test, if you schedule a job to run once within the test's timeframe, it will execute immediately during Test.stopTest() (similar to future/queueable behavior). For example:

Test.startTest();

String cron = '0 0 0 * * ? *'; // this cron is daily at midnight

System.schedule('TestJob', cron, new MySchedulableClass());

Test.stopTest();

When Test.stopTest() executes, the scheduled job will run once (immediately, not waiting until midnight). This is by design so you can test the outcome. Keep in mind, if your schedulable calls a batch or queueable, that subsequent async will also run during the stopTest phase. You'll then verify the results. **Tester Tip:** Write tests to ensure that the scheduled logic respects boundaries. For instance, if your schedulable is not supposed to start a batch if one is already running, simulate a "already running" condition and ensure it doesn't start another. You can simulate this by perhaps inserting a marker record or using a static flag in the class for testing indicating "simulate running state" and have the execute check that in tests. While this introduces test-specific logic, it's often necessary to simulate scenarios that are hard to create in a test context.

Special considerations for different roles:

For Product Owners/Stakeholders: ensure they understand that scheduled jobs run in the background and may not have a UI component. If the business needs visibility, plan for some form of reporting or logging. For example, a scheduled data clean-up could log how many records were cleaned each time, and that summary could be surfaced on a dashboard or sent in an email. A Product Owner might want to see a monthly summary of these background operations to be confident things are running smoothly.

For Testers: beyond unit tests, in UAT or staging environments testers might want to verify that scheduled jobs indeed fire. Since waiting in real-time is impractical, testers may manually execute the Apex (perhaps by calling an exposed method or using the developer console to run the schedulable immediately). It's useful to provide admins/testers with a manual trigger mechanism for scheduled logic, such as a checkbox "Run job now" on a custom setting that a simple admin script checks. This isn't a standard pattern, but it can be helpful for manual runs. At minimum, ensure thorough logs so that when a schedule triggers, testers can confirm via logs or results that it ran.

Now, let's see how **ChatGPT can further assist** with Scheduled Apex. Some possible interactions:

Generating Cron expressions: As mentioned, the cron syntax can be tricky. For example: *"Generate a Salesforce cron expression for the last day of each month at 11:45 PM."* ChatGPT can output something like "0 45 23 L * ? *" (where L means last day of month).

Optimizing schedule: "I have 5 scheduled jobs. How can I ensure they don't run at the same exact time?" – ChatGPT might suggest staggering times or combining jobs if possible.

Writing a schedulable class: *"Provide a simple example of a Scheduled Apex class that runs a batch job."* – It can produce a template showing the pattern (Schedulable class calling Database.executeBatch in execute).

Troubleshooting: *"My scheduled job didn't run as expected at 5 AM. What could be the reasons?"* – ChatGPT may enumerate possibilities (job wasn't scheduled properly, was removed, org was paused, etc., or time zone confusion).

Creating reminders: One creative use: ask ChatGPT to outline an email template or log format for after a scheduled run. e.g., "Draft an email text that a scheduled Apex job could send to admins summarizing its success or failure, including placeholders for number of records processed and errors." This could help you quickly implement a notification feature.

On using ChatGPT beyond typing: You might leverage **image generation** to visualize schedule timelines. For example, if explaining the schedule to stakeholders, you might want a diagram that shows a calendar with markings when jobs run (e.g., job A runs daily, job B weekly). ChatGPT itself doesn't draw, but you could ask for an ASCII art timeline or use an integrated tool. For instance: *"Draw a simple timeline for the weekly and daily jobs across one week."* It might respond with text like:

Mon: [DailyJob]

Tue: [DailyJob]

Wed: [DailyJob]

Thu: [DailyJob]

Fri: [DailyJob][WeeklyJob]

Sat: [DailyJob]

Sun: [DailyJob]

This is a simple representation that could then be turned into a more formal diagram if needed. Alternatively, if you have an image generation capability, you could ask it to create a chart image highlighting run times. Always review any AI-generated schedule visualization to ensure it matches reality.

Callout tips for Scheduled Apex

Developer Tip: Always give your scheduled jobs clear, descriptive names when scheduling (the first parameter of System.schedule). This name appears in the job list. For example, "Nightly_Data_Cleanup" is better than "Job1". It will make monitoring easier. Additionally, document the purpose of each scheduled job in the code or in a central wiki – including when it runs and what it does – so that if something goes wrong at 2 AM, on-call developers or admins can quickly identify the job and understand its function.

Product Owner Tip: Communicate to the development team any business constraints on scheduling. For example, "We cannot run this job on the first of the month because that's when we do a data export." Such information can be crucial. Also, ensure that any user-facing impact of scheduled jobs is known. If a scheduled job might lock records or consume significant API calls (maybe interacting with external systems), product owners should know if users might experience anything different during those times (generally they shouldn't if done correctly, but it's good to confirm).

Tester Tip: Use sandbox or lower environments to simulate production schedules. If possible, shorten the interval for testing (like schedule a job to run every hour in a test environment instead of daily, just for testing purposes, then change it to daily in production). This helps catch issues early. Ensure that test verification includes checking the results of the scheduled run; e.g., if the job is supposed to update stale records, intentionally create some stale records and check they get updated after the schedule triggers (or manually run it in sandbox).

By effectively using Scheduled Apex, you can automate routine tasks and ensure your Salesforce org stays up-to-date without manual intervention. Combined with Batch and Queueable Apex, scheduling forms a complete toolkit for asynchronous processing: you can schedule batches or queueables to run at needed times. This leads to a smoothly running system where nightly maintenance, data synchronization with external systems, and periodic computations happen reliably in the background.

MANAGING GOVERNOR LIMITS IN ASYNCHRONOUS CONTEXTS

Salesforce's **governor limits** are a fundamental aspect of Apex development. These limits (on CPU time, memory, SOQL queries, DML operations, etc.) protect the multi-tenant environment by preventing any single org or execution from consuming excessive resources. When working with asynchronous Apex (Batch, future, queueable, scheduled), we often have higher limits or separate limit pools compared to synchronous execution. For instance, as mentioned earlier, an asynchronous Apex transaction can run for up to 60 seconds of CPU time (versus 10 seconds for synchronous) and use up to 12 MB of heap memory (versus 6 MB sync) in a single transaction. Additionally, batch jobs allow you to process far more total records by splitting them into chunks, each with their own per-chunk limits. However, "higher" limits do not mean "infinite" limits. In this section, we'll clarify how governor limits apply in asynchronous contexts and why careful management is still essential. We'll provide strategies to proactively manage limits within asynchronous jobs, such as optimizing batch sizes, selective data processing, and iterative processing approaches. The goal is to ensure that just because something is running asynchronously, we don't become complacent – good design and coding practices are needed to stay within limits and avoid exceptions like LimitException errors.

How limits differ in asynchronous Apex

Heap size: As noted, you get 12 MB heap per transaction asynchronously. This means a batch execute can use up to 12 MB for its 200 records (or whatever batch size) of processing. If your code needs to handle a large data structure, doing it asynchronously may allow it to succeed where it would fail in a synchronous context. For example, if you had to build a very large JSON string from thousands of records, doing that in a future might be possible within 12 MB whereas it might exceed 6 MB in a trigger. Still, 12 MB is a hard cap – if you try to hold too much in memory, you'll hit "Apex heap size too large" error. Strategy: Free memory when possible. If working with large collections, after processing a portion, set references to null or use local scopes so that objects can be garbage-collected within the transaction. In batch, avoid using Database.Stateful to accumulate huge lists (as that stays in memory across executes). Instead, use each execute to persist results incrementally (e.g., insert into a temp object or update records as you go) so that memory doesn't snowball.

CPU time: 60 seconds of CPU per async execution is generous, but heavy computations or extremely large loops can still hit it. For instance, a poorly written batch that does an extremely nested loop or complex calculation on each record could conceivably use 60s for 200 records (which would be very slow per record, but not impossible if doing crypto-level math or something). Strategy: Watch out for inefficient algorithms. If something might be pushing CPU, consider splitting the work. Maybe use two queueables sequentially (where each gets a fresh 60s, effectively giving 120s combined, though in separate contexts). Or break a batch into smaller chunks (if 200 records took 60s, 100 records might take ~30s, giving more headroom). Also, leverage efficient data structures; for example, use Sets/Maps for lookups in Apex instead of double looping.

SOQL queries and DML: These limits (100 SOQL queries, 150 DML statements per transaction) do not increase for async; they are the same as sync. The main benefit in async is that if you need to process many records, you can do it across multiple transactions (each chunk of a batch gets its own 100 queries allowance). But within one execute, you still can't exceed 100 queries. So if your code design is such that it might run a query per record, it will break if batch size > 100. Strategy: Bulk queries – one query to get all needed data for all records in the batch. If you truly need more than 100 queries worth of data for 200 records, then the logic probably needs an overhaul or data model adjustment. Sometimes, hitting query limits is a sign you should do processing in smaller increments or use a different approach (like maybe use SOSL if searching across objects, or pre-aggregate data in a big query). Similarly, for DML: if you have logic that does DML in a loop for each record, you could hit 150 limit with more than 150 records in a batch if not bulkified. In practice, proper bulkification (one DML for all or chunks) should avoid that.

SOQL rows: In a synchronous context, you can fetch up to 50,000 rows in a single SOQL query. In asynchronous context (like batch), a single query in execute still has that 50k limit per query. But you rarely need to pull that many in one chunk, because typically you partition via batch. If you needed to handle, say, 1,000,000 records, you wouldn't query them all in one go; batch's QueryLocator will feed them to you chunk by chunk. However, one nuance: QueryLocator in batch can retrieve up to 50 million records total across all batches, which is the main way to handle extremely large volumes. If you tried to manually query 50 million records in Apex, it's impossible in one go – batch with QueryLocator is the mechanism. Strategy: Use QueryLocator (by returning a SOQL in the start of batch) for huge volumes so Salesforce handles the chunking and pagination automatically. If you have to manually query inside execute, limit your queries properly (e.g., using LIMIT or filters) such that each execution doesn't try to pull an excessive number of rows that could hit row limits or run too slowly.

Concurrent asynchronous limit: Salesforce has a limit on how many asynchronous Apex executions (future, queueable, batch chunks, scheduled runs) can run concurrently and how many can be queued. For instance, you can have at most 50 queued Batch jobs at once (beyond that they go into a "holding" queue up to 100). For future/queueable, there is a general limit of how many total async calls you can make in 24 hours (per license type, e.g., some thousands per user license in the org). Hitting that daily limit is rare but possible if you fire off a huge number of futures/queueables (for example, an org with a trigger launching 1000 queueables per hour could run into limits by sheer volume). Strategy: If you find you are consuming a large number of async jobs, reconsider design. Maybe you don't need to start a separate job per record action – could group them. Or maybe some logic can be done in a single batch nightly instead of many futures throughout the day. Logging the usage of async (like counting how many jobs launched per day) can help identify if you're nearing limits. If needed, an architect might throttle certain operations (for example, only allow a certain number of background jobs per hour by queuing them in a custom object and processing them gradually).

Selective processing and iterative approaches: One key strategy to manage limits is *doing less work*. Sounds obvious, but in design terms: filter out unnecessary processing. If a batch runs nightly to update records but 90% of those records haven't changed and don't need updating, it's wasteful to process them. Add a filter to only process relevant records (e.g., "WHERE LastModifiedDate = YESTERDAY" or some criteria). This reduces the data volume and thus lowers the chance of hitting limits (and speeds up completion). Another example: if you have to process a million records but can do 200k each night over five nights, that might be safer and within limits than trying all million in one go (even though theoretically batch can handle 50 million, you might not want to saturate your nightly window with one job). Architect Tip: Consider data partitioning strategies. For example, run multiple batch jobs in series, each on a subset of data (like partition by record type or by range of IDs). Each job deals with fewer records and thus lower risk of hitting cumulative limits or timing out. This is especially useful if each record's processing is heavy. Partitioning can ensure one job doesn't carry all the weight. However, balance this with the concurrency limits – don't run them all at once unless needed (which could itself cause limit issues). An iterative approach could be: schedule batch 1 at 1am, batch 2 at 2am, etc., or chain them via finish methods.

Using Apex Limits methods for insight: The Limits class in Apex provides methods like Limits.getCPUTime() and Limits.getLimitCPUTime(), which can tell you how much CPU has been used and the max CPU allowed. Similarly for queries and DML. In a long-running batch execute, you could periodically check these to see if you're close to limits and perhaps adjust behavior. For example, if your execute method processes records in a loop and you notice after processing 150 of them that you've used, say, 55 seconds of CPU (nearing 60), you might decide to stop processing further records in this chunk to avoid hitting the limit, and let the remaining records be processed in the next chunk. This is advanced and rarely needed, but it's a safeguard approach. Essentially, you'd catch the scenario by yourself rather than letting a LimitException occur (which would abort the whole chunk). By checking, you could say "if CPU time > 55000 (55 sec), exit loop early". Those unprocessed records might not be marked done; you'd need to handle them perhaps by re-queuing them via another job or marking them to do later. This is complex, so it's typically better to ensure by design that each chunk can complete within limits rather than dynamic mid-chunk adjustments. But it's good to know the tools exist.

Monitoring governor usage in logs: When asynchronous jobs run, if you have debug logging set appropriately (at least on error or finer for the user context that runs them), you can see in the logs how close to limits they got. For example, logs will show "Number of SOQL queries: 10 out of 100" etc. If you notice in a test or sandbox that a batch execute is consistently, say, "Queries: 100 out of 100" or "CPU: 59000 out of 60000 ms", that's a red flag. You should refactor before production to add breathing room. Running that close to the edge means any small data change could push it over the limit and cause failure. Tester Tip: As part of performance testing, consider creating data scenarios that push the boundaries – e.g., maximum expected data size – and then run your batch or queueable. Check logs to see how it fared. This is akin to a stress test for governor limits. If testers aren't familiar with reading debug logs, developers or admins might need to assist. Or testers could use ChatGPT: *"I have a debug log from a batch job. Can you help interpret the governor limit usage? Here's the log..."* (If the log is large, summarizing the relevant part or attaching a file is better.) ChatGPT can parse and highlight if any limits were nearly hit.

Now, how can ChatGPT assist with managing governor limits?

Code review for limit usage: If you provide ChatGPT with code (especially batch or queueable code), you can ask it to identify potential spots where limits might be exceeded. For example: *"Review this execute() method for any patterns that may hit governor limits when processing 200 records."* It might point out if it sees a SOQL inside a loop or multiple DML statements in succession that could be merged, etc. This is effectively a static analysis from an AI perspective.

Optimization suggestions: "How can I reduce the heap usage of this code?" or "I'm hitting CPU limits occasionally in this queueable – what general strategies can I use to optimize CPU?" ChatGPT can offer strategies like using more efficient loops, eliminating redundant calculations, or using built-in features (maybe formulas on fields instead of calculating in Apex, etc.).

Understanding specific limits: If you encounter an error like "Too many queueable jobs added to the queue: 51", you might ask: *"Why do I get this error and how to fix it?"* The answer should explain the 50-per-transaction limit and suggest combining jobs or restructuring the logic.

Advising on splitting work: *"I need to update 1 million records. Should I use one batch job or multiple? What batch size?"* ChatGPT can weigh in: perhaps recommending one batch with default batch size (since batch can handle 1 million fine, as 1,000,000/200=5000 executes which is okay as long as each execute is efficient). Or if context suggests heavy logic, maybe multiple batches. It might also mention using data loader vs Apex if appropriate (though context here is Apex best practices, so presumably using Apex).

Checking the limit usage across chain: You could describe an entire asynchronous flow and ask where limits might be a concern. E.g., *"I have a batch that calls a queueable in finish, then that queueable calls another. What limits should I watch out for across this chain?"* ChatGPT might mention the transaction boundaries (each is separate, so mostly just per transaction limits but also the number of total jobs counts for async limit).

Voice assistance for brainstorming: Sometimes explaining a problem out loud can clarify it. You could use voice input to quickly tell ChatGPT the scenario: *"I have an Apex batch that times out at 60s CPU with 200 records, how might I refactor it?"* and listen to its response. That way you can do a high-level brainstorming while maybe away from keyboard or multi-tasking.

One more thing: When an org grows, sometimes the code that once worked fine begins to hit limits because data volume increased. **Architect Tip:** Design with future growth in mind. If currently there are 10k records and a batch of 200 runs in 2s, that's fine. But if in 2 years there will be 10 million records, how will that scale? Will you run 50 batches a day or one batch of 50k records nightly? Plan architecture to either archive data or adjust logic when volumes grow. Perhaps implement a dynamic batch size that can be configured smaller as data grows and processing per record becomes heavier. Keep an eye on Salesforce release updates too; occasionally limits are adjusted (though rarely relaxed significantly).

In summary, while asynchronous Apex gives you more breathing room with governor limits, it's not a license to write inefficient code. You must still practice good Apex coding discipline: bulkify, avoid unnecessary work, and keep an eye on resource usage. The advantage is that you can tackle much bigger tasks by spreading work out. By proactively managing limits—using smaller batches, filtering data, optimizing queries and loops—you can prevent runtime limit exceptions and ensure your background jobs run smoothly. And by using tools like debug logs, the Limits API, and even ChatGPT's analysis, you can catch potential limit issues early in development or testing. *(Now that we've covered performance limits, our next focus is on making these asynchronous processes robust. In the following section, we'll explore error handling and recovery strategies specific to asynchronous Apex, so that when things go wrong (and eventually they will), our system can recover gracefully.)*

ERROR HANDLING AND RECOVERY IN ASYNCHRONOUS APEX

In any software system, things don't always go as planned – exceptions occur, external systems fail, data conditions surprise us. In asynchronous Apex, error handling deserves special attention because these processes run in the background. Unlike a user-triggered operation where an error can be immediately presented to a user (or rolled back such that the user knows nothing was done), asynchronous errors happen out of band. If not handled properly, they can go unnoticed or leave data in an inconsistent state. In this section, we'll highlight error handling considerations unique to asynchronous Apex and discuss best practices for building resilience: from structured error handling in code (try-catch, finally blocks, etc.), to robust logging and alerting mechanisms, to strategies for recovery and retries. We'll cover common error types (e.g., governor limit exceptions, unhandled exceptions in a batch chunk, callout failures) and how to approach them, as well as how to recover state if a job partially fails.

Challenges of asynchronous error handling

Lack of user context: Since async jobs are often triggered by some event and then run later, the user who initiated or the context of initiation might not be present when an error happens. For example, an Account trigger launches a future call to update some related data via callout. If that callout fails, the original user transaction already succeeded. The user might assume everything is fine. The error in the future method doesn't automatically notify them. This means communication of errors is non-trivial – you may need to inform someone (an admin or even the initiating user) through additional means.

Partial successes: Consider a batch Apex job processing 1000 records. It might successfully process 900 and fail on 100. In synchronous operations, typically everything is all-or-nothing (transaction rollback). In batch, by default, those 900 successes stay committed and 100 failed records might remain unprocessed or partially processed. It's up to the developer to decide what to do with those failures. This partial success nature is powerful (you get things done for most records) but requires planning to handle the ones that didn't succeed.

Complex flows and state recovery: With chained queueables or scheduled sequences, an error in one step might disrupt a whole flow. For instance, if Step 2 of 3 fails and you haven't designed for it, Step 3 might never run, leaving the overall process incomplete. Recovery might involve re-queuing the failed step or otherwise compensating (maybe rolling back what Step 1 did, if possible).

Given these challenges, here are some best practices for error handling in asynchronous Apex:

Use try-catch in every asynchronous execution block: Whether it's the execute of a batch, the execute of a queueable, or the body of a future method, wrap your logic in a try-catch. Catch the broad Exception if you must (to ensure no error escapes uncaught). Within the catch, implement appropriate handling:

Log the error details (see logging strategies below).

If it makes sense to retry, you might attempt the operation again, perhaps with a small delay or different approach, though in Apex you can't really pause a running job. A batch chunk can't easily "retry itself" except by reprocessing in the same execute which is risky. A queueable could catch an exception and decide to System.enqueueJob(new SameJob(params)) to try again.

If cascading failures are possible, consider failing fast vs continuing. For batch: if one record fails, you often want to continue with others (so catch exceptions around per-record processing rather than let one exception abort the whole execute). For a queueable that processes a list, similarly you can catch per item errors and continue.

Always ensure that the catch doesn't completely swallow an error silently. At minimum, increment some error counter or log something. Silent failure is the worst outcome (nothing happens and nobody knows).

Leverage the Database method result for partial successes in batch: In batch Apex, if doing DML on a list of records, prefer using the Database.insert/update/etc with allOrNone=false (or capturing the result of a normal DML which is effectively allOrNone true by default, which will throw exception on first failure). Using the Database methods returns a Database.SaveResult[] for inserts/updates or Database.UpsertResult[] for upserts, etc. These allow you to see which records failed and why, without throwing an exception that halts the execution. You can then handle each failed record in the code (e.g., log its ID and error message). This way, your batch can attempt all records in the scope and handle any that failed, instead of aborting on the first failure. For example:

```
Database.SaveResult[] results = Database.update(records, false);

for (Integer i = 0; i < results.size(); i++) {

  if (!results[i].isSuccess()) {

    Database.Error err = results[i].getErrors()[0];

    // Log error for records[i]: err.getStatusCode() and err.getMessage()

    // maybe add to an error list to process later

  }

}
```

Using this approach, your batch execute won't throw unless something outside the DML logic throws. You effectively handle record-level errors gracefully. Tester Tip: Write tests to simulate a failure in one record and ensure your batch doesn't stop processing the others. For instance, create one record that violates a validation rule so that when updated it fails, and others that succeed. See that the batch marked others as done and the bad one is logged appropriately.

Logging errors: Logging is crucial for async processes. Salesforce debug logs are one option, but they are ephemeral and require you to have set up trace flags. A more persistent approach is to use a Custom Object as an error log. Many implementations have an object (e.g., Async_Error__c) with fields like JobId, JobType (Batch/Future/Queueable name), ErrorMessage, StackTrace, RecordId (if applicable), Timestamp, maybe a Status (new, resolved). In your catch blocks, insert a record into this object capturing relevant info. This way, all errors are stored and can be reviewed at any time. Mark them as resolved once addressed. Ensure this logging itself doesn't fail; keep it simple (text fields for error, maybe truncated if too long). If you worry about performance, you could aggregate errors and insert once per execute or in finish for batch (to reduce DML overhead).

Another modern option is Platform Events. Salesforce has a standard Platform Event called BatchApexErrorEvent which, if a batch fails due to an unhandled exception (one that aborts the batch), Salesforce will publish this event with details (batch job ID, exception). You can subscribe to this via a Trigger or CometD client. If using that, you could automate some handling when a batch fails hard. For other async types, there isn't a built-in event, but you could create a custom Platform Event like Async_Error_Event__e and publish an event in your catch blocks. This has the advantage of real-time delivery to subscribers (maybe to an external monitoring system or to a Lightning component). But platform events have some limits and complexities, so a simpler approach is often just writing to a custom object or sending an email.

Alerting responsible parties: Logging is passive; alerting is active. Determine who should know when an async job fails. Often, system admins or a support email should receive notifications. You can use Apex Messaging.sendEmail in your catch block for critical failures. Be careful to avoid spamming if, say, 100 records fail in a batch – maybe send one summary email in finish instead of per record. For a future/queueable, maybe send an email right in the catch since it's one context. In a large org, an error logging system feeding into something like Splunk or Salesforce Shield Event Monitoring might be used. But for many cases, an email to admins or posting to a Chatter group can suffice. Architect Tip: Define an error handling policy. For example: "If any async job fails with an error, log it to X object. If the error is of type Y (e.g., an external system issue) and persists, notify the integration team. If error is data-related, open a case for data steward." It's good to classify errors so the right people act on them. ChatGPT can even help draft such a policy if you ask it for best practices on error handling roles.

Implementing retries or recovery workflows: Not every failure should be automatically retried – it depends on the cause. But some transient errors (like a callout to a service that times out or a record lock failure if two jobs try to update the same record concurrently) might succeed on a second attempt. Consider implementing a retry mechanism:

For **future methods**, retry logic is limited by Salesforce's execution rules. A future method cannot directly call another future method, including calling itself again, because Salesforce does not allow a future to be invoked from an existing future context. Attempting to do so will result in a runtime error indicating that a future method cannot be called from a future execution context. This makes simple "catch and retry" patterns impractical when you are already running asynchronously in a future.

In practice, this means that if a future method encounters a recoverable error, such as a transient callout failure, you cannot simply schedule a delayed retry from within the same future method. While you can catch the exception and handle it conditionally, any attempt to enqueue another future from that context will fail. Future methods are intentionally constrained to keep asynchronous execution predictable and to avoid runaway recursion.

If you need retry behavior, the usual workaround is to hand off responsibility to a different asynchronous mechanism that *can* be scheduled or chained. Queueable Apex is often the preferred option, as it allows limited chaining and better control over retry logic. Alternatively, you can use Scheduled Apex to retry the operation at a later time. In both cases, you can track retry attempts explicitly and apply backoff or termination logic as needed. The key takeaway is that future methods are best used for simple, fire-and-forget asynchronous work, not for workflows that require retries or orchestration.

For queueable: In the catch, you can enqueue a new instance of that queueable (perhaps with a flag incremented for attempt count). This effectively retries in a new job. You might also schedule a slight delay by enqueuing via a scheduler (like schedule it to run 5 minutes later via a Scheduled Apex that runs once). An easier approach is just immediately queue another job; it will run when resources free (which might be moments later). If the error is due to an external system down, immediate retry might also fail, but perhaps the second attempt might hit a different server or work if it was a glitch.

For batch: If one batch chunk fails entirely (uncaught exception aborts that chunk), the batch will abort overall and no automatic retry for that chunk. You'd have to rerun the batch for those records. A pattern could be: in the finish of the batch, detect if there were errors (maybe by checking a static or a stateful flag, or querying AsyncApexJob status). If errors occurred, you might automatically schedule another batch to handle the failed records. But since you might not easily know which records failed (unless you logged them), you'd need to have recorded them. If you did log failed record IDs in execute, you could retrieve them in finish and perhaps kick off a follow-up batch on those specific records. This essentially implements a retry for failed items. Keep an eye though: if they failed due to a consistent issue (like data problem), repeating might just loop. Perhaps put a cap (don't retry more than N times for the same record). This is quite advanced but is doable for critical processes that must eventually process all data.

State recovery considerations: If an async job partially completes, how do you continue later? For example, a scheduled job runs nightly; one night it fails halfway. Next night, you may want to pick up from where it left off. That means you need to know where it left off – some state must be stored. One approach: track a "last processed record ID or timestamp" somewhere. If the job fails, that value isn't updated to completion, so next run sees it and continues from there (maybe overlapping a bit but handling duplicates carefully). Or maintain a status field on records to mark done vs pending. Then each run processes those marked pending. That way, if it fails mid-run, the unprocessed ones remain pending and will be picked up next run. This is more of a design pattern for idempotence: design jobs to be restartable without negative side effects (no double-processing that causes harm, or if double-processing can happen, handle it gracefully). Architect Tip: Strive for idempotent asynchronous operations – meaning if the same job or record is processed twice, the result is the same as if processed once. This avoids issues if a retry happens. For example, if a batch is sending out emails and it fails halfway, a naive retry would resend some emails. To make it idempotent, you could record which records already had email sent, and check that to not resend, or design content such that double sending is acceptable (like in a log, mark emailSent = true per record, and the batch only sends where emailSent is false).

Testing error scenarios: It's important to test how your asynchronous code handles errors. Some ways:

Use test methods to simulate exceptions. You can write a subclass or use dependency injection to force an exception. For example, if your queueable calls an external service, in test you can set a static variable that makes it skip the callout and instead throw a CalloutException to simulate failure, then see if your catch caught it and logged it. Or if your batch processes records, include a dummy record that will cause a DML failure (like a required field missing) and see that it's handled not to crash the batch.

Ensure that error logging mechanisms themselves are tested. If using a custom object, verify in test that an error record is created under the simulated failure.

This might involve some creative coding, such as introducing a way to trigger exceptions in test mode. But ensure that any test-only logic is removed or inactive in production (you can use Test.isRunningTest() conditions to do things only in tests).

Now, let's consider how ChatGPT can assist with improving error handling:

Reviewing error handling logic: Provide a portion of code (like a batch execute) and ask: *"Do you see any potential unhandled exceptions in this code? How can I improve error handling?"* ChatGPT might spot something like an operation that isn't in a try-catch or a scenario like null pointer possibility.

Brainstorming recovery strategies: "If a batch job fails in the middle due to a governor limit, what are some ways to recover or retry remaining records?" ChatGPT can outline strategies like recording progress, splitting data.

Implementing a logging framework: You might ask for an example implementation: *"Show an example of logging errors from a queueable job to a custom object and sending an email to admins."* It could output a snippet illustrating inserting a custom object record and using SingleEmailMessage.

Dealing with specific error types: e.g. "In a future method that does an HTTP callout, how should I handle a CalloutException (like timeout)? Should I retry or catch and log?" ChatGPT might discuss exponential backoff strategies or at least logging and notifying. It might mention that automatic retry isn't built-in so you'd have to implement it.

Using voice or file upload for analysis: If you have a long log of an error and you're not sure where it came from, you could paste or upload the log and ask ChatGPT to help pinpoint. Or simply describe an error scenario verbally to ChatGPT: *"ChatGPT, yesterday my nightly batch job aborted because of a 'Too many SOQL queries' error. How do I investigate which part of the code caused it and fix it?"* ChatGPT could guide you to check logs or instrument the code.

Finally, consider the human aspect: when people (admins, developers, support) see error logs, they should be able to understand them. Make error messages clear and actionable. Instead of just logging "Exception in batch," include context like "Batch Job UpdateContacts failed on record Id=001xx...: FIELD_CUSTOM_VALIDATION_EXCEPTION, 'Email is invalid'". That way, someone can follow up on that record's data issue. Using ChatGPT's language abilities, you can even have it help format or improve error messages, making them more user-friendly.

E.g., *"Given an error code, how can I phrase a message for admin to understand what to do next?"*

With robust error handling and recovery strategies in place, your asynchronous Apex processes become much more reliable. Instead of mysterious failures in the night, you have logs and alerts that quickly pinpoint issues, and in some cases, self-healing mechanisms to retry or continue processing. This reduces downtime and ensures data integrity. It also builds trust with users and stakeholders, because even if something goes wrong behind the scenes, the system is designed to deal with it gracefully. *(Having addressed reliability and error recovery, our final topic in this chapter is maintaining visibility into these asynchronous processes. Next, we will discuss monitoring and auditing asynchronous Apex jobs to complete our best practices for enterprise-grade async operations.)*

MONITORING AND AUDITING ASYNCHRONOUS APEX JOBS

Once you have asynchronous Apex jobs running in your Salesforce org – batches processing data, queueables chaining, scheduled jobs firing off at intervals – it's crucial to have **visibility** into their execution. Monitoring and auditing are about knowing what's happening (or has happened) with your asynchronous processes, ensuring they are functioning as expected, and catching any issues or anomalies proactively. In this section, we'll highlight the importance of proactive monitoring for async Apex and discuss the tools and techniques to achieve it. We'll cover Salesforce's native tools like the Apex Jobs page, debug logs, and event monitoring, as well as custom monitoring solutions (dashboards, email reports, etc.). We'll also provide guidance for different roles – product owners, architects, developers, and testers – on how to use monitoring data to ensure the ongoing health of asynchronous processes.

Salesforce built-in monitoring tools:

Apex Jobs page: Salesforce provides a standard UI for monitoring asynchronous Apex. In Setup, under "Monitoring", there is "Apex Jobs". This page lists details of all asynchronous jobs (Batch Apex, Queueables, future methods, Scheduled Apex) that have been executed or are in queue. For each job, you can see:

Status (Queued, Holding, Processing, Completed, Failed, Aborted)

Job Type (Batch, Future, etc., and for batch it shows the class name)

Submitted By (the user who triggered it or whose context it runs under)

Last Processed and Total Batches (for batch, e.g., 10/20 if halfway done)

Started and Completed dates

Errors (for batch, number of errors in processing)

Admins and developers should regularly check this page, especially after deploying new async processes, to verify jobs are completing successfully. If you see a lot of Failures, that's a red flag to investigate. Product Owner Tip: For critical processes, a product owner or manager might want to occasionally review this as well (or have a simplified dashboard version of it) to gain confidence that, say, "the nightly sync job has run successfully for the past 7 days". They may not dive into logs, but a quick glance at a dashboard or report summarizing Apex job outcomes can be useful. You can't directly report on Apex Jobs via normal reports (since AsyncApexJob is not a reportable object by default), but you can query it via API or maybe use an Analytical Snapshot if needed. An easier approach: use a custom object for logging as mentioned, and report on that.

AsyncApexJob object (and BatchApexWorker): Under the hood, Salesforce stores job data in AsyncApexJob (each job) and for batch, each chunk is a BatchApexWorker record. Developers can query these objects in Apex or via API to get status. For example, you could write a Visualforce or Lightning Component to display current running jobs or the history of jobs of a certain type. This can be part of a custom monitoring console within a Salesforce app. For instance, an architect might commission a "Batch Monitor" LWC that shows all batches running today, highlighting any that failed, and maybe giving a button to resubmit failed ones if that's relevant. While the Apex Jobs page is fine for admins, a custom UI can filter specifically your business-critical jobs and show statuses in a friendly way (for example, show last run duration, records processed, etc.). You could use the job's Status and CompletedDate to compute runtime.

Email notifications for Apex job failures: Did you know Salesforce can send an email when a batch job fails due to an unhandled exception? By default, it sends to the user who submitted the job. If your integration user is running it, that user's email might get it (which could be an unattended mailbox). Ensure that these emails go to a monitored mailbox. Possibly, set the integration user's email to a group mailbox or an internal system that creates a ticket. Alternatively, you can catch errors and send emails in Apex as we discussed. But Salesforce's built-in is a backstop: any unhandled batch exception triggers an "Apex job failed" email with some info. Queueables and futures might not send these emails (I think future doesn't, queueable might not either; batch does for sure). Relying on these isn't enough, but be aware they exist.

Debug logs and Apex log levels: You can set up debug logging for specific users to capture their asynchronous operations. If an async job runs as a certain user (usually the user who triggered it, or the user who scheduled it), having a log level set for Apex code can capture the debug statements in your code. For instance, if you liberally use System.debug in your batch classes to log progress, you need an active trace flag to see those. In production, continuous debug logging can be impractical due to log limits and performance. Instead, consider using a toggle (like a Custom Setting "Debug Mode") your code checks to log only when needed. Or use a new platform feature: the Apex Replay Debugger and checkpoints can help debug after the fact, but those are more dev tools.

Event Monitoring (Salesforce Shield): If your org has Event Monitoring, there is an event type called ApexExecution. It logs details about Apex executions including CPU time, heap, number of SOQL queries, etc., for both synchronous and asynchronous executions. You could use these event logs to monitor performance trends of your async jobs. For instance, by analyzing ApexExecution events, you might catch that a particular batch's average CPU time is increasing over time (maybe data volume is growing). This can be an early warning that you need to tune the code or adjust batch size. Event Monitoring requires using the API or tools like Splunk (or Salesforce's own Analytics) to analyze logs. It's more of an advanced auditing approach, often used by performance engineers or IT departments to ensure everything is within healthy parameters.

Custom monitoring solutions

Dashboards & Reports: Create a dashboard for system operations that includes components like:

Number of async job failures in the past week (this could be sourced from your custom error log object, e.g., count of errors per day).

Average duration of key batch jobs (if you log start and end times or use AsyncApexJob data).

A list of any jobs that are currently in "Failed" status and not yet addressed.

Perhaps a trend of records processed per day by certain jobs (if you log that info).

This kind of dashboard can be reviewed by the development team or IT operations daily. It transforms raw technical data into something actionable.

Periodic audits: Have a scheduled job (meta, right?) that itself monitors other jobs. For example, a scheduled Apex that runs every day and checks: did all the expected jobs run in the last 24 hours? If not, alert. Or are there any jobs stuck in "Queued" for too long (maybe indicating a blockage)? Or for batches: if any batch job's error count > 0, flag it. This meta-monitor could create summary records or notifications. Essentially, automate the oversight to reduce manual checking. Architect Tip: In a large org, establishing a routine (automated or manual) to audit async processes ensures nothing slips through cracks. For mission-critical processes, consider having redundancy or at least high visibility when something fails. For instance, if a nightly job fails and no one notices until a user complains next afternoon that data is stale, that's not ideal. Proactive monitoring would have caught it in the morning.

Audit trails and compliance: If your org needs to keep a history of processing (for compliance, like proving that certain data was processed as required), ensure you store that info. For example, if you must demonstrate that every record was checked by a batch job, keep an audit log with job ID and date for each record (this could be heavy; maybe just a "last processed date" on the record itself is enough). Auditors might also want to know who ran a job and when. AsyncApexJob has created by, but if jobs are auto-scheduled, it's usually a generic user. If you allow users to manually trigger async jobs (like a "Recalculate Now" button that enqueues a job), you might log that user's action explicitly ("User X triggered recalculation batch at 2:05pm").

Using monitoring data to ensure health:

For Product Owners: Monitoring data can be presented in terms of business impact. For instance, "The daily order sync processed X orders, with Y failures that were retried." A product owner might not care about Apex jobs per se, but they care that all orders got synced. So tying the technical monitoring to business KPIs is useful. Maybe a report that says "Number of accounts processed by the nightly process" etc. If that number drops or spikes unexpectedly, it could indicate a problem or a change in usage.

For Architects: They should periodically review the logs and metrics to identify patterns: Are any jobs nearing limits regularly (time to refactor)? Is the volume of async calls growing as the org usage grows (maybe need to consider splitting or optimizing)? Are some jobs no longer necessary (maybe legacy that could be turned off)? Architects might also compare actual performance vs. initial design assumptions. If a job was expected to take 10 minutes and now it takes 50, figure out why – maybe data grown 5x, and plan to accommodate further growth.

For Developers: Monitoring helps in debugging issues that only appear in production data. It's often in logs that a developer finds that one weird record caused a null pointer, for example. Also, by monitoring, developers can verify that their error handling is working – e.g., they see error logs being created and can fix the underlying issues. It's a feedback loop for improving code quality.

For Testers: They can use the monitoring tools in testing phases to validate that scheduled jobs run when expected, and that batch jobs process the correct number of records. In UAT, testers might be given access to an Apex Jobs page or a custom monitor to ensure "Yes, the nightly job did what it should." In performance testing, testers/QA might use monitoring to ensure no limits were exceeded and to record how long jobs took with test data.

Now, how can ChatGPT help with monitoring and auditing tasks?

Interpreting AsyncApexJob records: If you pull some AsyncApexJob data (maybe via Workbench or API) and want to understand it, you could paste a few records to ChatGPT. *"These AsyncApexJob records show one job stuck in Holding, what does that mean and what should I do?"* ChatGPT can explain that Holding means in flex queue waiting for a slot, and that maybe you've hit the 5 concurrent batch limit.

Setting up queries or scripts: *"Give me a SOQL query to list all batch jobs in the last 7 days with their status and number of errors."* ChatGPT could produce a query: SELECT Id, ApexClass.Name, Status, JobItemsProcessed, TotalJobItems, NumberOfErrors, CompletedDate FROM AsyncApexJob WHERE CreatedDate = LAST_N_DAYS:7 AND JobType='BatchApex'.

Building a Lightning Component for monitoring: *"Outline how I could create a Lightning Component that displays currently running batch jobs and their progress."* ChatGPT might suggest using Lightning Data Service or Apex controller to query AsyncApexJob and display a table, perhaps even using Live updates (though polling might be needed).

Log analysis with ChatGPT's help: If you have a complex debug log from a failure, you can have ChatGPT summarize it (be mindful of sensitive info though). It might identify the error and which line of code (if debug logging has line numbers).

Image generation for trend charts: You could feed ChatGPT data like "jobs per day" and ask it to generate an ASCII chart. Or better yet, use a tool: since we're in a hypothetical scenario, if you had ChatGPT with a graphing plugin, you could give it a series of data and ask for a line chart image. This can help visualize trends from your monitoring data for reports or discussions.

Finally, remember auditing is not only technical but also about continuous improvement. If monitoring shows frequent small errors that don't break the system (maybe they auto-retry successfully), that's an opportunity to clean them up to reduce noise. Possibly discuss within the team: are our asynchronous processes meeting SLAs (if any)? For example, if a batch is supposed to finish by 6 AM and sometimes it finishes at 8 AM, that could affect users. Monitoring lets you catch that.

With monitoring and auditing in place, you close the loop on asynchronous Apex management. You not only build and run these processes, but you actively observe and refine them. This ensures high reliability and performance over time, even as data grows or requirements change. Moreover, strong monitoring provides confidence to stakeholders that the system is under control – issues will be caught, and successes are visible. For enterprise and SMB professionals alike, having this transparency is key to trust in automation. *(In the summary to follow, we will recap the key best practices covered in this chapter and prepare you for the advanced Apex performance optimization techniques in the next chapter.)*

SUMMARY

In this chapter, we explored the world of **Asynchronous Apex** and learned best practices to design, implement, and manage background processes in Salesforce effectively. We began with an introduction to asynchronous Apex – why it's crucial for scalability and user experience – and then looked at seven focused sections, each addressing a vital aspect of async processing:

Batch Apex Patterns and Practices: We learned how to handle large data sets by processing records in manageable chunks. Best practices like choosing appropriate batch sizes, chaining batches for multi-step processes, maintaining state cautiously (using Database.Stateful), and handling partial successes were discussed. We also saw how ChatGPT can help review batch code for efficiency and suggest improvements. Key takeaway: Batch Apex is powerful for bulk operations, but success lies in careful structuring and error handling to avoid governor limits and ensure as many records as possible are processed even if some fail.

Future Methods: Usage and Considerations: We covered the simple @future method for asynchronous tasks, particularly for callouts and quick one-off jobs. We highlighted guidelines like limiting the number of future calls in a transaction, using proper annotations (callout=true when needed), and testing future methods with Test.stopTest(). While future methods are easy to use, we noted their limitations (no chaining, limited parameter types) and emphasized using them for the right scenarios. The section reinforced that even with futures, good practices (bulkifying calls, catching exceptions) must be followed.

Effective Use of Queueable Apex: We introduced Queueable Apex as a more advanced and flexible async mechanism. Unique features like job chaining and passing complex data to jobs allow developers to create sophisticated asynchronous workflows. We discussed how to implement queueables, when to use them over futures, and how to handle errors and even chain retries in queueables. The guidance to architects and developers was to use queueables for structured background processing and to remain mindful of limits (like the 50-queueable-per-transaction limit and chain depth). We also considered how ChatGPT and even tool integrations ("operators") can assist in writing and reviewing queueable code.

Scheduled Apex for Recurring Processes: We looked at how to schedule Apex jobs for execution at specific times or intervals. This section provided best practices for writing schedulable classes and using cron expressions, as well as combining Scheduled Apex with batch/queueable to do actual work. We discussed strategies to avoid conflicts (e.g., staggering job start times) and testing scheduled jobs in a controlled way. A key point was that scheduling should be aligned with business needs — timing matters — and that one should monitor scheduled jobs to ensure they're running and not overlapping undesirably. Generating an image of a schedule or timeline was even suggested as a way to visualize and communicate the job timetable.

Managing Governor Limits in Asynchronous Contexts: Here we reaffirmed that "async doesn't mean no limits." We detailed how certain limits are higher (CPU, heap) in async contexts, but others remain the same, and new considerations (like total number of jobs) come into play. Strategies like adjusting batch scope, selective processing (filtering out unnecessary work), optimizing queries and loops, and splitting large tasks across multiple jobs were given to ensure that asynchronous jobs complete without hitting limits. The use of the Apex Limits class and proactive checks was mentioned for advanced control. The overarching idea is that maintainability and performance go hand in hand: efficient code that respects limits will be more maintainable and less likely to cause issues in production.

Error Handling and Recovery in Asynchronous Apex: Perhaps one of the most crucial sections, we focused on building resilience. We explored how to handle exceptions in async code by using try-catch, how to log errors to persistent stores, and how to alert stakeholders when things go wrong (since no user is watching these processes live). We discussed patterns for retries and recovery, acknowledging that sometimes reprocessing or manual intervention is needed. The best practices include designing idempotent operations (so that a retry doesn't harm) and ensuring partial failures are accounted for (e.g., a batch failing on some records should log those and possibly try them again or mark for review). By implementing robust error handling, we can greatly reduce the impact of inevitable issues in background jobs.

Monitoring and Auditing Asynchronous Apex Jobs: In the final section, we expanded on the need for visibility. We covered Salesforce's tools like the Apex Jobs page and AsyncApexJob object, and how to use them, as well as custom approaches like building dashboards or automated alerts. We emphasized that regular monitoring allows teams to catch failures early, understand performance trends, and continuously improve the asynchronous processes. Auditing these jobs not only helps in troubleshooting but can also be required for compliance or capacity planning. The advice was for each role to engage with monitoring: developers to debug, testers to validate behavior, architects to ensure system health, and product owners to gain confidence in the system's reliability.

Throughout the chapter, a recurring theme was **maintainability and performance.** Asynchronous Apex, by its nature, deals with potentially heavy or long-running tasks. To keep such processes maintainable:

We break the work into clear, modular units (batches or queueables) with single responsibilities.

We write clean code that is bulkified and well-documented (often using ChatGPT to assist in documentation or prompt creation).

We include logging and comments to make future debugging easier.

Performance-wise:

We take advantage of the platform's async capabilities (higher limits, parallel processing via batches) but still write efficient algorithms.

We monitor and test with realistic data volumes to ensure performance scales, adjusting where needed (e.g., reducing batch size as data grows).

We also wove in how **ChatGPT** can be a powerful assistant at every stage: from brainstorming solutions (e.g., how to handle a million records in a batch) to reviewing code for anti-patterns, from generating test scenarios to even helping format cron expressions or error messages. Moreover, we highlighted using **multiple input methods** with ChatGPT. Whether by speaking to explain a complex problem, uploading a code snippet or log file for analysis, or even asking for an illustrative diagram, these methods showcase that interacting with AI doesn't have to be just typing questions. Modern tools allow voice conversations and image generation, which can make understanding and solving problems more intuitive – all while reminding users to keep sensitive data private when sharing with AI.

By applying the best practices from each section, readers should now be equipped to implement asynchronous Apex solutions that are robust, efficient, and aligned with enterprise standards. We've seen how a combination of careful design, thorough testing, proactive monitoring, and leveraging tools like ChatGPT for support can significantly enhance the quality of asynchronous processes in Salesforce.

As Salesforce implementations grow, the importance of asynchronous processing only increases – more data, more integration, more automation. Thus, mastering these techniques is key for both Salesforce SMB professionals and enterprise architects. With these skills, you can ensure your background jobs run smoothly, your users remain happy (since heavy lifting is done behind the scenes), and your system can scale to meet future demands.

This wraps up our exploration of asynchronous Apex best practices. In the next chapter, we will advance to **Apex Performance Optimization** techniques. Building on the foundation of efficient and well-managed code, we'll review strategies to further tune Apex execution, both synchronous and asynchronous, to achieve the highest level of performance for enterprise-grade applications.

10

APEX PERFORMANCE OPTIMIZATION

Performance optimization in Apex is about more than just speed—it's about ensuring that your Salesforce applications run efficiently within the platform's governor limits and deliver a smooth user experience. Salesforce's multi-tenant architecture enforces strict limits (like CPU time and SOQL queries) to ensure no single app hogs resources. Poorly optimized Apex code can hit these limits or slow down under heavy load, leading to runtime exceptions or sluggish transactions that frustrate users and impede business operations. By investing time in performance tuning, teams can prevent such issues before they escalate.

Common symptoms of Apex performance issues include pages taking too long to save or load, asynchronous jobs (like batch processes) running slowly, or frequent errors such as *Apex CPU time limit exceeded* or *Too many SOQL queries: 101*. These warning signs indicate that the underlying code might be doing more work than necessary. For example, you might notice transactions nearing the 10-second CPU limit or see repeated queries fetching the same data. In large data volume environments, even minor inefficiencies become magnified—what's a slight delay with 100 records could turn into a serious problem with 10,000 records.

The benefits of optimizing Apex are far-reaching, positively impacting every stakeholder on a Salesforce project. **Developers** gain more reliable code that is easier to maintain and less prone to hitting limits in production. They spend less time firefighting performance bugs and more time building features. **Architects** appreciate a performant codebase because it aligns with scalable design principles and ensures the system can handle growth in users and data. Optimized code helps architects meet non-functional requirements like speed, scalability, and efficient use of resources, which are critical in enterprise environments.

Other team members also reap rewards from a well-tuned Salesforce implementation. **Testers** experience more predictable and faster execution of test suites; they can run bulk test scenarios without encountering random timeouts or governor limit failures, leading to more effective performance testing. **Product owners** and managers see tangible improvements in user satisfaction. A faster application means better user adoption and fewer support tickets about things being slow. It also gives product owners confidence that adding new features won't degrade the overall user experience or violate service-level agreements for response times.

Performance optimization is a continuous effort, not a one-time task. As new features are added or data volumes grow, Apex code that once ran smoothly might need revisiting. This is where **ChatGPT** becomes an invaluable co-pilot for Apex teams. With ChatGPT's assistance, you can quickly analyze debug logs, review code for inefficiencies, and brainstorm solutions to tricky performance problems. For instance, a developer might ask ChatGPT to examine an Apex snippet for potential bottlenecks, or an architect could prompt ChatGPT to outline a high-level approach for a caching mechanism. AI can also demystify complex Salesforce concepts or governor limits for team members, improving everyone's ability to reason about performance.

In this chapter, we'll explore seven major areas of Apex performance optimization—from effective profiling of Apex executions to advanced techniques like asynchronous processing and continuous monitoring. Each section offers practical guidance, real-world examples, and even ChatGPT prompts you can try to reinforce your learning. Let's begin with the foundation: understanding how to profile Apex execution and diagnose performance issues before attempting any fixes.

APEX PROFILING AND DIAGNOSTIC TOOLS

Before you can improve Apex performance, you need to identify where the bottlenecks are. Apex profiling is the process of gathering data about how your code executes – which operations are taking the most time or resources – so you know what to optimize. Salesforce provides robust **diagnostic tools** for this purpose, the most important being the Apex debug log and the Developer Console's performance analysis features. By using these tools, developers can visualize execution timelines, measure database query times, and see how close their code is to hitting governor limits.

Developer Console – Execution Logs and Timeline: The Salesforce Developer Console isn't just for writing code; it has dedicated panels for performance analysis. When you run Apex (for example, by executing a test or performing an action in the UI), you can open the resulting debug log in the Developer Console's Log Inspector. The console can display a **graphical timeline** of the request, show aggregated time spent in each operation (such as each SOQL query or DML action), and even summarize governor limit usage. This makes it much easier to spot, say, a particular SOQL query that took 300ms or a trigger that executed multiple DML statements. Instead of combing through raw text, the timeline view highlights gaps and spikes in execution time. Ensure your debug log levels (especially for Apex code execution) are set to a fine level, so that detailed timing information is captured. For example, setting logging to **FINER** or **FINEST** will include profiling details; without it, you might see "gaps" in the timeline where the system didn't log details for performance.

Interpreting Debug Logs: Even without the graphical interface, the raw debug log is a goldmine of information. Each log line is timestamped, allowing you to calculate how long each operation took by comparing start and end times. You'll also find sections in the log that show governor limit usage – look for the keyword LIMIT_USAGE_FOR_NS. For instance, a log might end with lines indicating **SOQL queries** used vs. the limit, CPU time consumed, heap size, etc. By searching for "LIMIT_USAGE" in a log, you can quickly see how close the execution came to critical limits. If you notice that a transaction used 98 out of 100 SOQL queries or nearly hit the 10,000ms CPU limit, that's a sign the logic needs optimization. Another flag in logs is the presence of too many repeated queries or DML operations; if you see the same query text appear multiple times, it could mean the code is querying inside a loop (a known anti-pattern we'll address in the next section). Additionally, pay attention to EXECUTION_THROWN events – an exception like *CPU time limit exceeded* clearly indicates a performance problem.

One handy trick is to use **checkpoints** in the Developer Console (or log System.debug statements strategically) to measure how long certain parts of your code take. For example, you might log the timestamp before and after a complex loop to calculate its duration. However, be careful: excessive logging can itself slow down execution and clutter your analysis. Use debug logs with a purpose – capture just enough detail to diagnose issues. Salesforce's documentation and tools like the Developer Console's "Analysis Perspective" can guide you on reading logs effectively. Over time, you'll learn common patterns: e.g., a surge in heap size usage might point to large collections being built, or a long gap between log lines might suggest waiting on a callout or a workflow processing step.

Beyond the Developer Console, you can leverage **Event Monitoring** to track Apex performance in production environments. Event Monitoring (part of Salesforce's Shield platform) can log detailed event data for Apex executions, including execution time, CPU time, and SOQL usage for every transaction. This is extremely useful for enterprise orgs where you want to continuously monitor performance. By pulling Apex Execution event logs, architects or DevOps engineers can identify slow-running transactions in production (for example, a nightly batch that is close to timing out or a user operation that consistently takes several seconds). While Event Monitoring might be more of an advanced tool requiring API access and analysis outside Salesforce (such as in a SIEM or monitoring tool), it's worth mentioning as part of your profiling toolkit for continuous improvement.

ChatGPT as a Log Analyst: Analyzing a 2 MB debug log file can be tedious. This is where ChatGPT can assist as your log analysis partner. For example, you can copy a portion of a debug log and ask ChatGPT to help interpret it. ChatGPT might not understand every Salesforce-specific detail, but it can summarize the sequence of events or even calculate time differences between log entries if you prompt it correctly. If your ChatGPT interface supports file uploads, you could upload a full debug log file for analysis. *(Always remember to remove or anonymize any sensitive information from logs before sharing them with any third-party service, including ChatGPT.)* By using AI in this way, a **tester** could quickly get a second opinion on where a script is spending most of its time, or a **developer** could verify if their understanding of the log matches the AI's summary. For instance, you might ask, "The following log excerpt shows a single transaction – can you identify any slow steps or potential bottlenecks?" and paste the log lines. ChatGPT could highlight that a particular SOQL query took longer than others, or that there were multiple invocations of the same trigger handler, etc.

Tips for Different Roles: Effective use of profiling tools should be a team effort. **Developers** should habitually run and inspect debug logs for complex use cases – not just when things go wrong, but also during development to catch inefficiencies early. It's far easier to fix a slow query or unbulkified loop in the development stage than to refactor after deployment. **Testers** can incorporate performance as part of testing by executing actions with debug logging on, then examining the logs (possibly with ChatGPT's help) to ensure no obvious inefficiencies. They might do this in a full sandbox with production-like data volumes to see how the code scales. **Architects** can define standards for what "good" looks like in a log (e.g., no single transaction should use more than 50% of any governor limit as a buffer) and can periodically review log samples from critical processes. Architects might also set up automated monitoring using event logs or custom Apex that pings if certain limits are exceeded. **Product owners** may not dive into logs themselves, but they benefit from understanding what these tools reveal. For instance, if a particular batch job is running close to a limit, a product owner can prioritize its optimization in the next sprint knowing it's a risk to uptime or user experience.

To make the most of these tools, start by profiling one end-to-end scenario in your org – for example, creating a key business record that triggers multiple automations – and see where time is spent. The next section will build on this by addressing a very common source of slowness uncovered by profiling: inefficient use of collections and loops in Apex logic.

Try it with ChatGPT: Consider experimenting with prompts like:

"Explain how to use the Salesforce Developer Console to identify performance bottlenecks in Apex code."

"Here is an excerpt from a debug log: [paste log lines]. What does it tell me about where my Apex code is spending time?"

EFFICIENT USE OF COLLECTIONS AND LOOPS

One of the most common performance killers in Apex is inefficient logic in loops. Salesforce applications often deal with collections of records, and Apex code that isn't written with bulk processing in mind can run into serious issues. Inefficient loops typically manifest as hitting the governor limit for SOQL queries or DML statements, or simply consuming excessive CPU time due to redundant operations. The mantra **"bulkify your code"** is core to Apex best practices: always write code to handle multiple records at once, and leverage collections (lists, maps, and sets) to minimize work inside loops.

Bulkify and Avoid SOQL/DML in Loops: Apex imposes a limit of 100 SOQL queries and 150 DML operations per transaction (and these limits are even lower in some contexts, like synchronous Lightning calls). A classic mistake by novice developers is running a SOQL query inside a for loop. For example, imagine a trigger iterating over 50 contacts and querying for each contact's account within the loop – that would perform 50 queries and likely hit the limit. The optimized approach is to first gather all needed Account IDs in a collection (e.g., a Set<Id>), do a single SOQL query outside the loop to retrieve all those Accounts in one go (using a WHERE Id IN :set), and then use a Map<Id, Account> to quickly lookup the account info within the loop. This way, no matter if you're processing 1 record or 200 records (the maximum in a trigger context), you stay well under limits with a constant number of queries. The same principle applies to DML: instead of inserting or updating records one by one in a loop, collect them in a list and perform one insert or update outside the loop. Not only does this avoid governor exceptions, but it's also much faster because Salesforce can bulk-process those operations.

Using Collections Efficiently: Collections in Apex (Lists, Sets, and Maps) are your allies for writing efficient code. Use **maps** to avoid expensive lookups. For instance, if you need to relate two sets of records (like matching contacts to accounts by AccountId), loading accounts into a Map<Id, Account> allows you to retrieve the relevant Account in O(1) time by key, instead of doing a nested loop (which would be O(n) per lookup). Likewise, use **sets** for membership checks instead of lists – mySet.contains(x) is generally faster than iterating through a list to find x. Also, prefer performing bulk operations on collections when possible: many Apex methods allow list inputs (for example, using Database.insert(listOfRecords). When writing triggers, always assume Trigger.new can have 200 records and code accordingly; use a single loop to process them and within that loop, avoid any operation that doesn't scale linearly.

Looping Patterns and Algorithmic Complexity: Think about the complexity of your code. A single loop that goes through 1000 records is fine, but a nested loop (loop inside a loop) that results in a million iterations (1000 x 1000) can become a CPU hog. Sometimes nested loops are unavoidable, but often you can refactor by using a map to index one of the lists. For example, instead of two nested loops to match children to parents, put the parents in a map keyed by ID, and then just loop over children and do a constant-time map lookup for each child's parent. This turns an O(n*m) operation into O(n+m). In Apex, this can be the difference between staying under the 10-second CPU limit or not. Another tip: if you have logic that can short-circuit, use it. Suppose you are searching a list for a particular record and you find it; break out of the loop rather than continuing to iterate unnecessarily. These micro-optimizations add up, especially in triggers which might run many times in bulk data loads.

Efficient use of collections also means being mindful of memory. If you load thousands of records into a list or map, you could hit heap size limits or slow down the Apex runtime with garbage collection. **Architects** should consider data volume patterns: do you really need to handle 10,000 records in one go, or can you batch or filter the data? Sometimes archiving old data or adding a selective filter (such as "only query active records") can dramatically reduce the work a loop needs to do. If extremely large loops are a necessity, you might be better off using batch Apex (covered later in the asynchronous section) which can process records in smaller chunks.

ChatGPT for Code Review: Using ChatGPT, you can perform a *pseudo-code review* of your Apex loops and collection usage. For example, a **developer** can paste a snippet of a trigger or class method into ChatGPT and ask, "Can you spot any performance issues with this code?" ChatGPT might catch things like a query in a loop or an inefficient use of a list where a map would work better. Because ChatGPT has knowledge of common Apex best practices, it can suggest bulkification where needed. This is particularly useful for less experienced developers to validate their approach. Additionally, you could describe the logic in natural language (perhaps using voice input via a ChatGPT-enabled mobile app) and ask if there's a more efficient way to achieve the same result. ChatGPT might respond with a recommendation to use a different data structure or to restructure the loops. Exploring these suggestions can be a great learning exercise.

Tips for Different Roles: When dealing with large collections and loops, **developers** should always test their code with the maximum expected batch of records – e.g., if an import job could bring 500 records into a trigger, simulate that in a test to see if it holds up. Following best practices like "no SOQL/DML in loops" is non-negotiable for enterprise-grade code. **Testers** should include a bulk data test case: for example, creating 200 records at once via data loader or a script, then verifying the operation succeeds and observing the execution time. If a bulk operation is significantly slower per record than a single record operation, that might indicate inefficiencies in loops. **Architects** must enforce design patterns that favor bulk operations – like using trigger frameworks that batch work – and should conduct code reviews focusing on how collections are used. They might also set standards such as using maps for any cross-object data fetching. **Product owners** can help by defining requirements that consider scale (e.g., "the system should handle importing 1000 contacts in one go"), prompting the team to think about performance from day one.

By efficiently utilizing collections and writing smart loops, you eliminate a huge class of performance problems. Code that is bulkified and well-structured not only runs faster, but also is resilient to growing loads. Next, we'll address another key strategy for speeding up Apex: reducing database interactions through strategic caching.

Try it with ChatGPT: You can also leverage ChatGPT by asking things like:

"Review the following Apex code for any performance issues: [paste your loop or trigger code]."

"What are some ways to optimize nested loops in Apex when dealing with thousands of records?"

REDUCING DATABASE CALLS WITH STRATEGIC CACHING

Reading from the Salesforce database (SOQL queries) is a relatively expensive operation. If your Apex code repeatedly fetches the same data, it incurs a cost each time – not just in CPU time, but also against governor limits. **Strategic caching** is the practice of temporarily storing frequently used data in memory or in a faster storage layer so that subsequent requests can retrieve it quickly without an extra database trip. In a platform with strict limits on queries, caching is a powerful technique to improve performance and stay within limits.

In-Transaction Caching (Static Variables): Apex transactions are short-lived, but you can cache data within the lifespan of a single transaction using static variables. For example, if you have a utility class that provides configuration data (say, a list of valid product codes from a custom object), you can implement it such that the first time it queries the database, it stores the results in a static variable. On subsequent calls in the same transaction, the method simply returns the cached list instead of querying again. This pattern is often used in triggers that might be invoked multiple times in one transaction (for instance, recursive triggers or multiple triggers on related objects). By using a static cache, a **developer** ensures that each unique piece of data is fetched from the database at most once per transaction. Static variables reset after the transaction, so this is not a cross-transaction cache – it's purely to avoid redundant work within one execution context.

Platform Cache for Cross-Transaction Caching: Salesforce offers Platform Cache as a feature for more persistent caching needs. Platform Cache is essentially a key-value storage that lives in Salesforce's infrastructure and can be accessed from Apex. There are two types: Session cache (which is per-user session) and Org cache (shared across the org). If you have data that many users or processes need frequently – for example, a set of configuration records or a result from a complex calculation – you can store it in the Org cache so all Apex requests can reuse it. Reading from Platform Cache is much faster than a SOQL query, and it doesn't count against SOQL governor limits. A typical use case might be caching reference data that changes infrequently (like exchange rates updated nightly, or a list of countries). The Apex Cache namespace provides methods like Cache.Org.put(key, value) and Cache.Org.get(key) to interact with the cache. Before using Platform Cache, you need to allocate cache space in Setup. Keep in mind that cached data can expire or be evicted (for example, when the cache is full, least-recently used data might be purged), so your code should always handle a cache miss by fetching from the database and then repopulating the cache.

Caching Strategies and Considerations: Caching introduces the classic trade-off between performance and data freshness. It's important to have an **invalidation strategy**. For instance, if you cache a list of Product__c records, and someone updates a product, how will your cache know to refresh? In some cases, you might use a short Time-to-Live (TTL) so that the cache automatically refreshes periodically. In others, your code can explicitly invalidate the cache when it detects a change (e.g., after a DML update to that object, remove or update the cached entry). A simple strategy is to use a naming convention for cache keys that includes a version number or date, so that deploying new code or changing a key can implicitly start fresh. Another consideration is **cache size and limits.** Salesforce imposes limits on Platform Cache (measured in MB allocated). If you try to cache too much data, you might exceed the space and start evicting entries constantly, which hurts the benefit of caching. Thus, cache only what provides value – typically data that is expensive to compute or query, and is reused often. **Architects** should identify these opportunities: for example, if a homepage component queries the same summary data on every page load for every user, that's an ideal candidate to cache once and reuse.

It's also worth noting that Salesforce has some built-in caching behavior for certain things. For example, if you use **Custom Settings** (especially the list type) or **Custom Metadata Types**, reading them in Apex is generally cached for the transaction and doesn't count against SOQL limits. That makes them a good choice for storing configuration data that you need to access in Apex frequently. However, those are more for small data (like config flags or constants). For larger data or query results, Platform Cache or custom static caching patterns are more appropriate.

Practical Example – Caching to Reduce SOQL: Suppose you have a trigger on Opportunity that needs to verify something about the Opportunity's Account each time. Without caching, if 100 opportunities are updated in bulk, the trigger might query the Account 100 times (worst case, if written poorly). A bulkified trigger would query all needed Accounts once, but we can do even better: if multiple opportunities share the same Account, you don't want to query that Account twice in one transaction. A static map cache can ensure you only query each Account once. Going further, if this check happens frequently across transactions (not just in one trigger execution), storing some Account info in the Platform Cache (Org scope) means subsequent transactions might not need to query the database at all. This can drastically reduce overall query count and latency for the operation. In enterprise scenarios, caching can be the difference between an application that scales to thousands of users versus one that starts hitting governor limits or response time SLA breaches under load.

ChatGPT as a Caching Coach: Designing a caching strategy can be tricky – you have to think about what data to cache, how long to cache it, and how to handle updates. ChatGPT can help brainstorm these decisions. For example, you can ask ChatGPT, "What's a good approach to caching frequently used reference data in Salesforce Apex?" and it may suggest using Platform Cache and discuss pros/cons of various approaches. You could also have it review a caching implementation. Perhaps you wrote an Apex class that caches query results in a static map; you could present that to ChatGPT and ask if there are any pitfalls (ChatGPT might remind you about static variables being per transaction, or warn about memory usage). This kind of design validation is useful for **developers** and **architects** alike – it's like having an architecture review buddy available anytime. Additionally, ChatGPT's capabilities aren't limited to text: with certain plugins or tools, you could even generate a simple diagram of your caching approach. For instance, using an image generation tool via ChatGPT, you might create a diagram that shows "User -> Apex -> Cache -> Database" data flow to help communicate the idea to stakeholders. This multi-modal use of ChatGPT (text and images) can help ensure everyone on the team, including **product owners**, understands how caching is improving performance.

Tips for Different Roles: Implementing caching requires cooperation across the team. **Developers** should be diligent in using existing cache utilities rather than re-querying data. If a common utility method for caching is provided, use it everywhere it applies so you get consistent benefits. They should also write logic to handle cache misses gracefully (fetch from DB and update cache). **Testers** need to test scenarios both with and without cache available – for example, test the first run (cold cache) and a subsequent run (warm cache) to see the performance difference and ensure correctness. They should also test cache invalidation; e.g., if data is updated, does the next transaction get the new data or is it erroneously serving old data from cache? **Architects** have to decide what data is worth caching and ensure proper governance of the cache (like allocating enough Org cache space, and monitoring cache hit rates if possible). They may set guidelines such as "cache any reference data that's used in more than 5% of transactions" or similar. **Product owners** should be informed about the caching strategies as well, because it might have functional implications (for instance, "Changes to exchange rates will take effect 15 minutes later in the app due to caching"). Knowing this, they can adjust requirements or at least set the right expectations with users and stakeholders.

By reducing database calls through thoughtful caching, Apex applications can achieve significant speed-ups and avoid hitting query limits. Caching is a powerful complement to efficient coding practices, ensuring that the expensive operations are done once and reused. Next, we turn to another aspect of performance: making sure our triggers and automation logic run as efficiently as possible.

Try it with ChatGPT: Some prompt ideas for exploring caching with AI:

"Provide an Apex code example that uses Platform Cache to store and retrieve a list of records."

"What are best practices for cache invalidation in Salesforce Apex?"

OPTIMIZING TRIGGER AND PROCESS AUTOMATION PERFORMANCE

Salesforce record-triggered automation – whether implemented as Apex triggers, record-triggered Flows (the newer declarative option), or older tools like Process Builder – is often a hotspot for performance issues. These automations run whenever data changes, so inefficiency here can impact every save operation and even chain into other processes. Optimizing triggers and processes involves ensuring they execute as quickly as possible, do only the necessary work, and handle bulk operations gracefully.

Best Practices for Apex Triggers: Apex triggers should be lean and efficient. This means a few things in practice. First, follow the one-trigger-per-object design principle: instead of multiple triggers (each possibly doing different things) on the same object, consolidate logic into a single trigger entry point. This avoids redundant overhead and gives you control over execution order. Use trigger handler classes to organize logic and make it testable. Second, as covered earlier, make the trigger logic **bulkified.** Trigger.new can contain up to 200 records, so any query or operation inside the trigger must handle a collection – never assume a single record context. Third, avoid heavy computations or callouts directly in a trigger. If a trigger needs to perform something costly (like making a web service call or processing a large dataset), consider offloading that to an asynchronous context (e.g., using a future method or Queueable, which we'll discuss in the next section). This keeps the trigger's own execution time short. Fourth, guard against recursive trigger executions. For example, if your trigger updates a record of the same type (causing the trigger to fire again), you can use a static boolean flag to ensure you don't enter an infinite loop. There are also trigger frameworks available (like the fflib Apex Common library) that help structure trigger logic efficiently and only run specific pieces when necessary.

Flows and Process Builders: Salesforce has been encouraging the use of Flows for declarative automation. From a performance standpoint, modern **record-triggered Flows** can be quite efficient, especially the "before save" flows for simple field updates (they are known to be faster and avoid extra DML, acting similarly to before triggers). However, complex Flows can still accumulate processing time and SOQL queries much like Apex. One best practice is to minimize the number of separate flows on the same object. If you have, say, five record-triggered flows all on Account update, each one incurs its own overhead. It may be better to combine logic into fewer flows (or even a single one with multiple branches) to reduce the per-transaction overhead. Also, just as with triggers, Flows should be built with bulk operations in mind: use collection processing elements (like Loop and Assignment to collections) rather than operating on one record at a time, and avoid SOQL queries inside loops in Flow formulas or actions. Salesforce has tools like the **Flow Execution logs** and debug details that can help identify which element of a Flow took the most time. If a Flow is getting too complex or hitting limits, that might be a sign to move part of the logic to Apex where you have more control over optimization (for instance, an after-save Flow could invoke an Apex @Invocable method that is optimized for bulk data).

Reducing Automation Redundancy: Sometimes performance issues come not from one trigger or flow being slow, but from having *too many things* happen on the same event. Consider an update to an Account that triggers a cascade: an Apex trigger updates related Contacts, which in turn fire their own triggers, plus a Process Builder process sends an email, and a Flow does a field update. All of this happens in one transaction. To optimize, assess if all those steps are necessary or if they can be combined. Maybe the Contact update could be done in the Account trigger itself to avoid firing separate Contact triggers, or vice versa. Or perhaps some of those actions could be moved to an asynchronous context or scheduled job if they don't need to happen immediately. **Architects** play a key role here – they need to have a birds-eye view of all automation on an object and ensure there is a clear strategy (e.g., "All logic on Case updates is handled in a single Apex trigger and helper class; we don't use Process Builder on Case to avoid duplication"). Also, prefer configuration over customization only when it doesn't compromise performance: for instance, an admin might add multiple declarative automations because it's easy, but a developer could potentially merge those into one efficient Apex solution if needed for speed.

Tooling to Analyze Trigger Performance: How do you know if your triggers or flows are slow? Aside from using debug logs (which will show how much CPU time a trigger took, or how many queries it ran), Salesforce's event monitoring can capture **Apex Trigger events** and **Flow Execution events** with timing info. There are also third-party tools and Salesforce features like the Apex Performance Collector (in the Tooling API) that can measure CPU time usage. As a quick check, if you suspect a particular object's automation is causing slow saves, try doing a simple update of a record of that object with debugging on – the log's timeline will reveal if the trigger or flow took significant time. Additionally, the Health Check might flag if you have too many automations on one object. Within the Developer Console, the Execution Overview panel will list each trigger execution and you can see their cumulative resource usage.

ChatGPT for Automation Optimization: You can describe your trigger or flow logic to ChatGPT and ask for optimization tips. For example, a **developer** might say, "I have a trigger that does X, Y, and Z; it sometimes hits CPU limits when 200 records are processed. How can I improve it?" ChatGPT could suggest combining queries, using maps, or moving part of the work to a future method. If you have a Flow, you could even outline its steps in text and ask ChatGPT if there are ways to simplify the logic. In the near future, we might even see AI tools integrated directly with Salesforce ("Operator" modes where an AI agent can analyze your org's setup). For instance, imagine a tool that inspects all your flows and triggers and then advises where you have overlap or inefficiencies – ChatGPT's underlying technology could power such an **operator.** In the current state, you can still manually provide details to ChatGPT. Another use case: ask ChatGPT to convert a complex Process Builder flow into Apex code for better performance, or vice versa, to ensure you're using the right tool for the job. By iterating with ChatGPT on designs, both **architects** and developers can validate whether an approach might cause unintended performance consequences.

Tips for Different Roles: For **developers**, always keep the "bulk and efficient" mindset when writing triggers. Use frameworks or templates so that every trigger in the org has a consistent, optimized structure (e.g., all do bulk processing, all avoid recursive loops, etc.). For **admins or low-code developers** building flows, get familiar with the Salesforce provided best practice: e.g., use one before-save flow for trivial field updates (for speed), and limit the number of after-save flows. They should test flows with bulk records in debug mode to ensure they don't blow past limits. **Testers** should include scenarios that cover combined automations – for instance, testing an update that will fire both a trigger and a flow, to see the end-to-end effect. They should measure the time it takes and monitor for any limit exceptions in the log. **Product owners** should be aware that sometimes simplifying requirements (or accepting certain technical solutions) can vastly improve performance. For example, if a requirement can be met either with four different automations or one combined automation that's a bit more complex, it might be worth opting for the one combined path for the sake of maintainability and speed. Having open conversations about trade-offs (like immediate vs. slightly delayed processing of a non-critical task) allows the team to decide on approaches that balance functionality and performance.

By streamlining triggers and process automations, you ensure that when data changes in Salesforce, it happens swiftly and efficiently. Users will notice faster save times and fewer weird errors from limits. Now that we've optimized internal logic, let's focus on improving performance when Apex interacts with external systems through web services and callouts.

Try it with ChatGPT: Prompts to consider for optimizing automations:

"How can I refactor multiple Process Builders on the same object into a single efficient flow or trigger?"

"My Apex trigger is hitting CPU limits with large batches – what strategies can I use to optimize it?"

IMPROVING APEX WEB SERVICE AND CALLOUT PERFORMANCE

Many Salesforce implementations involve integration with external systems – whether calling out from Apex to a web service (REST/SOAP API calls) or exposing Apex methods as services for others to call. These integrations can become performance bottlenecks if not designed carefully. Unlike purely internal operations, web service interactions introduce network latency and external processing time, which can significantly slow down an Apex transaction.

Optimizing Outbound Callouts: When Apex calls an external service (using classes like HttpRequest and Http for REST, or generated Apex stubs for SOAP), it's important to minimize both the number of callouts and the waiting time per callout. Each callout has overhead: establishing connections, transmitting data, etc. If your code makes many small callouts in a loop, the delays add up. A better strategy is often to batch requests if the external API supports it – for instance, sending one request to get information about 50 records rather than 50 requests for one record each. Additionally, Apex enforces a limit of 100 callouts per transaction and a cumulative 120 seconds of callout time. This means if you make a lot of callouts or ones that take a long time, you risk hitting those limits. To stay safe, set reasonable timeouts on your HttpRequest (for example, if the external service usually responds in 2 seconds, don't leave the timeout at the default 120 seconds; maybe set it to 5-10 seconds to fail fast and recover). Also, handle exceptions gracefully – a callout that fails or times out should not crash your whole trigger; catch the exception and decide if you can retry later or log an error for follow-up.

If the logic requires calling out for each record (for instance, validating an address via an external API for each contact in a batch), consider using **asynchronous techniques.** One approach is using future methods or Queueable Apex for callouts: these run in a separate transaction where you have more breathing room (and they won't hold up the user's transaction). Another approach, if in a user-facing context like a Lightning page, is to use the **Continuation** class (available for Visualforce, Aura, and LWC) which allows Apex to make a callout without blocking the user interface – the response is handled asynchronously via a callback, and the user isn't stuck waiting. Continuations can even run multiple callouts in parallel (up to three) and then return when all are done, which can speed up composite requests. The key idea is not to do long callouts in the middle of a synchronous process that the user is waiting on. By decoupling them, either via Continuation or by processing after a save (async), the overall user experience improves.

Another technique for performance is **caching external data** (which overlaps with the caching discussion earlier). If your Apex regularly calls out to get data that doesn't change often (like an exchange rate, or a list of holidays), you might store that data in Salesforce (maybe in Custom Metadata or Platform Cache) so you don't call out every time. This reduces dependency on the external service's latency and improves consistency. Just remember to refresh the data periodically.

Optimizing Inbound Calls (Apex as a Web Service): When external systems call into Salesforce (using Apex REST/SOAP endpoints or the standard API), performance optimizations largely mirror what we do for internal triggers. Make sure the Apex methods behind the API are bulkified if they accept lists of records. Encourage clients to batch requests – for example, design your REST API to accept up to 200 records in one call, instead of requiring one call per record. This way, the overhead of API calls (which includes Salesforce authentication, etc.) is amortized over more data. On the Salesforce side, treat inbound requests like triggers: avoid redundant queries, and if the logic is complex, consider processing asynchronously. For instance, a REST API call that kicks off a long process might simply enqueue a Batch job and immediately return a status ID, rather than doing all the work while the HTTP connection is open.

Managing External System Latency and Limits: A challenge with integrations is that you might be at the mercy of external system performance. To mitigate this, design for timeout and retries. A common pattern is an **exponential backoff** for retries: if a callout fails, retry after a short delay (maybe via a scheduled retry or storing a record to try again) rather than instantly, to give the external system time to recover. Also be mindful of API rate limits on the other side – hammering an external API can throttle that side and ironically slow your calls more. If you need to make a high volume of calls, work with the external system's team to find a solution (maybe a bulk API or an asynchronous callback model).

From Salesforce's perspective, one special limit to note is the **Concurrent Long-Running Request limit.** If more than 10 Apex requests (such as Visualforce/Aura actions or API calls) are running longer than 5 seconds at the same time, Salesforce will start queuing additional requests. Heavy callouts can trigger this if many users perform an action that calls out simultaneously. Using the asynchronous patterns mentioned can help avoid hitting this, since future/queueable calls won't count against the synchronous limit that affects user-facing actions.

ChatGPT for Integration Design: ChatGPT can be very helpful when working through integration performance questions. For example, you might ask, "How can I speed up a Salesforce callout to an external API that is taking too long?" and it could suggest some of the strategies above (batching, caching, async). If you have the interface details of an API (like a snippet of a WSDL or JSON response), you could even show it to ChatGPT and ask for ideas to optimize the usage of that API. Perhaps the response includes a lot of data and ChatGPT points out you could request a smaller subset. For more hands-on help, you could use ChatGPT to generate a template for a Continuation-based callout, saving you time looking up the syntax. Think beyond just text as well: if you have a sequence diagram of how data flows between Salesforce and an external system, you could describe it or even sketch it and have ChatGPT analyze it (some advanced tools might even let you upload the diagram image for analysis). This multi-modal interaction means even if you're not a developer, as a **product owner** or **architect** you could ask high-level "what-if" questions about integration performance and get understandable answers. For instance, "What if the external service is down? How should our Salesforce app respond?" – ChatGPT might recommend a graceful degradation or user message.

Tips for Different Roles: Developers implementing callouts should pay attention to governor limits specific to callouts (100 callouts, 120s total time) and code accordingly. They should also use tools like Named Credentials for authentication (which can simplify token management and avoid unnecessary callouts for login). **Testers** should include integration scenarios in performance testing – for example, test how long an end-to-end process takes with the external service actually being called, and simulate the service being slow or returning large payloads. If the external API has a sandbox or test mode, testers can use that to measure performance. They should also test error paths (timeouts, service down). **Architects** must consider integration design in the overall system architecture: sometimes the solution might be to redesign the process (e.g., instead of making Salesforce call an external service for each request, have the external system push data to Salesforce periodically, which can be more efficient). Architects should also make sure there's monitoring on integration performance, possibly by logging callout durations or using event monitoring (Salesforce provides an Apex Callout event log type) so that any slowdown can be detected. **Product owners** should set realistic expectations in requirements: if an external system has a known response time of 3 seconds, expecting a real-time Salesforce save that includes that call to be sub-second isn't realistic. They should also be looped in on decisions like caching external data or doing work asynchronously, since those can have user-facing implications (e.g., "Data will show up a few minutes later"). Clear communication ensures that performance optimizations align with business needs.

By making integrations efficient and resilient, you ensure that external calls don't become the Achilles' heel of your Salesforce app. With both internal and external performance considerations covered, we can now discuss how to leverage asynchronous processing to handle large workloads and keep things snappy.

Try it with ChatGPT: You might find these prompts useful for integrations:

"Give an example of using Continuation in Apex for a long-running callout."

"My Apex callout to an external API is timing out occasionally. What are some strategies to make it more robust and faster?"

LEVERAGING ASYNCHRONOUS PROCESSING FOR PERFORMANCE

Asynchronous Apex is a crucial tool in the performance optimization toolbox. By running code in the background, outside of the immediate transaction that a user initiates, you can handle larger workloads and longer processes without making the user wait or risking governor limit failures in a synchronous context. Salesforce provides several asynchronous mechanisms – including Future methods, Queueable Apex, Batch Apex, Scheduled Apex, and Platform Events – each suited to particular scenarios.

Why Asynchronous = Better Performance (When Used Right): In a synchronous context (like a trigger or a Visualforce/Aura action), Apex has stricter limits: for instance, a 10-second CPU time limit and lower heap size limit. In an asynchronous context, some limits are higher (e.g., 60-second CPU time limit for Batch jobs, larger heap, more SOQL queries allowed in batch per 200 records processed). Asynchronous processes also don't count against the concurrent long-running request limit that affects user threads. This means if you have a chunk of work that is heavy, offloading it to async not only avoids frustrating the user with a spinner, but also gives that work more breathing room to complete. The trade-off is that the user doesn't get the result immediately – but for many tasks (like sending confirmation emails, syncing data to another system, recalculating a big report) that is an acceptable or even preferred approach.

Future vs Queueable vs Batch: *Future methods* (annotated with @future) are the simplest form of async – fire-and-forget with some limitations (no return values, not chainable, and limited to primitive parameters). They're great for quick things like calling an external service or doing an update in the background. *Queueable Apex* is a more advanced form that allows more complex data types, chaining (one queueable can enqueue another), and is generally replacing many future use cases. If you have a sequence of tasks or need to process records progressively, queueables are helpful. *Batch Apex* is designed for processing very large data sets by breaking them into chunks (batches of up to 2000 records, though 200 is a typical chunk). Batch jobs can process up to millions of records by iterating chunk by chunk, committing along the way to stay within limits. They also run with higher limits as noted. You'd use Batch Apex for things like nightly recalculations, large data migrations, or any scenario where you need to handle more than a few thousand records in one go. *Scheduled Apex* simply allows running Apex at specified times (which under the hood runs as a batch or queueable at that time), useful for off-peak processing. *Platform Events* and the newer *Change Data Capture* are another way to handle things asynchronously: a record change can fire a platform event, which an Apex trigger can subscribe to and handle in a separate transaction, often used to decouple processing from the main transaction.

Using Async to Improve User Experience: A common pattern is to move non-critical post-save logic to async. For example, an Opportunity trigger might need to update related objects and send some notifications. The critical part is maybe updating a couple of fields on the Opportunity itself (which can be done in the trigger quickly), but the rest – updating related records and sending notifications – could be done in a Queueable. The trigger can enqueue the Queueable and finish fast. The queueable then executes independently, perhaps a few seconds later, handling the heavier tasks. From the end-user's perspective, their record saved quickly. The updates and notifications happen shortly after, which in many cases is fine. Of course, this requires designing for eventual consistency; the system should tolerate that these related updates aren't immediate. Another example: processing uploaded data. If a user uploads 5000 records via a custom Visualforce or Lightning component, instead of inserting all 5000 in one go (which would likely fail or be slow), you might accept the file, record a reference, and then start an asynchronous Batch job to process in the background. The user could be notified when it's done via a notification or they might check back later.

Async Considerations and Limits: Asynchronous Apex has its own limits. For instance, you can only enqueue 50 queueable or future calls per transaction (so you shouldn't, for example, in a trigger launch a future for each of 200 records – that would hit 200 futures which is not allowed). If you have a scenario where many async jobs need to be spawned, consider merging tasks or using Batch Apex which can inherently handle chunks of records. There's also a limit of 250,000 batch executions per 24 hours (that's plenty in most cases, but keep in mind for extreme scale). Also, chaining too many queueables sequentially could delay overall completion (since each queueable might wait in line to run). **Architects** need to design an async architecture that scales – sometimes using a combination of patterns. For example, a common pattern for extremely large jobs is to use a Scheduled Apex to kick off batch jobs in serial or even a stateful batch that re-enqueues itself until done.

Monitoring Async Performance: Just because something is async doesn't mean you can ignore its performance. Long-running async jobs could pile up. Salesforce provides some tools: for batch jobs, you can get logs for each batch execution. Event monitoring also has an event type for Batch Apex executions. You might build custom logging into your batch/queueable classes to record start and end times, and any key stats (like records processed). In unit tests, you can use Test.startTest() and Test.stopTest() to actually run queued jobs and batches to verify they do what you expect within limits. It's a good idea to test with near-governor-limit scenarios. For instance, in a test, generate 200 records and run your batch against it to ensure it stays within CPU time. If it doesn't, you know to optimize the batch's execute logic (maybe process fewer records per batch or optimize queries).

ChatGPT for Async Patterns: If you're unsure which async mechanism to use, you can ask ChatGPT something like, "I need to process thousands of records nightly and update related data, should I use batch Apex or queueable, and why?" ChatGPT can compare the options and help you make a decision. It can also assist with code examples – for example, "Show me a simple Apex Batch class structure for processing Contacts" or "How do I chain queueable jobs and pass data between them?" Often, copying a code snippet of your batch or queueable and asking for improvements can yield suggestions (maybe ChatGPT will point out that you are querying inside the execute method instead of in the start method of the batch, which is a common mistake). This kind of review is helpful for developers getting used to asynchronous Apex. On the **tester** side, you could even ask ChatGPT how to write a unit test for a batch class (covering the start, execute, finish). Additionally, forward-looking uses of AI might include an "operator" that monitors your async job queue and predicts bottlenecks – as an analogy, ChatGPT could be thought of as that advisor if you feed it the right context (like descriptions of your scheduled jobs and their runtimes).

Tips for Different Roles: Developers should become comfortable with at least the basics of future, queueable, and batch Apex. Knowing when to use each is key: e.g., use batch for bulk data processing, queueable for sequential tasks or when you need more complex types than future allows, etc. They should also handle exceptions in async code carefully (since it might not have a user watching it, perhaps implement a retry mechanism or alerting when a job fails). **Testers** must ensure that asynchronous processes eventually do what they're supposed to – often by checking final outcomes in tests or using Test.stopTest to force execution. In integration testing (like UAT), testers might trigger processes that spawn async jobs and then verify outcomes after those jobs have had time to run. **Architects** should set guidelines for async usage: for example, "Any operation expected to take more than 2 seconds or 50k records must be async." They also need to coordinate schedules – if you have 10 batch jobs all scheduled at midnight, maybe stagger them to avoid contention. Salesforce has an Async Apex Flex Queue for managing batch job scheduling, which architects/admins can use to prioritize jobs. **Product owners** should be aware that some features will be implemented to run in the background. This might affect how requirements are phrased (instead of "user clicks and data is updated immediately," it might be "user clicks and data will be updated within 5 minutes"). They should communicate to end-users when processes are asynchronous, perhaps by showing notifications or status updates ("Your data is being processed, you'll get an email when it's done"). Managing expectations helps ensure that the benefits of async (throughput and reliability) are appreciated rather than seen as a delay.

Asynchronous processing is a powerful way to maintain performance as workloads grow. By smartly scheduling and offloading work, you keep the system responsive and avoid running into limits in the middle of business operations. Finally, let's discuss how to keep an eye on performance over time and continuously optimize even after your code is deployed.

Try it with ChatGPT: To dig deeper into asynchronous Apex, try asking:

"When should I use a Queueable Apex job instead of a Batch Apex job, and vice versa, for performance?"

"Help me convert this synchronous Apex process into an asynchronous one to improve performance: [describe or paste pseudocode]."

CONTINUOUS PERFORMANCE MONITORING AND OPTIMIZATION

Performance tuning isn't a "set it and forget it" task – it requires ongoing vigilance. Salesforce orgs are dynamic: data grows, user behavior changes, new features are added, and Salesforce itself evolves with seasonal releases. A piece of Apex that runs fine today might become a bottleneck a year from now due to a tenfold increase in data or a change in how it's used. That's why establishing a culture of continuous performance monitoring and periodic optimization is essential for long-term success.

Monitoring in Production: In production, you likely won't have developers staring at debug logs all day. Instead, you should leverage monitoring tools and analytics. Salesforce's **Event Monitoring**, as mentioned earlier, is a powerful source of data. By capturing events like Apex Execution, Apex CPU Time, and Apex Callout, you can feed this data into an analytics tool (even Salesforce's own Einstein Analytics or external systems like Splunk or New Relic) to identify trends. For example, you might track the average execution time of a certain Apex API or the 95th percentile CPU usage of all transactions on a weekly basis. If you see those numbers creeping up over time, it's a sign that something needs attention. Another approach is to build **custom lightweight monitoring:** for critical processes, you can instrument the code to log performance metrics to a custom object. For instance, after a batch job finishes, have it record how many records it processed and how long it took. Over time, this record can show if the processing time per record is increasing.

Periodic Performance Audits: It's a good practice for **architects** or leads to schedule periodic performance reviews, perhaps every few sprints or each quarter. In these audits, the team can review recent changes: Did any new feature push the limits? Are there any new warnings or limit exceptions in the logs? Often, testers or admins can help gather anecdotal evidence too – maybe users started complaining that a certain page is slower. Use that as a trigger to investigate. Salesforce's multi-tenant environment means sometimes performance issues surface as limit exceptions (which are obvious), but other times they may simply be slow execution that doesn't break a limit but still degrades experience. By regularly reviewing both hard data (event logs, custom logs) and soft signals (user feedback), you can catch issues early.

Adapting to Change: Continuous optimization also means being ready to refactor code as better techniques or platform improvements come out. For example, if Salesforce releases a new indexing feature or a more efficient way to query big objects, it might be worth revisiting an old query that was slow and adjusting it. Likewise, keep an eye on Apex enhancements – perhaps a future Salesforce release introduces a more efficient asynchronous mechanism or increases certain limits (occasionally Salesforce does raise limits or provide opt-in higher limits for specific use cases). A recent example is the introduction of tools like Scale Center and ApexGuru (in 2024) that help identify org-wide performance issues and recommend optimizations. Being up-to-date on platform updates (through release notes, developer blogs, and community discussions) ensures you're not missing opportunities to improve performance with new features.

Automating Performance Checks: In a mature DevOps setup, you can incorporate performance checks into your pipeline. For instance, before deploying to production, run a suite of Apex tests that not only verify correctness but also measure execution time for critical operations. While Apex tests don't directly give you timing metrics, you can infer performance by using large data volumes in tests and ensuring they complete within an acceptable time. There are also third-party tools that can run load tests against Salesforce (simulate many users or repeated actions) – these can be part of staging environment testing. If something starts taking 2x time compared to last release's benchmark, it should flag a warning. In an automated way, you might use scripts to fetch event log data nightly and if CPU times spike, alert the team. This is where an **AI assistant** could come in handy: one could imagine using ChatGPT's capabilities to summarize daily performance anomalies from logs or even to analyze code changes that might have caused them. For example, feeding a diff of Apex code to ChatGPT and asking, "Could any of these changes impact performance significantly?" can give a quick sanity check.

ChatGPT as a Continuous Assistant: Performance optimization can benefit from an extra pair of eyes – or in this case, an AI's perspective – over time. As new team members join or new code gets written, you can leverage ChatGPT to educate and enforce best practices. For instance, a new developer might draft a trigger and then use ChatGPT to review it for any obvious performance anti-patterns, catching issues before they ever hit production. Team-wide, you might set up a knowledge base (with ChatGPT Custom Instructions or an enterprise AI assistant) that knows your company's performance guidelines and can answer questions like "What's our standard approach for handling large data updates?" ensuring consistency. Some organizations integrate AI bots into their Slack or Teams; you could have a "Salesforce Performance Bot" (powered by something like ChatGPT) where anyone can ask, "Have we had any Apex CPU limit errors this week?" or "What was the slowest transaction today?" and it could retrieve that info from monitoring systems and respond. While this is an advanced setup, it highlights that ChatGPT and similar AI can be woven into the monitoring process, not just used ad-hoc when someone remembers to ask.

Fostering a Performance Culture: At the end of the day, tools and AI aside, maintaining peak Apex performance is about instilling it into the team's mindset. **Developers** should feel a sense of ownership about the efficiency of their code, and code reviews should include performance considerations. **Testers** should incorporate performance scenarios in test plans. **Architects** should set clear non-functional requirements (e.g., "all save operations must complete in under 2 seconds for up to 500 records") and ensure designs meet them. **Product owners and**

managers should allocate time in sprints not just for feature development but also for performance improvements and tech debt reduction. It's much cheaper to address an inefficiency early than to overhaul a module under fire from production issues.

Salesforce itself is introducing more support for performance management. Tools like **Scale Center** and **ApexGuru** (announced in late 2024) aim to help admins and developers identify performance issues such as frequent record locking, high concurrent usage, or CPU spikes, and even suggest optimizations. This trend means the platform is providing more built-in guidance, but your team's own attentiveness remains crucial.

One often overlooked aspect of continuous optimization is data management. As data sets grow, certain operations naturally slow down. **Archiving old data** or purging unnecessary records can improve query performance dramatically. Ensure that reports or batch jobs are scoped to only the data that matters (for example, active records or last N years of data). If an object is expected to accumulate millions of records (a "large data volume"), work with Salesforce support to add appropriate indexes or consider design strategies like skinny tables or partitioning data by criteria (using separate record types or custom objects for historical data). By managing data growth, you prevent slowdowns that no amount of code tweaking can fix.

Finally, remember that performance mindset when building new features. It's easier to build it right from the start than to retrofit performance later. Use the lessons from this chapter as a checklist during design and code review: Are we bulkified? Are we caching? Did we consider async processing? Asking these questions upfront is the best form of continuous optimization, because it embeds best practices into the DNA of your application development process.

By continuously monitoring and proactively optimizing, you'll keep your Salesforce application running smoothly as it evolves. This ongoing effort goes hand-in-hand with understanding and respecting Salesforce's platform limits. A deep understanding of Salesforce's platform limits is the next cornerstone to address. Knowing and managing those limits is closely related to performance, and it will further empower you to build high-performing Apex solutions as your journey continues.

Try it with ChatGPT: As you continue optimizing, you might ask:

"What metrics should I monitor to ensure my Salesforce Apex code maintains good performance over time?"

"Create a checklist for an Apex performance review that our team can use every quarter."

SUMMARY

Throughout this chapter, we explored a comprehensive toolkit for Apex performance optimization. We began with **profiling and diagnostic tools**, emphasizing that you can't improve what you don't measure. Using debug logs and the Developer Console's analysis features helps pinpoint exactly where your code spends time or approaches limits. Next, we discussed writing efficient **collections and loops**, underscoring the importance of bulkification – handling many records at once and eliminating wasteful queries or nested loops. We then moved to **strategic caching**, where we saw how storing frequently needed data can cut down on repetitive database calls and significantly speed up transactions.

Optimizing **triggers and process automation** was another focal point – ensuring that the business logic running on data changes is streamlined and avoids doing unnecessary work (or doing necessary work in a clumsy way). We looked at how combining or refactoring automations can reduce overhead and how declarative tools like Flow should be used with the same care for bulk and efficiency as code. For **web service callouts**, we learned that integrating with external systems requires special attention: combining requests when possible, using asynchronous callout patterns, and guarding against slow or unreliable networks to keep Salesforce responsive. We also covered the breadth of **asynchronous Apex** options (future methods, queueables, batch jobs, etc.) and how they allow us to handle big jobs or complex logic without bogging down user-facing transactions. Finally, we highlighted the need for **continuous monitoring and re-optimization** – performance isn't a one-time project but an ongoing concern as conditions change.

A few key takeaways emerge from these topics. One is that efficiency in Apex often comes from **reducing needless work**: whether that's avoiding duplicate queries, avoiding doing things one-by-one when they can be done in bulk, or skipping work entirely by caching a result. Another takeaway is the value of **governor limits as guides** – instead of seeing limits purely as restrictions, we can view them as design parameters that shape efficient code (for example, knowing the 100 SOQL limit encourages us to architect for fewer queries). We also saw that **architecture and planning** play a huge role: making decisions early about data model, automation approach, and integration design can either set you up for success or leave you with inherent performance challenges. Using asynchronous processing and caching are architectural choices that pay dividends when used appropriately.

Crucially, a culture of performance-awareness is as important as any single technique. Developers, architects, testers, and product owners all have a part to play in keeping a Salesforce implementation running quickly and smoothly. If everyone considers performance implications during design, development, and testing, issues can be caught before they impact users. And with the help of AI tools like ChatGPT acting as a smart assistant – from reviewing code for inefficiencies to brainstorming better designs – teams can enhance their capabilities and catch subtle issues that might be overlooked.

By applying the best practices covered in this chapter, you'll be well-equipped to build Apex solutions that scale and perform in enterprise environments. Performance optimization is an ongoing journey of improvement and learning. As you continue this journey, an equally important focus is mastering Salesforce's platform limits. Understanding and navigating those limits will ensure your applications remain robust under Salesforce's multi-tenant constraints – a topic that naturally complements the performance techniques we've discussed, and a focus for the next stage of our journey.

11

APEX AND SALESFORCE PLATFORM LIMITS

In the Salesforce ecosystem, every architect and developer quickly encounters the reality of *platform limits* – the rules that govern how much data and processing an Apex transaction can consume. Salesforce's multi-tenant cloud architecture means that many customers share the same computing resources, so the platform enforces strict limits to guarantee fair resource allocation and stable operations for everyone. These limits (often called *governor limits*) span various aspects of Apex execution, from how many database queries you can run to how much memory your code can use. They are not arbitrary restrictions; rather, they ensure that no single piece of code can "hog" databases or processors at the expense of other users. In practice, this means Apex developers must always be mindful of how their code uses resources, writing efficient and scalable logic that stays within these guardrails.

Understanding and respecting Salesforce platform limits is not just an academic exercise – it has direct, real-world impacts on application quality. Improper handling of limits can lead to runtime exceptions that crash processes, slow page response times, or even complete transaction failures. For example, code that runs too long or queries too much data will be halted by the runtime, resulting in errors that users or admins eventually see. Such failures disrupt the user experience and can undermine confidence in the system. In extreme cases, a poorly designed feature that continuously hits governor limits might make a business process unusable. As one analysis noted, *failing to manage these limits effectively can lead to performance bottlenecks, transaction failures, and disrupted business processes*. In short, ignoring limits is not an option if you want a reliable, scalable Salesforce implementation.

This chapter sets the expectation that writing Apex code *within platform limits* is a foundational skill for any Salesforce professional. We will explore the key Salesforce limits that developers and architects must keep in mind, strategies for anticipating and handling those limits, and best practices for writing limit-conscious Apex code from the outset. By the end, you should appreciate how limit-aware design benefits all stakeholders: architects can ensure system designs remain robust under load, developers can avoid firefighting late-stage issues, testers can better simulate real-world conditions, and product owners (as well as enterprise managers) gain confidence that new features won't break at scale. Effectively managing limits means smoother deployments and fewer unpleasant surprises in production.

Another theme throughout this chapter is the supportive role that modern AI tools like ChatGPT can play in this arena. ChatGPT can act as a proactive assistant in identifying potential limit issues early. For instance, a developer might use ChatGPT to review a snippet of Apex and highlight any patterns that could risk hitting a governor limit. Or a tester could ask ChatGPT, *"What's the maximum number of records I should create to test for SOQL query limit breaches?"* and get guidance on designing a stress test. We'll demonstrate multiple practical ChatGPT prompt examples – including using voice input and file uploads – that encourage you to adapt and extend these techniques. As you'll see, ChatGPT can recommend code optimizations, explain complex limit concepts in simpler terms, and even help brainstorm approaches to refactor logic to be more efficient. In essence, AI assistance can augment your team's ability to adhere to Salesforce's constraints by providing on-demand advice and analysis.

It's worth noting that Salesforce itself continues to evolve in ways that emphasize governance and best practices around limits. As of Winter '25, for example, the Flow Builder tool in Salesforce will proactively warn admins and developers if a flow they are building could hit Apex governor limits. This kind of in-app guidance underscores how critical it is to consider limits during design – it's no longer just a back-end concern for coders, but a first-class consideration even in declarative development. Additionally, Salesforce's Winter '25 release introduced free monitoring for Apex unhandled exceptions, which include governor limit errors, making it easier to detect and troubleshoot limit issues in production (a topic we will revisit in a later section). The take-away is that limit awareness is becoming ever more integrated into the platform's tools and processes, and professionals should keep up with these developments.

With that background in mind, the following sections will explain specific areas of Apex and platform limits. We'll start by clearly defining the different kinds of Salesforce limits and why they exist. Then, we'll move into strategies for *proactive governor limit management* – essentially, coding patterns that help avoid hitting limits in the first place. After that, we'll dedicate individual sections to two of the most commonly problematic limit areas: CPU time & heap memory, and database operations (SOQL queries and DML). We'll also examine how working in *asynchronous Apex* contexts (like Batch Apex and Queueables) can change the limits landscape, sometimes offering higher limits but bringing their own considerations. Next, we'll discuss what happens when things go wrong – how to anticipate and respond to limit exceptions to minimize disruption. Finally, we'll emphasize the importance of *continuous monitoring and optimization* to keep your org running smoothly over time. Each section includes role-specific tips – whether you're an architect planning a big implementation or a product owner defining requirements – and each transitions smoothly into the next, so you can see how all these pieces fit together in an enterprise Salesforce strategy.

Let's begin by *understanding Salesforce platform limits* at a high level, and why these constraints are the backbone of developing robust Apex solutions.

UNDERSTANDING SALESFORCE PLATFORM LIMITS

Salesforce platform limits are essentially guardrails that ensure no single piece of custom code monopolizes shared resources. They apply to every *transaction* – a set of operations that Salesforce executes as a single unit (for example, all the logic that runs when a record is saved, including triggers and flows, is one transaction). Within a transaction, the Apex runtime strictly governs resource consumption. The reasoning is straightforward: because Salesforce is cloud-based and *multi-tenant*, many customers share the same infrastructure, and Salesforce must prevent any one org (or one rogue script) from degrading overall performance. Governor limits protect the platform's stability by capping what each transaction can do. From an engineering standpoint, they also encourage us to write efficient code. If our Apex code tries to do too much – whether that's querying an enormous dataset or executing complex logic without yielding – the platform will stop it in its tracks to protect everyone else.

What kinds of limits are we talking about? Salesforce imposes limits on nearly every resource an Apex transaction might use. Some of the most important categories include:

SOQL queries – e.g. a transaction can issue at most 100 SOQL database queries (synchronous limit) and retrieve at most 50,000 total records from those queries. (SOSL text searches are capped at 20 queries per transaction, with up to 2,000 records returned per SOSL query.)

DML operations – e.g. you can perform at most 150 DML statements (inserts, updates, deletes, etc.) in one transaction, and those statements can affect no more than 10,000 records in total. This ensures bulk updates are handled in manageable chunks.

Heap memory – the transaction can use up to 6 MB of memory (e.g. for storing object variables or collections in Apex) in a standard context. In asynchronous jobs, this is doubled to 12 MB. If you attempt to load too much data into memory, you'll hit a heap size limit and get an error.

Callouts – if your Apex makes HTTP callouts to external services, you're limited to 100 callout requests per transaction and a total of 120 seconds of callout time. This prevents excessive waiting on external systems within a single transaction.

Additional limits – there are limits on the depth of recursive logic (to prevent infinite recursion), on the number of static Apex classes you can have, on email sending, and more. Salesforce also sets *org-wide limits* (for example, no more than 250,000 asynchronous Apex executions per 24 hours for most orgs , and a limit of 10 concurrent long-running transactions over 5 seconds each). But the majority of day-to-day Apex coding concerns the per-transaction limits listed above.

Most of these limits are documented as absolute "hard" limits – they cannot be increased by Salesforce support (with a few exceptions like some email limits or if you purchase additional capacity). Certified managed packages from the AppExchange get a separate set of governor counters, effectively granting them their own allowance beyond your org's limits. But for your custom code, assume the published limits are immutable boundaries. Hitting one typically results in a *runtime exception* (specifically a System.LimitException) and an error message like "Too many SOQL queries: 101" or "Apex CPU time limit exceeded." Such errors cannot be caught by your code – once a governor limit is violated, the Salesforce engine stops the transaction immediately. That makes it crucial to design within limits rather than expecting to handle limit errors after the fact.

In summary, platform limits are a fundamental aspect of Salesforce development. They shape how we write Apex and design solutions. Far from being hindrances, they enforce good behavior: efficient queries, bulk processing, and mindful use of resources. By clearly understanding the limits up front, architects can design scalable systems and developers can implement solutions that work within these bounds. Now that we've covered the what and why of Salesforce's limits, we can move on to strategies for *proactively managing those limits* in our Apex code – essentially, techniques to avoid hitting the ceiling in the first place.

PROACTIVE GOVERNOR LIMIT MANAGEMENT IN APEX

The best way to deal with governor limits is to *not hit them in the first place*. This section is about coding techniques and patterns that make your Apex inherently efficient, so that even under heavy workloads it stays within Salesforce's limits. In practice, this comes down to writing *bulkified, streamlined, and thoughtful* Apex code. Many governor limit issues can be traced back to inefficient code that does the same thing over and over when it could be done once, or that handles records one at a time instead of in batches. By fixing these patterns, you "manage" the limits proactively – your code plays nicely within the limits rather than constantly flirting with them.

Bulkification is the cornerstone of limit-aware Apex. Salesforce is designed to process multiple records at a time, and it rewards code that handles bulk data. A bulkified Apex trigger or method means that it can process a list of records with almost the same logic as it would a single record, just using loops and collections internally. Bulkification avoids blowing limits because it minimizes the number of expensive operations (like queries or DML) per record. For example, consider an unbulkified trigger that runs a SOQL query inside a loop over 200 records. That would execute 200 queries – far above the limit of 100 – and throw a governor limit error. A bulkified version of the same trigger would collect all needed record IDs first and then perform a single SOQL query outside the loop to retrieve the data for all 200 records at once. The loop then iterates over the query results held in memory. This bulk approach uses maybe 1-2 queries total, instead of 200, and stays well under the limit. The same principle applies to DML: rather than inserting or updating records one by one in a loop (which would count toward the 150 DML statements limit for each operation), you add those records to a list and perform one insert operation for the whole list. Bulk DML can handle up to 10,000 records in a single statement, so it's extremely efficient compared to 10,000 individual inserts. In short, bulkify every trigger and method – always assume your code could be invoked with a batch of records rather than a single record.

Another proactive pattern is to *reduce redundant work*. This might mean caching information within a transaction to avoid duplicate queries. For instance, if you need data about a parent Account for each of 50 Contact records, don't query the Account 50 times. Query all required Accounts once (using WHERE Id IN (...) with the set of unique parent IDs) and store them in a map for quick lookup by ID. That way, you use just one SOQL query regardless of how many Contacts are in play. Similarly, if the same computation or logic can be reused, calculate it once and reuse the result instead of recalculating in every iteration. Efficient *looping constructs* also matter: avoid nested loops over large collections whenever possible, because a nested loop (especially if paired with queries) can lead to a combinatorial explosion of work. Sometimes you can replace a double loop with a single loop that uses a map for lookups – achieving the same result with far fewer iterations.

Using ChatGPT to review and improve code is a powerful way to enforce proactive limit management. You can paste a code snippet or even upload a file (for example, an Apex class or trigger) into ChatGPT and ask something like, *"Do you see any governor limit issues in this code?"* This essentially performs an automated code review focused on limits. ChatGPT might point out, for instance, that there's a Database.update call inside a loop, or that a recursive trigger could be firing multiple times. It often explains why that's a problem (e.g., "You could exceed the DML statements limit if a large number of records are processed") and can suggest a solution (such as collecting records in a list and updating them outside the loop). Many developers have started to include AI-based checks like this in their workflow – it's like having a pair programming partner who has the governor limits reference guide memorized.

Proactive limit management should also be part of your *testing strategy*. *Tester Tip:* When QA engineers or developers write test classes for Apex, they should include tests that invoke the code with a bulk set of records (for example, insert 200 test records to simulate a batch trigger) to ensure it still succeeds and doesn't throw any LimitException. If the test with 200 records fails, that's a red flag that the code isn't bulk-safe. Testers can leverage ChatGPT as well – for instance, to generate test methods or test data. A tester might ask, *"Can you write a test case that will verify this trigger doesn't hit governor limits?"* ChatGPT can outline a test method that inserts a large list of sample records and perhaps uses Limits.getQueries() after the operation to assert that the query count stayed below the threshold. This not only speeds up writing the test, it also codifies the expectation that the code should run within limits.

Finally, consider some *architectural patterns* that help with governor limits. One is the "one trigger per object" recommendation: by consolidating all logic for a given object into a single trigger (and controlling the order of execution within handler methods), you avoid scenarios where multiple triggers on the same object each do their own queries or DML. Multiple triggers on one object can inadvertently perform redundant work and make it harder to govern overall behavior. A single, well-structured trigger framework (using helper classes) can ensure that queries and DML are batched together optimally. Another pattern is to offload certain heavy computations to asynchronous Apex (like Queueable or future methods) where appropriate, thereby using the expanded limits in those contexts. We'll talk more about asynchronous Apex soon, but the point is that proactively deciding *when* and *how* to execute logic (synchronously vs asynchronously, or in batches) is itself a tool for staying within limits.

MANAGING CPU AND HEAP SIZE LIMITS

Even with perfectly bulkified, SOQL-efficient code, you still have to pay attention to two critical resources during Apex execution: *CPU time* (how long your code runs on the Salesforce servers) and *heap memory* (how much data your code stores in memory at once). Hitting the CPU time limit or the heap size limit will also result in a LimitException that halts your execution. These limits ensure that any single transaction doesn't tie up the shared server for too long or consume excessive memory. Managing CPU and heap is about writing fast, lean code – not just in terms of database operations, but overall algorithmic efficiency and data management.

CPU time tends to be exceeded when code performs an enormous amount of processing or inefficiently iterates over data. To reduce CPU usage, start by examining the *algorithmic complexity* of your Apex code. If your code's runtime grows very quickly as the data volume grows (for example, a quadratic *O(n^2)* loop that compares every record to every other record), it may run fine for 100 records but time out for 10,000. Optimizing algorithms – using maps for quick lookups, avoiding nested loops when possible, and moving heavy computations out of tight loops – directly cuts down CPU time. Consider a scenario where you have two lists and you need to find matches between them. A naive approach might use two nested loops, which is *O(n·m)* and could be catastrophic if both lists are large. A better approach is to put one list's values into a Set or Map (which is an *O(n)* operation) and then loop through the other list (*O(m)*), checking for inclusion in the set – an overall *O(n + m)* approach. This kind of refactoring can mean the difference between a transaction finishing in under a second versus one that drags on until the 10-second CPU limit is hit.

Another way to manage CPU is to *perform work asynchronously* when appropriate. Certain tasks that are very CPU-intensive (for example, recalculating a complex roll-up on 10,000 records) might be better done in a Batch Apex job. Batch Apex not only gives you 60,000 ms of CPU per batch execution, but it also lets you spread the workload across multiple transactions. If you find that a particular operation consistently runs up against the 10,000 ms limit in testing, that's a good candidate to turn into a batch or queueable job. That way, you leverage the asynchronous processing limits and free up the synchronous transaction (say, a user's save operation) to finish quickly. *Architect Tip:* Be cautious about doing massive recalculations or data transformations in a trigger context – if it's something that can be deferred to run after the fact (and doesn't need to block the user's action), consider using a future method or queueable. This not only avoids CPU limits in the immediate transaction, but it also improves perceived performance for the end user.

Now let's talk about *heap size*. The heap is essentially the memory allocated to store objects and variables during execution. 6 MB (in a synchronous context) is plenty for most scenarios, but certain operations can use it up quickly. Large collections of sObject records, especially if they have many fields, can bloat memory usage. If you query 50,000 records and each record has dozens of fields, you might hit the heap limit when storing all those objects. To manage heap, you want to *limit how much data you load into memory at once*. Only query the fields you truly need – for instance, if you're just checking for the existence of records, querying a single ID field is enough instead of selecting all fields. If you need to process a very large data set, consider processing it in smaller chunks rather than one giant list. Batch Apex is again a friend here: it can process records in batches (e.g., 2,000 at a time by default), which implicitly limits heap usage per batch. If you are dealing with extremely large strings or blobs (for example, parsing a huge JSON payload or processing file contents), be mindful of those sizes – you might need to break the string into pieces or stream the processing.

Another technique for heap management is to *release references* to objects that you no longer need. In Java or other languages, you might explicitly null out variables or use other memory management tricks; in Apex, once an object goes out of scope (for example, outside the loop it was defined in, or after a function ends), it becomes eligible for garbage collection. More commonly, you would refactor code to process records in streams. For instance, instead of building up a list of 10,000 records in memory and then processing them, you might process each batch of 200 records as you retrieve it (then let that batch go out of scope before moving on to the next batch, keeping only a smaller subset in memory at any time).

How can ChatGPT help optimize CPU or heap usage? One way is by acting as a *performance analyst* on a piece of code. You can prompt ChatGPT with a code excerpt and say, *"This code is running slowly – do you see any way to make it more efficient?"* If the issue is algorithmic, ChatGPT might point out the nested loop or the repeated computation and suggest using a collection or simplifying the logic. For example, a developer might paste a method that builds a big list via multiple loops and ask for suggestions – ChatGPT could identify that some loops can be merged, or that a map could cut down a lot of lookups. For heap issues, you could describe a situation like *"I have an in-memory list of 50,000 Account records and I'm hitting heap limits"*, and ChatGPT might advise to paginate the processing or use Batch Apex to handle the records in chunks.

Another creative use of ChatGPT is generating test scenarios to *probe the limits*. As a tester or developer, you could ask, *"Generate Apex pseudo-code that would intentionally hit the CPU limit."* By studying the generated example (say it creates a huge loop with complex logic), you gain insight into what not to do. Conversely, you could ask for an example of memory-efficient processing of a large data set, and use that as a model. ChatGPT can also assist in explaining Apex debug log output. If you have a log snippet showing "Maximum heap size: 6,000,000 bytes" or a "CPU time limit exceeded" error, you can feed that to ChatGPT and ask, *"What does this imply and how might I fix it?"* – it will likely explain that the code ran out of memory or time and reiterate some of the strategies to mitigate it.

HANDLING SOQL QUERY AND DML OPERATION LIMITS

Database interactions are often where Salesforce limits are most immediately felt. Exceeding the limit of 100 SOQL queries or 150 DML statements in a single transaction will cause an error, so effective Apex code must use these operations judiciously. Fortunately, the platform gives us tools to work with data in bulk and to retrieve exactly what we need with minimal calls. The guiding principle here is to do more with fewer queries and fewer DML statements.

We've already touched on the most important tactic: bulk operations. When inserting, updating, or deleting records, always try to group records and perform one bulk DML operation, rather than many individual ones. Not only does this keep you safely under the 150 DML statements limit, it's also far more efficient – Salesforce can optimize a single bulk operation internally in ways that it can't when handling many separate calls. For example, if you need to insert 500 new records, a single insert recordsList is ideal. But what if those 500 records come from different parts of your code? It may be worth accumulating them into a list (perhaps using a static variable or a stateful helper class) and inserting them together at a later point. This kind of pattern (sometimes called a buffer or accumulator) ensures that even if multiple pieces of code want to do DML, they coordinate to do a single combined operation. It's also important to avoid unnecessary DML. Ask yourself: does a record really need updating? If you query a record, change a field, and update it, that counts as one DML. But if the field value you're writing is the same as it was before, that DML was effectively wasted. It still counts against the limit and even consumes CPU time for no net change. A proactive check – only perform the DML if something actually changed – can save you from burning through DML statements on no-ops.

On the SOQL side, *minimizing query count* and *maximizing each query's productivity* are key. This often means writing queries that retrieve *all the data you need* in one go, rather than multiple queries for related data. Suppose you have a scenario where for each Account, you need to get some related Contacts. A poorly written approach might iterate over Accounts and run a separate [SELECT... FROM Contact WHERE AccountId = :accountId] for each – this is the dreaded query-in-loop pattern that could easily exceed 100 queries. A better solution is to perform a single SOQL query to get all Contacts for all relevant Accounts, using a WHERE AccountId IN :accountIds filter, or even to leverage a parent-child subquery (for example, SELECT Id, Name, (SELECT Id, Name FROM Contacts) FROM Account WHERE...). This way, you replace dozens of small queries with one or two comprehensive queries. Likewise, if you need to retrieve different types of data, see if you can combine them. Salesforce allows some flexible query patterns – for example, you might fetch multiple object types via a polymorphic relationship or use a single SOQL query on a parent object with selective fields instead of querying the parent and child separately.

It's also crucial to *use filters and indexes* to keep SOQL queries efficient and within limits. While governor limits like "50,000 records retrieved" are a hard cap, in practice you typically want to fetch far fewer records than that. Writing selective queries (queries that filter on indexed fields or otherwise greatly narrow the result set) not only improves performance, it reduces the chance you'll bump into the 50,000 record return limit. For example, if an org has a million Contact records, a query with a broad filter could try to return hundreds of thousands of rows – you'd hit the limit at 50k and get an exception. By filtering on things like Contact.Active__c = TRUE and maybe a date range or specific criteria, you ensure that the query returns only the subset you need, ideally well under the limit. From a design standpoint, if you anticipate needing to frequently query large data sets, consider working with your architect to create custom indexes on fields (Salesforce supports indexing on external IDs, unique fields, and you can ask for indexes on other fields via support). An indexed filter can make a query more selective and prevent full-table scans that might otherwise time out or return too much data. *Product Owner Tip:* When defining requirements, try to include criteria for narrowing down data operations (e.g. "update customers in a certain region" instead of "update all customers") – this not only aligns with business needs but also helps the development team implement queries that won't exceed limits.

Common pitfalls and how ChatGPT can help: One common mistake is not realizing that multiple triggers or processes in the same transaction cumulatively use the 100 SOQL queries. For instance, if you have a trigger on Account and a trigger on Contact, and an operation causes both to fire (say, inserting an Account and also creating Contacts in the same transaction), the SOQL queries used by both triggers count toward the single limit. It's important for developers to be aware of not just their piece of code, but the "big picture" of what else runs. ChatGPT can assist by analyzing a set of Apex triggers or classes together – you might provide summaries of two triggers and ask, *"Could these interfere with each other's governor limits?"* ChatGPT might recognize that both are doing a lot of queries and warn that combined they could exceed limits when run in one transaction.

Another scenario: you suspect that a particular operation is close to the SOQL limit, say 95 queries, which is dangerous. You could ask ChatGPT, *"This execution is using 95 SOQL queries, where are they all coming from?"* By giving it relevant snippets (or even a debug log excerpt), you might get insight. Perhaps it will identify that a helper method is being called repeatedly without caching its results, leading to duplicate queries. ChatGPT could suggest caching the results in a Map or using a static variable to ensure the query runs only once.

For DML, if you have complex business logic, you might inadvertently call many separate DML statements. For example, maybe you have a loop that occasionally inserts a related record for each item in a list – if that loop runs 200 times, that could be up to 200 inserts. A solution would be to collect those to-insert records in a list and do one insert at the end. If you share such code with ChatGPT, it can usually spot the issue and rewrite the code in a bulk manner. **Developer Prompt Example:** *"Rewrite this code to reduce the number of DML statements"* (after showing a snippet). ChatGPT might come back with a version that uses one insert outside the loop, with commentary like "moved the insert outside the loop to avoid hitting the 150 DML limit."

From a testing perspective, it's useful to validate that your code handles data volume without limit errors. **Tester Tip:** Create tests that perform bulk operations – for example, simulate a user uploading 200 records via Data Loader, which will fire your Apex in one transaction. After the test, assert that no LimitException was thrown and even inspect Limits.getQueries() to ensure its below 100. Testers can work with product owners to understand typical and peak data volumes so the tests can cover those scenarios. They can also use ChatGPT to generate or manipulate large test datasets (for instance, ask for an Apex code snippet that creates 200 dummy records of an object with necessary field values) to use in volume testing.

Lastly, *efficient database design* can preempt a lot of limit issues. If your data model and indexing strategy are sound, you won't need to write crazy workarounds in code. Sometimes denormalizing data (duplicating a piece of info on a child record) can save you from an extra query to the parent, but be careful – that can introduce data consistency challenges. Architects should weigh those trade-offs. The good news is that with careful use of relationships, queries, and DML, you can almost always stay within the limits. Salesforce's limits for queries and DML are actually quite generous when code is bulkified. By writing queries that do more and DML that handles batches of records, you maximize each transaction's efficiency and avoid hitting the governor ceiling.

At this point, we've looked at limits in synchronous Apex code. But what about asynchronous contexts like Batch Apex, Queueables, and scheduled jobs? Those have different limits (often higher in some respects) and their own best practices. In the next section, we'll explore the special considerations when working with *asynchronous Apex* and how to handle limits in those scenarios.

LIMITS IN ASYNCHRONOUS APEX CONTEXTS

Asynchronous Apex (such as Batch Apex, Queueable, Scheduled Apex, and future methods) runs in a different mode with its own set of governor limits. In general, Salesforce relaxes certain limits for asynchronous execution to allow larger data volumes to be processed – but it also imposes some new limits to keep background processing under control. When designing async processes, it's important to understand both the *expanded per-transaction limits* and the *global limits on async usage*.

First, the good news: many per-transaction limits are higher for asynchronous Apex. For example, a Batch Apex execute method or a Queueable job can perform up to 200 SOQL queries (double the 100 of a synchronous transaction) and up to 100 callouts, and it gets a full 60,000 milliseconds of CPU time (6× the sync limit). The heap limit is also doubled to 12 MB in async contexts. This means asynchronous jobs are well-suited for heavy lifting – they can query more data and run more complex logic without immediately hitting limits. Additionally, Batch Apex in particular is designed to handle large numbers of records by splitting them into manageable chunks (each chunk or "batch execute" is its own transaction with fresh governor limits). So if you need to process, say, a million records, a batch job could break that into 5,000 chunks of 200 records each, and each chunk gets its own 10 seconds of CPU and 100 SOQL queries. This is how Salesforce can scale data processing beyond what a single transaction could do.

Now, the trade-offs: asynchronous Apex comes with *governance of its own*. Salesforce limits how many async processes you can spawn and run, to protect the system from being overwhelmed by background jobs. Key async limits to be aware of include:

You can invoke at most 50 future calls per synchronous transaction (and note: you *cannot* call a future method from another future or batch context). This prevents recursive chains of future calls. If you attempt to call more than 50 futures in one go, you'll hit a limit error.

For Queueable jobs, you can enqueue up to 50 jobs in a single synchronous transaction, but if you're already in an async context (like inside a Queueable or batch execute), you can only enqueue one additional Queueable. In other words, you can chain Queueable jobs, but only one step at a time – a Queueable can't spawn an unlimited number of other queueables in parallel.

Only a certain number of batch jobs can run at the same time. Salesforce allows up to 5 batch executions to be running concurrently by default. Any batches beyond that will wait in the Flex Queue (which can hold up to 100 queued batch jobs). If you try to start a 106th batch job while 5 are running and 100 are already queued, you'll get an error attempting to enqueue the batch. This scenario is rare, but it can happen in orgs that schedule many batch processes or trigger batches from data loads.

There is a *24-hour rolling limit* on the total number of asynchronous Apex executions. The limit is quite high (currently 250,000 or 200× the number of user licenses, whichever is greater) , but it's not infinite. A very busy org with many scheduled jobs, platform events, or batch processes could approach this limit. Hitting it means new async jobs (batch, queueable, future, etc.) will be delayed or refused until the usage falls below the threshold.

Scheduled Apex has its own limit: you can only have 100 scheduled Apex jobs at one time (per org). If an org tries to schedule more (for example, scheduling 200 jobs into the future), the 101st will fail to schedule. Usually, 100 scheduled jobs is plenty – but large enterprises sometimes bump into this, especially if admins schedule a lot of maintenance jobs. A workaround in those cases is to have one scheduled job that itself kicks off multiple batch jobs (within the 100 limit, since the scheduler counts 1, and the rest are batch executions).

Asynchronous jobs have limits on chaining depth as well. A batch job can call another batch in its finish method, but you wouldn't want an infinite chain. Salesforce doesn't allow a batch to reschedule itself more than a certain number of times in a 24-hour period (to prevent runaway recursion). Similarly, queueable chaining (one queueable calling another) should be used judiciously – you can chain one after another, but if you try to build a long linear chain dynamically, consider using a batch instead or chunking the work differently.

Strategies for async limit management: One of the first decisions when writing a batch job is choosing an appropriate batch size (the number of records per chunk). The default is 200, but you can specify anywhere from 1 to 2,000. If each record is relatively light to process, a larger batch size (say 500 or 1,000) can reduce the overhead of context switching and finish the job with fewer execute calls. However, if processing each record is heavy – maybe it involves complex calculations or calls out to an external service – you might intentionally use a smaller batch size (perhaps 50 or 100) to ensure you don't hit CPU or heap limits within an execute. It's a balance. The nice part is you can adjust this if you notice limit issues: if a batch job is failing due to CPU time in execute, try reducing the batch size and running again.

Error handling in asynchronous Apex is also important because of the *partial processing* nature. In a synchronous transaction, hitting a governor limit or unhandled exception rolls back the whole thing. In a batch job, if one execute chunk fails (say it throws an exception because one record had bad data), that chunk's transaction is rolled back but the batch will continue with the next chunks. You might end up with 950,000 records processed successfully and 50,000 that failed in that one chunk. If you don't account for that, those failed records might be silently skipped. To handle this, you can use Database.executeBatch with a scope parameter and handle errors record-by-record. For example, using the Database.insert(records, false) method will allow some records to insert and report which ones failed. In your batch execute, you could collect the errors and perhaps log them or store them in a custom object. Another approach is to implement Database.Stateful in your batch class, which allows you to maintain state between executes – you could keep a list of failed record IDs or error messages in an instance variable and then process it in the finish method (for example, sending a summary email or creating a report record of which items failed and why). *Tester Tip:* Write tests for your batch classes that simulate errors in some records (e.g., create data that will violate a validation rule for one record in the batch) and ensure your batch's error handling logic catches it appropriately rather than letting the exception abort the entire job.

For queueables and future methods, one challenge is *orchestrating complex workflows* without hitting limits. Since a queueable can only chain one deep, if you have a scenario like "after this job, start these two other jobs," you can't do that directly from one queueable (because that'd be two enqueue calls). Instead, you might use a scheduler or a batch as a supervisor to launch multiple queueables. Or use Platform

Events: one pattern is to have a queueable publish a platform event upon completion, and have multiple subscribers (in Apex triggers on that event) respond by launching their respective jobs. This gets around the one-chain limitation because the logic is fanned out via the event bus. Of course, you then need to manage the event limits (there are limits on event publishing too), but they're quite high for non-Volume events.

ChatGPT in asynchronous design: Architects and developers can use ChatGPT to sanity-check their approach to async processes. For instance, you could describe your plan: *"I need to process 1 million records. I plan to use a batch with batch size 1,000. Each execute will do X, Y, Z. Is this a good approach?"* ChatGPT might respond with considerations about CPU time and suggest maybe a smaller batch size if X, Y, Z are complex. Or it could confirm that the approach sounds fine. You might also ask, *"How can I chain batch jobs safely?"* and ChatGPT would likely explain the technique of calling one batch from the finish of another. It might even caution you about the 5 concurrent batch limit and advise queuing subsequent batches only after prior ones finish. Another helpful prompt: *"What are best practices to avoid hitting limits with many Queueable jobs?"* ChatGPT could enumerate tips like grouping work into a single batch instead of many queueables, or using the event-driven approach mentioned above.

From a monitoring perspective, remember that async jobs have their own tracking. Salesforce provides the AsyncApexJob object (which you can query) to see the status of batch and queueable jobs, how many batches remain, etc. It also logs errors for batch jobs in the Apex Jobs page. Make sure to review those logs or set up alerting (for instance, an email when a batch fails) so that limit issues in async processes don't go unnoticed.

In summary, asynchronous Apex is a powerful tool to handle large-scale processing under Salesforce's limits, but it introduces another layer of limits that govern how you can start and chain these processes. By understanding those rules – like how many jobs can run or be enqueued, and how to design chunked processing – you can leverage async Apex to do heavy lifting that would be impossible in a single transaction. Our final step in this chapter is to look at how to *anticipate and handle limit exceptions* when they do occur, and more importantly, how to continuously monitor and optimize your Apex code so that you catch limit issues before they impact users.

ANTICIPATING AND RESPONDING TO LIMIT EXCEPTIONS

Despite our best efforts, there may be times when governor limits are exceeded. When this happens, Salesforce will throw a runtime exception (usually System.LimitException) and halt the execution. *By design, you cannot catch these exceptions in your Apex code* – if a limit is violated, the error is unhandleable and the transaction is rolled back. This means the only real cure is prevention. However, we can still build safeguards and fallback strategies to *anticipate* limits before we hit them, and to respond gracefully in the aftermath so that users and administrators are informed about what went wrong.

Defensive coding is the primary way to anticipate limit issues. This involves checking the context and inputs and deciding not to proceed with an operation if it's likely to exceed limits. For example, if your code is about to perform a query based on user input, and the input criteria are so broad that you estimate the query could return hundreds of thousands of records, you might preemptively refuse to run and instead prompt the user to narrow their criteria. It's better to stop upfront with a polite message than to try and fail mid-way with a LimitException. You can also use the Limits class to gauge consumption. For instance, if you're processing records in a loop, you might periodically check Limits.getCPUTime() or Limits.getHeapSize(). Apex doesn't allow you to yield or sleep, but you could implement a pattern where if you detect you're close to a limit, you abort the operation deliberately (perhaps by throwing a custom exception that you *do* catch in a higher context) and reschedule the remaining work to continue later. One pattern for long-running processes is to use a *continuation or stateful retry*: process as much as you can, then store a progress marker (maybe in a custom object or static variable that a follow-up transaction can read) and trigger the next chunk of processing asynchronously. This way, you never actually hit the limit – you bow out before that point. It requires careful design, but it's a strategy some advanced implementations use to handle massive jobs without hitting the wall.

For limits like the SOQL 50,000 records limit, you can anticipate by adding a LIMIT 50000 clause to your query if there's a risk it might return too many rows. That way, Salesforce will stop at 50k and not throw an error (you can then handle the scenario of "partial results" gracefully in your code). Or you might break up the query into multiple smaller queries by adding conditions (like query 10k at a time using some indexed criteria or an ORDER BY with range). These are forms of *graceful degradation* – the code detects it can't do everything in one go, so it does as much as it can and either informs the user or prepares to continue later.

User-friendly messaging is essential when a limit exception does occur (or is about to occur). If an operation is halted because it ran out of CPU time or performed too many DML operations, the user will otherwise just see a generic error ("Apex CPU time limit exceeded" or "Too many DML statements: 151"). We can do better by catching foreseeable issues and translating them. While we cannot catch the actual LimitException, we can catch some "upstream" indicators. For example, if the user initiated something that would clearly violate limits, you can throw a custom exception (like BusinessLogicException) with a message like "The data set is too large to process at once. Please refine your criteria." and catch that in the calling context (e.g., the Aura component or Visualforce controller), then show that message to the user. In a trigger context, you can use addError on a record to prevent an operation from committing and give a specific message (for instance, if saving a record would trigger an inordinate update on thousands of related records, you might proactively addError on that record with a message "Cannot complete this operation – it would affect too many related records."). This stops the transaction gracefully before hitting the limit and informs the user in a controlled way.

Logging and monitoring play a huge role in responding to limit issues. Since our code can't trap a governor exception, we rely on Salesforce's logging mechanisms to tell us when one occurs. Traditionally, this meant combing through debug logs or receiving automated error emails (Salesforce sends an email to developers or admins when a batch job fails or an uncaught exception occurs in scheduled Apex, for example). But as of the Winter '25 release, we have a much better tool: *Apex Exception Event monitoring.* Salesforce now provides an event log for "Apex Unexpected Exception" that is accessible even with the free Event Monitoring feature. This log captures details about exceptions, including governor limit errors like CPU timeouts and too many SOQL queries. By regularly reviewing these event logs (or hooking them into a monitoring system), developers and support teams can quickly identify limit-related failures in production. For example, if a batch job hits the CPU limit on a certain batch of records, an event will be logged with that info. An administrator or monitoring script could catch that event and alert the team, or even automatically create a case or feed item for the issue. *Architect Tip:* Set up a routine (maybe using a monitoring tool or a scheduled Apex job) to scan for LimitException events in your logs. Early detection will allow the team to address performance problems before they escalate. In the past, a lot of limit issues were discovered only after users complained or data got stuck – with proper monitoring, you can be proactive.

Within Apex, one form of "logging" you can implement is to use a custom object to record occurrences of certain conditions. For instance, you might have a catch for a DML exception (not a governor limit, but say a validation failure) that logs an entry like "Tried to process 500 records, 20 failed validation." This isn't a governor limit issue per se, but the idea is to build a trail of what your code is doing. While you can't catch a governor LimitException, you might catch other exceptions and infer that limits were near. For example, a QueryException might indicate too many records (though again, Salesforce usually throws a LimitException for too many query rows). Another trick: if you use the Limits class at end of a transaction, you could record if certain thresholds were exceeded (like "transaction used 98 out of 100 queries" – that's dangerously close, worth investigating). In testing, you can even assert that your code stays well under limits (e.g., assert that Limits.getQueries() < 100 to catch any test that unexpectedly uses too many queries).

When it comes to ChatGPT's help: you can use ChatGPT to simulate or reason about worst-case scenarios. For example, you might describe a piece of code and ask, *"What could cause this code to throw a governor limit exception?"* ChatGPT might analyze the code and respond with something like, "If the list records is very large, the loop will execute many times and could hit the CPU time limit" or "If more than 100 SOQL queries are executed due to this recursive call, it will throw an error." This is useful as a thought exercise – it can confirm your suspicions or reveal a scenario you didn't consider. You can then ask for suggestions: *"How could I modify the design to handle that case?"* ChatGPT might suggest processing in chunks or using a batch. Essentially, you can have an interactive code review with a focus on limit-exception scenarios. Another use case: providing ChatGPT with an error log or exception message and asking for interpretation. For instance, *"I got a `Too many DML statements: 151` error in this context, what are some possible causes and solutions?"* It will likely explain that you exceeded 150 DML statements, possibly due to a DML in a loop or multiple triggers firing, and recommend combining DML operations or refactoring the logic.

For **developers and testers**, building resiliency means testing not just the happy path but also the edge cases where limits might be hit. A tester could create a scenario with extremely large data input to see how the system behaves – ideally, the system should refuse or defer the work rather than just break. If the system is supposed to handle it asynchronously, verify that the hand-off to async happens. If it's supposed to show an error message when something is too large, ensure that message appears and is understandable. Testers might use tools or even ChatGPT to generate those extreme test conditions (e.g., "generate 201 test records to insert to see if the trigger prevents the DML limit from being hit").

In summary, while we can't *catch* a governor limit exception at runtime, we can do a lot around it: we can avoid it through defensive checks, provide graceful fallbacks or messages when we detect a looming issue, and monitor/log any incidents that do occur so we can fix the root cause. By anticipating where things could go wrong, we ensure that hitting a limit is a rare, managed event rather than a chaotic failure. This proactive stance on errors ties directly into the next (and final) section: continuously monitoring our applications and refining them so that limit-related issues are caught and addressed early, keeping our Salesforce org healthy as it grows.

CONTINUOUS MONITORING AND LIMIT OPTIMIZATION

Managing Salesforce limits isn't a one-and-done task – it requires ongoing vigilance. As your org grows (more data, more users, more customizations), Apex code that once ran comfortably may start to strain against limits. This is why *continuous monitoring* and periodic optimization are crucial. By regularly reviewing how close your system is to various limits, you can proactively make adjustments before an issue turns into a user-facing incident.

Salesforce provides a few native tools to help with monitoring. The Apex Exception event log we discussed in the previous section is a goldmine for detecting limit issues in production. If you set up a process to capture and review these logs, you'll have a near real-time view of any LimitException that occurred. Many teams integrate these event logs with external monitoring solutions – for instance, exporting them to Splunk or New Relic to generate alerts. If an Apex CPU timeout happens, the team can be notified immediately via email or Slack. Over time, you can even track metrics like "How many SOQL queries on average does this transaction use?" by parsing debug logs or event logs. A rising trend might signal that a particular piece of code is doing more work as data grows, hinting that an optimization (like adding an index or archiving old data) will soon be needed.

Aside from reactive monitoring, *regular performance audits* are a best practice. An architect or tech lead might schedule quarterly reviews of key Apex classes and triggers: Are they still efficient given current data volumes? This could involve re-reading the code or even rerunning certain test scenarios with present-day data sizes. If you have a test environment with a copy of production data, it can reveal a lot – perhaps a process that took 2 seconds last year takes 8 seconds now (maybe it's not hitting the 10-second limit yet, but it's trending upward). Catching that trend allows the team to optimize the code (or beef up filters, etc.) in advance. *Architect Tip:* Include governor limit considerations in your architecture review checklist. For example, when a new integration or feature is proposed, ask "How will we monitor the impact of this on our org's limits? Do we have logging in place to measure how many queries or how much CPU this operation uses in production?" By planning for monitoring upfront, you ensure that every major piece of Apex has a way to verify its health in real usage.

Teams should also take advantage of *static analysis tools* for continuous improvement. Tools like PMD or SonarQube can be configured with Salesforce rules (e.g., flag any SOQL inside loops, flag any method that calls more than, say, 50 queries as a warning). Running these tools as part of your CI/CD pipeline or on a regular schedule can catch anti-patterns introduced over time. Perhaps a well-meaning developer wrote a quick fix that isn't bulkified – the static analysis can call it out for refactoring. Some organizations also perform **peer code reviews** for all Apex changes, with one of the review criteria being "Is this code limit-safe and efficient?" That ensures no new code introduces a slow or limit-risky approach.

Embedding limit awareness into team culture is the human side of continuous optimization. Developers should share tips with each other when they find a more efficient way to do something. Testers should include performance and limit tests in their test plans, not just functional tests. Product owners and managers can help by allocating time for "tech debt" sprints where the team specifically focuses on optimizing or refactoring code that is getting close to limits. It's much better to spend a few days cleaning up a highly utilized trigger now than to deal with an emergency later because it started hitting governor limits during peak business hours. In the context of Scrum or Agile, consider adding non-functional requirements to user stories – e.g., "the solution should handle up to 10,000 records without timing out." This sets clear expectations that are testable.

ChatGPT can be a partner in ongoing optimization efforts too. For example, you might periodically ask it open-ended questions like, "*What are some new best practices for Apex optimization as of the latest Salesforce release?*" This could reveal platform enhancements you weren't leveraging (maybe a new SOQL feature or a new way to schedule jobs). With Salesforce releasing updates thrice yearly, it's possible that what was a limit last year has a mitigation this year. (As a hypothetical example, if Salesforce were to raise the SOQL query limit in the future or introduce a more efficient query mechanism, you'd want to know and adjust your code to take advantage.) ChatGPT, staying up-to-date, might mention a Winter '25 or Spring '26 change that affects governor limits or relevant features.

Another idea is using ChatGPT to analyze trends from your logs. Suppose you export a summary of governor limit usage (like a CSV with columns: date, avg CPU used, max CPU used, avg queries used, etc. for certain processes). You could feed that to ChatGPT (or a tool like the Code Interpreter plugin) and ask for an analysis – maybe it will spot that "Process X's CPU usage has been increasing 5% each month" or "Process Y occasionally spikes to near the limit when large records are processed". While you could find these manually, having an AI point them out can save time or catch subtle patterns.

Product Owner Tip: Keep the lines of communication open with the technical team regarding limits. If users are asking for a new report that pulls "all data from the past 10 years", that could raise red flags for limits – looping in an architect early can lead to a solution like a scheduled job that pre-aggregates data, rather than a huge on-demand query. By understanding that limits exist, product owners can help set realistic expectations (for example, knowing that a data-intensive operation might take a few minutes via a batch job, rather than being instant). This prevents turning limit workarounds into last-minute surprises.

In practice, a continuous improvement cycle might look like this: monitor your org (through logs, event monitoring, user feedback) → identify a hotspot (say, a particular batch job frequently runs at 80% of the CPU limit) → investigate and optimize that code or process (maybe rewrite part of it in a more efficient way) → deploy the improvement → continue monitoring to verify the issue is resolved and to catch the next hotspot. Over time, this reduces incidents and also often improves overall performance, which users will certainly appreciate. As one Salesforce blog put it, *efficient governor limit management ensures scalable and reliable performance for the long haul.*

By making limit monitoring part of your maintenance routine, you ensure that your Salesforce org remains healthy and scalable. You catch issues when they're still minor and easier to fix. Plus, your team builds a deep knowledge of the system's behavior. This foundation of awareness and agility will serve you well as you move into more advanced development topics. Our next chapter will deep dive into Apex design patterns and reusability strategies – themes that go hand-in-hand with writing efficient, limit-conscious code, and will further help in building robust Salesforce applications for the enterprise.

SUMMARY

In this chapter, we explored how *limit awareness* underpins every aspect of Apex development. We began with an overview of Salesforce's platform limits and why they exist – the multi-tenant architecture demands fair use of resources, so limits on things like queries, DML, CPU time, and memory keep all orgs running smoothly. Next, we discussed proactive governor limit management, emphasizing patterns like bulkification and efficient looping that ensure our code avoids hitting limits even under heavy load. We then focused on optimizing CPU and heap usage, learning to write faster, leaner Apex by reducing algorithm complexity and processing data in memory-smart ways. After that, we reviewed best practices for SOQL queries and DML operations, seeing how to maximize each transaction's work by doing more in fewer database calls (and always bulk handling our records).

We also examined the special case of *asynchronous Apex*. We saw that while Batch Apex and Queueables give us higher limits per transaction, they introduce their own considerations – like limits on how many jobs we can enqueue or run concurrently. Designing asynchronous processes involves tuning batch sizes and managing chained jobs to stay within those meta-limits. From there, we moved into anticipating and handling limit exceptions. We acknowledged that governor limit errors can't be caught in code, so the emphasis must be on defensive checks and graceful degradation (informing the user or splitting work into smaller chunks) and on robust logging/monitoring to catch any exceptions that do occur. Finally, we highlighted the importance of continuous monitoring and optimization: using tools, logs, and team processes to regularly review Apex performance and adjust it, ensuring the application remains scalable as data and usage grow.

Throughout these sections, a common theme emerged: *governor limits are not obstacles but guides* for building scalable, high-quality applications. By coding with limits in mind, we inherently write better, more efficient Apex. We also saw how ChatGPT can serve as a valuable ally in this journey – from suggesting bulkification approaches and spotting potential inefficiencies, to explaining error messages and recommending architectural adjustments. Using ChatGPT's assistance proactively can shorten the debugging cycle and provide insights drawn from a vast range of Salesforce knowledge, effectively acting like a pair programmer specialized in best practices.

Being disciplined about governor limits pays off in the long run. Applications that respect limits tend to exhibit fewer runtime exceptions, better performance, and are easier to maintain. When you enforce limit-aware practices (bulk processing, selective queries, optimized algorithms, etc.), you're essentially future-proofing your code against large data volumes and surges in activity. This foundation is critical for any enterprise or growing org – it means your Salesforce implementation can scale without a complete rewrite of your Apex logic.

As we conclude our deep dive into Apex and platform limits, remember that limit management and good design go hand-in-hand. In the next chapter, we will build on this foundation by exploring Apex design patterns and reusability strategies. These patterns will further help ensure that your code is not only efficient but also modular, maintainable, and adaptable to change – key qualities for long-term success in any Salesforce project.

12

APEX DESIGN PATTERNS AND REUSABILITY

Design patterns in software are **reusable solutions to common design problems**, serving as high-level templates that developers can follow instead of reinventing the wheel for every challenge. In the context of Salesforce Apex, applying design patterns means encoding proven architectural paradigms into your code, making it more robust and easier to maintain. Reusability, on the other hand, refers to writing code in a modular, generic way so that it can be leveraged in multiple scenarios with minimal or no modification. Together, design patterns and reusability form the foundation of sustainable Apex code: they help us build solutions that are easier to understand, extend, and scale over time.

Why are design patterns and reusability so important in Apex? Apex runs in a multi-tenant cloud environment with strict governor limits and a shared runtime. Poorly structured code can become difficult to maintain as an org grows and requirements evolve. Design patterns address this by providing **structured solutions** and a **shared vocabulary** for tackling recurring problems. By using patterns, development teams can communicate solutions more clearly ("let's implement a singleton for this config" or "use a factory to create these records") and ensure consistency across projects. Patterns also encourage best practices like loose coupling (minimal dependencies between classes) and separation of concerns, which directly boost **maintainability and scalability**. Reusable code reduces duplication; when a piece of logic exists in one well-designed place (say, a utility class or a service method), any changes to that logic only need to be made once. This dramatically cuts down on errors and effort when updating functionality.

In the Salesforce ecosystem, these principles translate to tangible benefits. **Enterprise architects** appreciate that a patterned, modular codebase can better accommodate new business requirements without extensive rework. **Developers** find that code guided by patterns is easier to debug and extend, since each class or module has a clear role. **Testers** benefit because patterns like Dependency Injection make it simpler to isolate components and create repeatable tests. Even the **product owners and managers** see advantages: consistent use of patterns often means faster development (since developers are following known approaches) and lower risk of regressions, leading to more predictable project timelines.

It's worth noting that Apex, unlike some languages, doesn't have an expansive ecosystem of frameworks to impose a project structure. For example, there's no built-in equivalent of a Spring or.NET Core framework that dictates how you organize Apex logic. This is why patterns become even more critical – they act as the building blocks of your Apex architecture. With the platform's continued evolution (Salesforce Winter '25 and beyond), Apex has gained features like improved runtime performance and refined syntax, but the core need for good design principles remains. In fact, as orgs accumulate more Apex code over the years, using design patterns is key to **long-term code health**. A well-structured Apex codebase is easier to split into modules or packages, easier to review for security and compliance, and more resilient to changes in Salesforce platform features.

Before diving in, let's clarify what we mean by "design patterns" in this chapter. We'll be focusing on classic **Gang of Four (GoF)** object-oriented patterns adapted to Apex where relevant, as well as some Apex-specific patterns and practices for code reuse. These patterns are typically grouped into three categories :

Creational Patterns – Handle the flexible creation of objects (we'll explore Factory and Singleton).

Structural Patterns – Deal with class and object composition, forming larger structures (Facade and Adapter are our focus).

Behavioral Patterns – Manage effective communication and assignment of responsibilities between objects (we'll cover Strategy and Command).

In addition, we'll discuss the principle of **Dependency Injection** (a cornerstone of decoupled design) and practical approaches to **code reusability** such as utility classes and libraries. Finally, we'll look at **refactoring** existing Apex code to introduce patterns, tying everything together for real-world use.

Throughout the chapter, we will also illustrate how to leverage ChatGPT as a copilot in your Apex journey. This goes beyond just typing queries; you can use voice input for hands-free brainstorming, upload code files to analyze and refactor, employ the new ChatGPT "Operator" mode to perform multi-step tasks (like searching documentation or refactoring code autonomously), and even use image generation tools to create UML diagrams or flowcharts for better understanding patterns. Each section will include multiple practical ChatGPT prompt examples to encourage you to apply what you learn. Think of these as interactive exercises – ways you can involve ChatGPT to solidify concepts, experiment with pattern implementations, or get guidance tailored to your own Apex scenarios.

With expectations set, let's begin our exploration of Apex design patterns. In the next section, we'll introduce the concept of design patterns within Apex in more detail, setting the stage for the specific patterns to come and demonstrating how ChatGPT can assist in recognizing and applying these patterns effectively.

INTRODUCTION TO APEX DESIGN PATTERNS

Design patterns originated in traditional software engineering, but they are highly applicable to Apex development despite the platform's unique constraints. In essence, **an Apex design pattern is a tried-and-true solution model for a common problem in Apex code design.**

Instead of solving a familiar problem from scratch (e.g., how to ensure only one instance of a class is used, or how to handle a variety of actions in a uniform way), developers can apply a pattern that provides a template for the solution. This section lays the groundwork by reviewing key design pattern concepts and how they map to Apex, plus the role reusability plays in Apex code quality. We'll also introduce some ways to use ChatGPT as a learning and coding assistant in this context.

First, recall the three broad categories of design patterns and their relevance:

Creational patterns concern how objects are created. They provide flexibility in object creation by abstracting the instantiation process. For example, rather than calling new directly throughout your code (which scatters knowledge of specific classes), a creational pattern might centralize object creation. This is particularly useful in Apex given the limits on object instantiation within governor limits – you want to reuse objects where possible and only create what you need. In Apex, creational patterns help enforce efficient use of resources and simplify switching out implementations.

Structural patterns deal with how classes and objects are composed to form larger structures. They often help to simplify complex relationships or integrate systems with incompatible interfaces. Given that Salesforce often involves integrating Apex with external services (via callouts) or with other org components, structural patterns can provide clear interfaces and reduce integration complexity. They promote organize-and-conquer: breaking a big problem (or complex subsystem) into approachable pieces and then providing a simple way to use them.

Behavioral patterns focus on communication and the delegation of responsibilities between objects. In Apex, these patterns can manage how different parts of your code interact, especially in asynchronous processes or trigger-based logic. Behavioral patterns often increase flexibility by allowing a system to choose behavior at runtime or by decoupling the sender of a request from the receiver. This is valuable in Apex trigger frameworks or service layers where different actions might occur based on context.

Apex being an **enterprise cloud platform language** means that design patterns often take on an added dimension: they need to coexist with Salesforce's declarative features and multi-tenant safeguards. For instance, in a Java application you might freely use threads or complex memory structures in a pattern implementation; in Apex, you're bounded by governor limits (CPU time, SOQL limits, etc.) and by the one-transaction-at-a-time model. The good news is that most classic patterns can be implemented in Apex with some adaptation, and doing so yields the same benefits of clarity and robustness. In fact, many Salesforce architects encourage using these patterns as part of "Apex Enterprise Architecture" – a collection of patterns and best practices adapted for Salesforce by experts like Andrew Fawcett.

Let's consider a concrete example of why a design pattern might be needed in Apex. Suppose you have a set of complex business rules that determine how to calculate discounts on opportunities. A novice approach might hard-code these rules in a giant if-else or switch block. That works initially, but as rules multiply, this approach becomes unwieldy and hard to maintain (a classic code smell). A seasoned Apex developer might recognize this as an opportunity to apply the **Strategy pattern** – encapsulating each discount algorithm in its own class and making the selection of which to use configurable at runtime. The result is code that's easier to extend (add a new discount type by just adding a class) and simpler to test (each strategy can be tested in isolation). We will explain more about the Strategy pattern in the behavioral patterns section, but this illustrates the thought process: look for recurring problems and resolve them with known patterns.

Reusability goes hand-in-hand with design patterns. Patterns often inherently promote reusability because they encourage separating generic logic from specific. For example, using a pattern, you might create a general purpose class that can be used by many different parts of your application, rather than duplicating logic. A core Apex best practice is *"always write scalable and reusable code so that it can be reused in another functionality or module"*. Achieving this means writing classes that are generic where possible, using configurations to specialize behavior instead of hard-coding, and leveraging inheritance or interfaces to allow one piece of code to serve multiple purposes. We'll see this theme in patterns like Factory (one factory can create many types of objects) and Facade (one facade unifies many operations).

Now, how can ChatGPT help you in understanding and applying design patterns in Apex? A conversational AI can accelerate your learning curve dramatically:

Explaining Concepts: If you're unsure about a pattern, you can ask ChatGPT in natural language. For example, *"Explain the Facade pattern in Apex and give me a simple use case"*. ChatGPT can summarize the pattern's intent (e.g., *Facade provides a simple interface to a complex subsystem*) and even relate it to Salesforce (such as a facade class managing multiple callouts). This is like having an on-demand tutor.

Interactive Brainstorming via Voice: Sometimes you might not even know what to ask. Using voice input, you could verbally describe your problem to ChatGPT. For instance, you could say, "I have three different ways to calculate a score in my Apex code and I keep using if-else to pick one – is there a design pattern that could clean this up?" Speaking out loud (using voice input on a ChatGPT mobile app or enabled device) can make the problem description more fluid, and ChatGPT might recognize this scenario as a candidate for the Strategy pattern and advise accordingly. This beyond-typing method can be useful when you want to brainstorm hands-free or think through a problem collaboratively with the AI as if it were a colleague.

Identifying Patterns in Code: You can upload a snippet of your Apex code (using ChatGPT's file upload or code interpreter capabilities) and ask, "What patterns (or anti-patterns) do you see here?" ChatGPT might point out, for example, that you have a lot of static utility methods that could be organized into a Singleton or that you're tightly coupling your trigger logic with helper classes (suggesting a need for a better separation, maybe a Command pattern or a trigger framework). This kind of code review can reveal opportunities to refactor for better reusability.

Step-by-Step Guidance: Patterns can be abstract, so seeing an example helps. You could prompt ChatGPT for a step-by-step implementation: *"How do I implement a Singleton pattern in Apex?"*. It might respond with a structured explanation: first make the constructor private, then add a static instance and a public getInstance method, etc., possibly even providing a code template. You can take that template and adapt it to your class.

Developer Tip: When learning a new pattern, practice by applying it to a trivial problem. For example, implement a Singleton logger class in Apex that just collects messages. Use ChatGPT to check your work: paste your Apex class and ask if it truly follows the Singleton pattern. This can validate your understanding in a low-risk way before you use the pattern in a mission-critical part of your org.

As we move into each pattern-specific section, we'll include **practical ChatGPT prompts** you can try. These will range from asking for examples or analogies, to getting help with refactoring your own code. By actively engaging with these prompts, you'll not only read about the patterns but also experience how an AI assistant can support your learning and implementation process.

With this solid introduction in mind, let's transition to our first category: creational patterns. In the next section, we'll explore the Factory and Singleton patterns, which deal with controlled object creation and global access – fundamental techniques for managing Apex object lifecycles.

CREATIONAL PATTERNS: FACTORY AND SINGLETON

Creational design patterns are all about flexible object creation – ensuring that your code isn't littered with direct class instantiations that can make maintenance and scaling harder. In Apex, where efficient use of resources is crucial, creational patterns can help centralize and optimize how objects are spawned. We'll focus on two of the most widely used creational patterns: **Factory** and **Singleton.**

Factory Pattern in Apex

The **Factory pattern** provides an interface or method for creating objects, but leaves the decision of which class to instantiate to that factory method, rather than the calling code. In simpler terms, a Factory is like a dedicated object-creation service: you ask it for an instance of something, and it delivers the appropriate one based on input or context, without the caller needing to know the exact class being instantiated. This indirection adds a layer of flexibility. Why is that good? Because if the way an object is created changes (maybe a new subclass is introduced, or the criteria for choosing a class changes), you update the factory in one place, and all code that relies on it automatically gets the new behavior.

In Apex, a common use case for the Factory pattern is when you have logic that needs to handle multiple record types or variations of a process. For instance, imagine an application that processes different kinds of support cases: technical issues, billing issues, and general inquiries. Each type might require creating a different helper object to handle the case (perhaps TechnicalCaseProcessor, BillingCaseProcessor, etc., all implementing a common interface like ICaseProcessor). Without a factory, your code might do something like:

if(case.Type == 'Technical') processor = new TechnicalCaseProcessor();

else if(case.Type == 'Billing') processor = new BillingCaseProcessor();

...

This logic scattered everywhere is hard to maintain. Instead, a Factory class (say CaseProcessorFactory) can encapsulate that decision:

ICaseProcessor processor = CaseProcessorFactory.createProcessor(case.Type);

Inside the factory's createProcessor method would be the if/else or mapping logic to return the right instance. Now the calling code is decoupled from the concrete classes. If a new case type "Legal" is added with a new LegalCaseProcessor, you only modify CaseProcessorFactory to handle it, and none of the call sites need changing.

The Factory pattern is extensively useful in Salesforce scenarios where object creation needs to be dynamic or based on configuration. Some examples include:

Dynamic Record Creation: Choosing which object type to create at runtime based on criteria (e.g., create a Lead vs. Contact vs. Account depending on input data). A factory can abstract the logic of deciding the object type.

Integration Adapters: When integrating with external systems, factories can help create appropriate Apex wrapper objects for data coming from different sources. For example, if data comes in JSON vs XML, you might have factory methods that return the correct parser object.

Configurable Processes: Using Custom Metadata or Custom Settings, you might let admins configure which "strategy" or class to use for a certain process. A factory can read that config and instantiate the class named in a metadata record. (This overlaps with Dependency Injection, which we'll cover later, but a simple approach is a factory that does something like: Type t = Type.forName(className); return (BaseClass) t.newInstance(); in Apex.)

Implementing a factory in Apex usually involves a static method or a dedicated class with methods to create instances. Apex doesn't support method overloading by return type alone, so often you'll have distinct methods or use parameters to distinguish what's being created. For example, you might have InvoiceFactory.newInvoice(String region) that returns an Invoice subclass based on region. Or use an enum to indicate which type of object is needed.

One thing to watch out for is that Apex has a governor limit on the number of script statements and queries, so while factories add an extra call, that overhead is negligible compared to the clarity and flexibility gained. Just ensure your factory methods themselves are efficient (for instance, if they need to query some config, maybe cache that info statically to avoid repeated queries).

ChatGPT Prompt – Using a Factory in Apex: Try describing a scenario to ChatGPT and let it design a factory. For example, "I have a base class *NotificationSender* with subclasses *EmailSender*, *SMSSender*, and *PushSender*. Can you write a factory method in Apex that returns the correct *NotificationSender* subclass based on an input string 'Email', 'SMS', or 'Push'?". ChatGPT can produce the Apex code for a NotificationFactory with a method like createSender(String type) that returns the appropriate object. This helps you see a concrete implementation. You can even follow up with, "How would I use this factory in my trigger handler? Provide an example."

Architect Tip: Standardize the use of factories in your Apex architecture. For instance, if you adopt a factory for creating service classes or domain objects, document this in your engineering playbook. This ensures all developers know where to look for object instantiation logic and prevents ad-hoc new calls scattered in code. ChatGPT can help generate that documentation or examples for your team, ensuring consistency in how patterns are applied across the org.

A variant of the factory concept in Salesforce is the **Factory Method pattern**, where the factory is not a separate class but a method in an existing class (often using inheritance). Apex classes can have factory methods that subclasses override to create their own instances. However, common practice in Apex is to use simple static factory methods in a utility class for most cases, as it's straightforward.

Before moving on, let's mention testing: When you use factories, writing unit tests for object-creation logic becomes easier. You test the factory in isolation: give it input X, assert that the returned object is of type Y. In other parts of your code, you can even **mock the factory** if needed (though if the factory is simple, you might not need to). This isolated testing improves reliability. **Testers** might not directly care about the factory pattern, but they will appreciate that the code using factories is less prone to weird issues — since a factory funnels the creation logic, there's less chance of something forgotten in one branch of code.

Singleton Pattern in Apex

The **Singleton pattern** ensures that a class has only one instance and provides a global point of access to that instance. In other words, no matter how or where in your code you ask for an instance of the Singleton class, you get back the same object. This is useful for managing **centralized state or resources** – things like configuration settings, caches, or coordinators that logically should be singular throughout your application.

In Salesforce Apex, the Singleton pattern is typically implemented using a static variable to hold the single instance. Here's the common recipe in Apex:

Private constructor – Make the class constructor private so no outside code can do new YourClass().

Static instance field – Have a private static variable of the class type, initially null.

Public static accessor (getInstance) – Provide a static method (often named getInstance) that creates the object on first call and then always returns the same object on subsequent calls.

This pattern in Apex looks like:

```
public class SettingsManager {
  private static SettingsManager instance;
  private SettingsManager() {
    // possibly load settings from Custom Metadata or Custom Setting
  }
  public static SettingsManager getInstance() {
    if (instance == null) {
      instance = new SettingsManager();
    }
    return instance;
  }
  // other methods to use the settings
}
```

In this example, SettingsManager could be a class that holds configuration data (say from a Custom Metadata record) for easy access. The first time getInstance() is called, it constructs the object and perhaps loads the config. Subsequent calls just reuse that in-memory data. This ensures consistency (everyone is using the same settings values) and efficiency (you don't repeat the loading logic, which might involve a SOQL query).

However, it's important to understand the scope of "single instance" in Apex. Apex runs in a stateless environment across transactions. A static in Apex persists only for the duration of a single transaction (for example, one execution of a trigger or a single web service call). Once the transaction is over, the memory is gone. So an Apex Singleton is not a singleton *across the entire org or all time*, but rather "singleton per transaction." That is usually fine: if you want something truly application-wide persistent, you'd use a different mechanism (like a Custom Setting record, Platform Cache, or a singleton held in an @AuraEnabled cache for a user's session). But for most cases, per-transaction is what we need — e.g., you want to load config values once per trigger execution, not every time a utility method is called in that execution.

Common use cases in Apex for Singletons include:

Configuration and Settings: As described, load a bunch of config data once. Perhaps you have a Custom Metadata type defining business rules; a singleton can fetch them and store them so all classes can use it without each querying the data separately.

Governor Limits Tracker or Logger: You might design a logger that collects log messages during a transaction and at the end (maybe in an after trigger or in finally block) it dumps them somewhere. Making it a singleton (e.g., LogService.getInstance().log('msg')) means everyone logs to the same place. This is especially handy for complex transactions involving multiple triggers or classes – the singleton logger accumulates context from all parts.

Shared Resources: If you have a class that manages a connection or a big data structure (like an in-memory cache of records queried early in the transaction), a singleton can ensure all parts of the code use the same instance instead of duplicating work.

Keep in mind Apex's limits: if your singleton holds onto a lot of data, you could run into heap size limits. So use the pattern judiciously for cases that truly need one instance. Also, because static state is not preserved across transactions, don't rely on an Apex singleton to cache data across trigger contexts or between separate invocations (for that, consider Platform Cache or storing in a Custom Setting record which persists).

In Apex, implementing the Singleton correctly means also considering test class behavior. Static variables do *not reset between test methods by default*. Salesforce runs all test methods in a class in one transaction by default (though each test method is isolated for DML etc., static state carries over if not cleared). To avoid interference, tests should either call a utility to reset that singleton (you could add a reset() method marked @TestVisible that sets the static instance to null, and call that in test tearDown or at start of each test), or use the Test.startTest() / Test.stopTest() pair which provides some isolation.

Despite these nuances, Singletons in Apex are straightforward and powerful. Many Apex frameworks or enterprise patterns use singletons under the hood. For example, you might encounter a "Service Locator" in Salesforce (like the FFLib_SObjectUnitOfWork in FinancialForce's Apex Commons library) which is often implemented as a singleton to register and provide services across the transaction.

ChatGPT Prompt – Implementing a Singleton: A good exercise is to let ChatGPT walk you through it. For instance: *"Explain how to implement a Singleton pattern in Apex and show an example with a class that provides global configuration."* This should yield an explanation and a code sample similar to our SettingsManager. You can follow up by asking, *"Why is the constructor private?"* or *"What happens if two users call getInstance at the same time?"* to clarify any doubts. (The answer to the latter in Apex: two users in different contexts have their own transaction, so they each get their own instance – which is expected; the pattern's intent is within one execution context.)

Tester Tip: When you see a Singleton in Apex, it often means the code is managing some state or resource that all parts of a process use. From a testing perspective, be aware that a singleton might retain state between operations in a single test method. Always write tests such that one test method doesn't rely on another (which is a general rule, but static state can subtly violate it). Use techniques like resetting static members in test setup. You can ask ChatGPT to generate a test class template for your singleton – it might show how to initialize and verify the singleton's behavior, which can guide you in writing robust tests.

To illustrate the benefit of singletons, consider an Apex Batch job that processes millions of records in chunks. If each execute chunk needs some reference data (say a set of ZIP code to region mappings), you could have a Singleton loader for that data. The first batch execute calls ZipCodeCache.getInstance().getRegionMap() which triggers loading from a Custom Object. That data is stored in a static variable. For subsequent execute calls *within the same batch invocation*, the data is already in memory – saving repeated queries. (Batch Apex runs each execute in a fresh transaction by default, so a singleton wouldn't persist across executes unless you use the rarely-used Database.Stateful interface – but that's another story. The point is, in contexts where multiple operations happen sequentially in one context, singletons help.)

By using Factory and Singleton patterns, we control object creation in Apex in two ways: **Factory** gives us flexibility in *what* subclass or type of object to create (and encapsulates that choice), while **Singleton** controls *how many* instances of a class exist (exactly one, when needed). These patterns often complement each other; for example, a Factory might itself be implemented as a Singleton if you only ever need one factory instance (though factories are often just static methods, not needing an instance at all).

Now that we've covered creation, our next step is to look at organizing complexity in Apex. In the following section, we'll move to structural patterns, examining how **Facade** and **Adapter** can help simplify interactions with complex systems and external interfaces, making our Apex code cleaner and more modular.

STRUCTURAL PATTERNS: FACADE AND ADAPTER

Structural design patterns help us **compose classes and objects into larger structures** while keeping these structures flexible and efficient. In Salesforce Apex, structural patterns are particularly useful when dealing with complex systems (like multiple Apex classes coordinating or external integrations). We will discuss two key structural patterns here:

Facade, which provides a simplified interface to a complex system, and **Adapter**, which allows classes with incompatible interfaces to work together. Both patterns can make your Apex code more modular, readable, and easier to integrate with other systems.

Facade Pattern in Apex

The **Facade pattern** is about providing a unified, easy-to-use interface to a set of interfaces in a subsystem. Think of it as a front-facing simplified API that hides the complexity of underlying operations. In Apex, a Facade is often implemented as a class with high-level methods that internally call various lower-level methods or services. The callers of the Facade don't need to know about the details – they just call a single method to accomplish a task that might involve multiple steps behind the scenes.

Consider a scenario in a Salesforce org: When a new customer is onboarded, you need to perform several actions – create an Account, create a Contact, set up some default Opportunities or Cases, and perhaps call out to an external system to register this customer. Without a facade, the onboarding code (maybe a trigger or a controller) would have to call each of these services in turn, handling any ordering or dependencies between them. That can lead to cluttered code at the call site and duplicated logic if you need to onboard customers from different entry points.

A **CustomerOnboardingFacade** class could simplify this. It might have a method onboardCustomer(CustomerInfo info) that inside creates the account, contact, calls external systems, etc., by delegating to various service classes (AccountService, ExternalIntegrationService, etc.). The trigger or UI controller just calls CustomerOnboardingFacade.onboardCustomer(info) – one call, high-level. The facade hides the orchestrations needed to fulfill that request.

In Apex, facades are useful to **encapsulate business processes** that involve multiple objects or external calls. They can improve consistency (everyone uses the same process) and reduce errors (fewer places where the orchestration logic can go wrong). If later the process changes (say we no longer auto-create an Opportunity on onboarding), you update the facade's code, and all callers automatically use the new behavior.

Another real-world Apex example of Facade might be in working with complex APIs. For instance, Salesforce has the Metadata API, Tooling API, and so on – if you've ever done callouts to those, they involve many steps (login, set up session, construct SOAP messages, etc.). A facade class could provide simple methods like MetadataApiFacade.deploy(metadata) which internally handles all the SOAP details and login. This way, your business logic doesn't get polluted with those complexities.

One more from within Salesforce: think of complex calculations. Suppose to calculate a price, you need to do tax computation, discount lookup, and apply currency conversion. You could write a PriceCalculatorFacade that exposes a calculatePrice(productId, customerId) method. Internally it calls a TaxService, DiscountService, CurrencyService in the right order and returns the final number. The caller doesn't need to know those steps.

Benefits of Facade (in Apex context):

Simplified interface: One place to call to get something done, rather than multiple calls. Makes code using the facade more readable.

Decoupling: The calling code is decoupled from the intricate workings of the subsystem. You can change subsystem classes without breaking callers, as long as the facade interface stays consistent.

Consistent usage: If many parts of the application need to do the same multi-step process, a facade ensures they all do it the same way (reducing bugs).

Limit Governor Exposure: By funneling operations through a facade, you might better manage governor limits. For example, the facade could ensure that the overall process is bulkified or uses a single SOQL query for what otherwise might be multiple queries spread around.

Testing: A facade can be tested as one unit (does it correctly perform the series of actions?). Also, because it's decoupled, you could replace a facade with a mock in tests for code that uses it, if needed.

Implementing a facade in Apex is straightforward: create a class, add methods that represent the high-level operations, and within those methods, create and invoke instances of the lower-level classes. There's no special language feature needed, it's more of a design decision.

For instance:

```
public with sharing class OrderFacade {
  private OrderService orderService;
  private PaymentService paymentService;
  private ShippingService shipService;
  public OrderFacade() {
    orderService = new OrderService();
    paymentService = new PaymentService();
    shipService = new ShippingService();
  }
  public void placeOrder(Order__c order, PaymentDetails payInfo) {
    orderService.createOrder(order);
    Payment p = paymentService.processPayment(payInfo);
    shipService.scheduleDelivery(order, p);
  }
}
```

This facade placeOrder method wraps the complexity of creating an order, processing payment, and scheduling delivery. A calling class (maybe an AuraEnabled class called from LWC, or a trigger after insert of Order__c) can just do new OrderFacade().placeOrder(order, paymentInfo); and not worry about the internals.

ChatGPT Prompt – Building a Facade: You can have a bit of fun by asking ChatGPT to simulate building a facade. For example, "I have classes EmailService, SMSService, and PushService each with a send-

Notification() method. Write an Apex facade class NotificationFacade with a method sendAlert(message) that sends the message via all three services." ChatGPT should produce a facade that internally calls each service's method. This demonstrates how multiple actions get wrapped in one call. You might follow up: "Modify the facade to only send certain types of alerts via certain channels (e.g., critical alerts via all, normal alerts via email only)." This would show how adding logic to the facade centralizes control of a process.

Product Owner Tip: When discussing requirements with your development team, if you hear them talking about needing to call several different parts of the system to accomplish a user story, ask if they plan to encapsulate that in a single module. You don't need to know the term "facade," but by ensuring the team thinks about simplifying interfaces, you're indirectly encouraging this best practice. This matters because down the line, if that process changes, having it centralized will mean less cost to modify. Product owners can thus advocate for such patterns by allocating time for "tech design" and emphasizing long-term maintainability as an acceptance criterion.

From a testing perspective, facades mean you can test the complex process in one go. But do remember to test edge cases (for example, in the OrderFacade, what if payment fails? The facade might need to handle that – maybe by rolling back the order creation or marking an order as pending payment). These decisions are part of design: sometimes the facade handles exceptions and compensating actions, other times it might throw an exception back to the caller.

Adapter Pattern in Apex

The **Adapter pattern** is about **allowing two incompatible interfaces to work together** by introducing an adapter class that translates between them. In other words, if you have class A expecting some interface X, but you have class B that offers similar functionality with interface Y, an Adapter can wrap B to present an interface X to A. The classic analogy is a power adapter: your laptop expects a certain plug shape (interface), the wall outlet is a different shape; the adapter lets them connect.

In Apex, adapter pattern usage often comes up with **external systems or legacy code**. For instance, maybe you have an Apex interface IPayment-Processor used in your code (with methods like processPayment()), but you want to integrate a new payment gateway that has a REST API or a different signature for its methods. You could write an adapter class NewGatewayAdapter that implements IPaymentProcessor, but internally makes the callout to the new gateway's API. The rest of your code can remain unchanged (it calls IPaymentProcessor.processPayment() and now via the adapter it works with the new system).

Another scenario: You have old code that works with a class, but you're introducing a new class that should be interchangeable. If you cannot change the old code easily, you might adapt the new class to the old interface.

In trigger development, sometimes adapters are used to allow reuse of code between trigger handlers and batch processes. For example, imagine you wrote a trigger handler that expects to be given a List<Account> and does something. Later you want to use the same logic in a batch context (where you have a Database.Batchable that deals with one record at a time). You could write an adapter that implements the batchable execute by calling the trigger handler's method on a single-item list. This way the trigger handler's logic is reused even though the batch execute has a different method signature (execute(Database.Batchable-Context, List<SObject>) vs a custom handleAfterUpdate(List<Account> newList) in your handler class). This is a simplistic example, but it shows adapting one calling pattern to another.

Adapters in Apex can also be used when working with different API versions or libraries. For instance, you have some library code (perhaps from an unlocked package or managed package) that expects data in a certain format or a certain class type. You can't or don't want to change that library (especially if it's managed, you can't). But you can write an Apex class that takes your data and converts it to what the library expects, then calls the library, and if needed converts results back. That class is acting as an adapter between your code and the library.

Apex's strongly-typed nature means adapters often involve implementing an interface or extending a class. For example, if class A expects interface X, your adapter will implement X. Inside those interface methods, it will call the real methods of the class B it's adapting.

Let's illustrate with a simple pseudo-code scenario:

Suppose Salesforce added a new Geolocation service you want to use, but your codebase uses an older interface ICoordinateFinder { Location getCoordinates(String address); }. The new service has a class GeoService with a method fetchLocation(String addr) returning a GeoPoint. You could adapt like this:

```
public class GeoServiceAdapter implements ICoordinateFinder {

  private GeoService service = new GeoService();

  public Location getCoordinates(String address) {

    GeoPoint point = service.fetchLocation(address);

    // convert GeoPoint to Location (assuming Location is some existing
class/struct)

    return new Location(point.lat, point.lon);

  }

}
```

Now wherever your code expects an ICoordinateFinder, you can pass an instance of GeoServiceAdapter and it will use the new service under the hood. The rest of the system remains unaware of the change, fulfilling **Dependency Inversion Principle** ("depend on abstractions, not concretions") which was mentioned earlier.

Key point: The Adapter pattern preserves existing interfaces and code, leveraging a wrapper to bridge difference. It's extremely valuable in integration contexts. In fact, one of the bullet points in that Medium article on design patterns by Mila Mirović was using the Factory pattern for **Integration Adapters** – conceptually, you could have a factory provide an appropriate adapter for a given integration at runtime.

Salesforce-specific example: Suppose you use a third-party package for sending emails (with its own classes/methods) but you also have a custom email sending service. You could create an interface Email-

Sender that your code uses. Then create two adapters: ThirdPartyEmail-Adapter (wraps calls to the package) and CustomEmailAdapter (wraps your custom email logic) – both implement EmailSender. A factory could choose one based on an org setting. This way, your code can switch between the third-party and custom implementations easily via configuration, because an adapter makes the third-party package conform to your EmailSender interface.

ChatGPT Prompt – Adapter in Action: Ask ChatGPT to act as if it's integrating two systems. For example: "I have an interface IWeatherService with method getForecast(city) that my code uses. I want to use a new API which provides weather via a class NewWeatherAPI with a method retrieveForecast(cityName). Show how to implement an adapter in Apex." ChatGPT should produce a class NewWeatherAdapter that implements IWeatherService.getForecast and internally calls NewWeatherAPI.retrieveForecast. It might show creating an instance of NewWeatherAPI inside and returning the data in the expected format. This exercise clarifies how the adapter pattern works by example.

Architect Tip: Adapters are especially handy when dealing with managed packages or external APIs. As an architect, mandate that any external system integration in Apex uses an adapter layer. This isolates external dependencies and makes future changes easier. For instance, if you swap one geocoding API for another, only the adapter needs to change if both adhere to your internal interface. You can use ChatGPT to generate skeleton adapters for new integrations, ensuring developers have a starting point that adheres to the pattern.

One caution in Apex: Adapters (and patterns in general) add extra classes and layers. In extreme governor-limit scenarios (like a loop that calls something thousands of times), the indirection of an adapter or facade could add tiny overhead (a few more function calls). Usually, this is negligible and should not dissuade you from using patterns, but be mindful to avoid unnecessary abstraction in performance-critical code. That said, clear structure often *enables* performance improvements (e.g., it's easier to add caching in an adapter or facade since it's a central point of control).

To test an adapter, you essentially test that given certain input, it produces the same outcome as the adapted class would have if called directly. If possible, test the adapter in isolation by mocking the underlying class. In Apex, mocking might mean you use a fake version of the external service if you can (if it's your own class, you can substitute; if it's a callout, maybe use HttpCalloutMock to simulate the external response). The adapter itself should have minimal logic (just conversion), so most of your confidence comes from integration testing it with the real service and unit testing that it correctly maps inputs/outputs.

With structural patterns like Facade and Adapter in your toolkit, you can significantly clean up how different parts of your Apex system connect to each other and to external systems. They both strive to make interactions simpler: Facade by hiding complexity, Adapter by reconciling differences.

Now that we've covered creation and structure, the next challenge is behavior: how do we make the interactions and algorithms in Apex more flexible? We will now move to **Behavioral Patterns**, covering Strategy and Command, which will show us ways to inject different behaviors and encapsulate actions in Apex for greater flexibility and reusability.

BEHAVIORAL PATTERNS: STRATEGY AND COMMAND

Behavioral design patterns focus on **how objects interact and distribute responsibility**. In Apex, behavioral patterns can help us write code that is more flexible and adaptable to changing business rules or complex workflows. We'll discuss two such patterns: **Strategy**, which allows selecting an algorithm's implementation at runtime, and **Command**, which encapsulates a request or action as an object. These patterns might sound abstract, but they have concrete benefits for Apex: Strategy can replace large conditional logic with clean, interchangeable modules, and Command can queue up operations, enable undo functionality, or simply decouple the sender of a request from the executor.

Strategy Pattern in Apex

The **Strategy pattern** encapsulates interchangeable algorithms (or strategies) inside classes and enables selecting one of these algorithms at runtime. Essentially, you define a family of behaviors, put each in a separate class, and make them all adhere to a common interface. The code that needs to perform the behavior (the context) holds a reference to the interface, but the concrete implementation can be swapped easily.

In Apex, Strategy is particularly useful for implementing **business rules that vary** based on some criteria. A classic Salesforce example: lead assignment or scoring strategies. Suppose you have different ways to score a Lead (one strategy for leads from webinars, one for leads from cold calls, etc.). You could implement this with a bunch of if-else statements in one big method, but that becomes messy as strategies proliferate. Instead, you define an interface LeadScoringStrategy with a method score(Lead l), then implement classes like WebinarLeadScoring, ColdCallLeadScoring, each with their own score logic. The main code (maybe a trigger or batch) can decide which strategy to use, perhaps by checking Lead.Source or using a Custom Metadata setting that indicates the active strategy, then call selectedStrategy.score(lead).

The benefit is that adding a new lead scoring method doesn't require altering a giant if-else; you just add a new class. The selection logic might need an update (especially if it's by code), but even that can be abstracted (you could have a factory as discussed, or use dependency injection to supply the strategy).

Another common use: **Discount calculation, tax calculation, or validation rules** in Apex logic. Salesforce admins often ask for changing logic ("Now we have a new discount rule for VIP customers"). If the code is strategy-oriented, you perhaps just write a new class VipDiscountStrategy implementing an IDiscountStrategy interface and adjust configuration for VIP accounts to use that strategy.

Apex's limited support for reflection means that to select a strategy by name (string) you might use Type.forName or if you use the fflib pattern, a registry of strategies. However, for many cases, a simple if in one place or using Custom Metadata to map conditions to class names is sufficient.

Let's give a tangible Apex scenario using Strategy:

Suppose we're building a validation framework for Contacts before insert. Some contacts might need email validation, others phone validation, maybe based on country or source system. We define:

```
public interface ContactValidationStrategy {
  Boolean isValid(Contact c);
}
```

We then have classes EmailValidationStrategy, PhoneValidationStrategy implementing that interface. Each checks the respective field on the contact (maybe ensure email contains "@" or phone matches a regex).

We have a ContactValidator (the context) that uses a strategy:

```
public class ContactValidator {
  private ContactValidationStrategy strategy;
  public ContactValidator(ContactValidationStrategy s) {
    this.strategy = s;
  }
  public Boolean validate(Contact c) {
    return strategy.isValid(c);
  }
}
```

At runtime, we decide which strategy to use. Perhaps:

```
ContactValidator validator;
if(contact.LeadSource == 'Web') {
  validator = new ContactValidator(new EmailValidationStrategy());
} else if(contact.LeadSource == 'Call') {
  validator = new ContactValidator(new PhoneValidationStrategy());
```

```
}

Boolean ok = validator.validate(contact);
```

This is a trivial example, but it demonstrates separating *what to do* (validate contact) from *how to do it* (which validation rule to apply).

Strategy pattern promotes open/closed principle: your classes for strategies are open for extension (you can add new ones) but the context doesn't need to change for new ones (closed for modification). Well, selection logic might need slight adjustment unless you externalize it, but the heavy lifting is separated.

In Salesforce's context, strategy can also be used to abstract different implementations of an interface depending on context – for example, maybe in a managed package you offer a default strategy but allow subscribers to override by plugging in a custom subclass via configuration (with careful design, maybe using Type.forName on a class name stored in Custom Metadata to instantiate a custom strategy).

ChatGPT Prompt – Exploring Strategy: A great way to see Strategy in Apex is to ask ChatGPT for an analogy or diagram. For instance, *"Explain the Strategy pattern with an Apex example of different algorithms for calculating shipping cost (standard vs express shipping)."* ChatGPT might describe an IShippingStrategy with classes StandardShipping and ExpressShipping and how an Order class might use one or the other based on user choice. You can also ask, *"How would I choose a strategy based on a Custom Metadata record in Apex?"*. This brings in a practical twist: maybe you have a Custom Metadata that has a field "ActiveStrategy" = "ExpressShipping", and you use that to decide which class to instantiate. ChatGPT could outline how to query that metadata and then use Type.forName to instantiate the class by name (since in Winter '25, Apex allows dynamic instantiation of classes by name, as long as it's in the codebase).

Developer Tip: Use Strategy to simplify complex conditional logic. If you find yourself writing a method with a lot of branching based on some type or status, consider refactoring that to use strategies. For example, a "calculateCommission" method that varies by role (sales rep vs partner vs reseller) can be refactored so each role has its own commission calculation class. This not only makes the code cleaner but also isolates the logic for easier testing. You can have separate test methods for each strategy, feeding in sample inputs and verifying outputs.

Testing strategies is straightforward: each concrete strategy is a small class to unit test, and then you test the context selecting the right strategy in integration tests. This modularity often results in more tests that cover each scenario more directly rather than one monolithic test that tries to cover all branches.

One thing to note: In Apex triggers, you often see something akin to Strategy when implementing trigger frameworks. For instance, a trigger might delegate to different handler methods depending on the type of operation or record attributes — that's not exactly the classic Strategy pattern, but a related concept (some frameworks implement something like a Strategy for handling different record types or events).

Command Pattern in Apex

The Command pattern encapsulates a request or an action as an object. This means that rather than calling a method directly to do something, you package the details of that action (what needs to be done, to which receiver, with what parameters) into a command object. This decouples the sender of the request from the object that will execute it. It also allows for interesting capabilities: you can queue commands, log them, undo them, or send them to be executed elsewhere (like asynchronously).

In a Salesforce context, think of a Command object as a little box that knows how to perform one thing. For example, you might have a command object called CreateCaseCommand that, when executed, creates a Case record and maybe sends an email. Or ApproveOrderCommand that approves an order and updates related records. The key is that the logic to do that action resides in the command object's execute() method, not in the caller. The caller might be something like a framework or an orchestrator that simply says "okay, command, do your thing now."

Why use Command in Apex? A few reasons:

Queuing and Batch: Salesforce has async processing (Queueable, Batch, Future). If you design parts of your system as commands, you can queue them for later execution. For instance, user clicks a button to perform a complex action that might take time – instead of doing it immediately, you package the action as a command and enqueue it to a Queueable Apex. The Queueable's execute just calls command.execute(). This is a form of command pattern usage.

Undo/Cancel: Perhaps in an order management system, you want to allow cancellation of operations. If each operation is a command and you implement an undo() method in each (not always needed, but Command pattern literature often mentions it), then you can programmatically reverse actions. For example, CreateCaseCommand.undo() might delete the case it created if possible. Not all scenarios allow perfect undo (especially across transactions), but the pattern encourages thinking in those terms.

Logging and Audit: If each important user action is a Command object, you could record them (say, serialize the commands or at least log the fact that "XCommand" was executed with these parameters at time T). This can help with audit trails or debugging, as you have a standardized way of describing actions.

Macro Commands: The pattern allows grouping commands into a composite command. In Apex, you could implement an ICommand interface with execute() and possibly undo(), and then have an AllOrNothingCommand that in execute just calls a bunch of others. This could correspond to, say, a scenario where a business transaction requires multiple steps (create Opportunity, create Task, update Account). Instead of writing a separate method to do all, you compose it from smaller commands. This is similar to Facade but with the difference that each step is a self-contained command object possibly known at runtime rather than a fixed sequence in code.

An illustrative scenario for Command pattern in Apex could be a simple task scheduler within Apex. Imagine you want to build a mini workflow engine: based on certain conditions, you want to execute a series of actions. You could define various Command classes like SendEmailCommand, UpdateFieldCommand, CreateRecordCommand, etc. Then your engine can pick and execute commands in order. If later you need a new kind of action, you add a new Command class rather than change the engine. You might even store commands in a Custom Metadata or Custom Object (though storing actual Apex objects isn't directly possible, you might store identifiers that your code maps to command instances).

One concrete Apex example: the Apex Database.Batchable interface's finish method could use Command. Let's say after all batch chunks are processed, you want to optionally perform some action (maybe send summary email or call an external system). You could design it so that different batches have different finish behavior. If you treat finish actions as commands, you could pass a command into the batch job (maybe via the constructor). Then in finish() you just call finishCommand.execute(). For one batch it might send an email, for another it might do nothing (a NoOpCommand pattern, an empty command), for another it might chain another batch (which itself can be a command). This decouples the batch logic from what to do at the end.

Implementing Command in Apex usually involves:

An interface, e.g., ICommand with an execute() method (and an undo() method if needed).

Multiple classes implementing that interface, each with specific execute logic.

Optionally, an invoker class or piece of code that holds a command and triggers it. Often the invoker can be the piece of code that decides when to run the command (like our Queueable, or a loop processing commands).

Possibly a receiver: in classic pattern terms, the command delegates to a receiver to do the actual work. In Apex, this might just be the command itself doing DML or calling a service class (so the service class would be the receiver).

Consider a snippet from the earlier found code: there was an ICommand with execute() and undo(), and classes ApplicationCreated, ApplicationProcessing, ApplicationApproved that set a status field on an Application record. They are concrete commands. A separate class had a

MacroCommand to execute a sequence. The use case likely was moving a custom object through stages (like a state machine) with ability to undo a step by reverting to the previous status.

In Apex triggers, one could imagine using Command to represent different operations that should happen after a record is created, and then deciding which commands to run. However, in practice triggers are often straightforward enough that we don't over-engineer them with Command. But in complex automations, it's a consideration.

ChatGPT Prompt – Generating Commands: Try asking ChatGPT for a simple command example. "Give me an Apex interface ICommand with execute() and two implementing classes: TurnLightOnCommand and TurnLightOffCommand that print 'Light on' or 'Light off' (simulate with debug logs)." It will likely produce something illustrative. While "printing" isn't what we do in Apex, debug logs can simulate the action. This shows the skeleton clearly. Then you can ask, "How would I use a list of ICommand to execute a sequence of operations?" This would demonstrate the idea of an invoker iterating over a list of commands and calling execute on each — analogous to MacroCommand.

A step further: *"Show how a Queueable Apex could execute an ICommand passed to it."* This merges patterns with platform features. ChatGPT might show a Queueable class that takes an ICommand in its constructor (maybe as an Object and cast to ICommand, since interface serialization might not be straightforward), and in execute() does command.execute(). This would be a cool demonstration of combining Command pattern with Salesforce's async processing.

Tester Tip: When commands are used to encapsulate business actions, testers can leverage this by creating sequences of commands to simulate user workflows. For instance, if a user process triggers a CreateCaseCommand then an AssignAgentCommand, a tester can in Apex test context instantiate those commands, execute them in order, and then verify the final state (case exists, assigned to an agent). This is effectively integration testing through the command objects. Also, if an undo feature is supported via commands, testers should rigorously test it: ensure that executing a command then undoing it truly restores the prior state. ChatGPT can help by generating testing scenarios or even suggesting what to verify for each command's effects.

Both Strategy and Command patterns might appear to add complexity by introducing extra classes. However, they significantly improve clarity and flexibility. Strategy reduces complex conditional logic and makes adding new behaviors easier. Command turns an action into a standalone object that can be manipulated (queued, logged, undone) like any other data, which is powerful in a system like Salesforce where you often need to schedule or audit actions.

By applying Strategy, you get a more pluggable Apex codebase (choose behavior A or B by just selecting a different class), and by applying Command, you get a more transactional and decoupled codebase (send a command to be executed without needing to know how or where it's executed).

As we conclude our look at behavioral patterns, consider how these might interplay: you could have a Command that internally uses a Strategy (imagine a "PerformCalculationCommand" that has a Strategy for which calculation to do). These patterns are building blocks that can mix and match.

Now that we have covered creation, structure, and behavior patterns, we have a rich toolkit of approaches. But applying them to an existing org's codebase may require changes. This is where Dependency Injection comes in as a facilitative pattern for using strategies or other patterns, and where refactoring skills are needed to introduce patterns into legacy code. In the next section, we'll tackle how to implement Dependency Injection in Apex to further decouple our code, making it easier to swap out implementations (like those strategies) and improving testability.

IMPLEMENTING DEPENDENCY INJECTION IN APEX

If you've ever heard the principle "Depend on abstractions, not concretions," you're already familiar with the core idea behind Dependency Injection (DI). Dependency Injection is a design technique (often considered a pattern or a set of patterns) where an object's required dependencies are provided (injected) to it from the outside, rather than the object instantiating them itself. This leads to a loosely coupled system – classes are not tightly bound to specific implementations of their collaborators, which improves flexibility and testability.

In simpler terms: when using DI, if Class A needs Class B to do its work, Class A doesn't create a new B() itself. Instead, it accepts a B (often via the constructor or a setter) that someone else created and handed in. That someone else could be orchestrating code or a factory. This way, Class A is not tied to *which* B it uses; it could be a real B, a subclass of B, or a mock B in tests.

Why is this important in Apex? Apex, being a strongly-typed OO language, benefits from DI just like Java or C#. It helps especially in writing unit tests where you might want to substitute a fake implementation to simulate certain scenarios or to avoid hitting external systems. It also makes large applications more maintainable by separating object creation from object usage.

Let's break down how you might implement DI in Apex:

Techniques for DI in Apex:

Constructor Injection – This is the most common approach. You provide needed dependencies via the class's constructor. For example:

```
public class OpportunityService {

  private PricingStrategy pricingStrategy;

  public OpportunityService(PricingStrategy strat) {

    this.pricingStrategy = strat;

  }

  public Decimal calculateTotal(Opportunity opp) {

    return pricingStrategy.calculatePrice(opp);

  }

}
```

Here, OpportunityService depends on a PricingStrategy (which is an interface or abstract class). Instead of instantiating a concrete PricingStrategy, it expects one to be provided when it's constructed. The calling code might do new OpportunityService(new StandardPricing()) or inject a different strategy.

Constructor injection makes it clear that the class cannot work without that dependency. By listing it as a parameter, you enforce that the dependency is provided.

Setter Injection (Property Injection) – You create the object, then set its dependency via a setter method or directly setting a public property. In Apex:

```
public class OpportunityService {

  public PricingStrategy pricingStrategy { get; set; }

  public Decimal calculateTotal(Opportunity opp) {

    return pricingStrategy.calculatePrice(opp);

  }

}
```

And elsewhere:

```
OpportunityService svc = new OpportunityService();

svc.pricingStrategy = new DiscountPricing(); // inject dependency
```

Setter injection is more flexible if the dependency is optional or if you need to change it during the object's life, but it also means the object could be used before the dependency is set (so you have to be careful to document or enforce the usage).

Apex has a twist: you can use the @TestVisible annotation to allow tests to set private properties. This is sometimes used to inject dependencies only in test code (making it somewhat like property injection, but not used in production code paths). For instance, if pricingStrategy was private but annotated @TestVisible, test methods could set it to a mock implementation.

Interface-based injection and Factory – Apex doesn't have a dedicated DI container, but you can achieve a kind of manual DI using factories or registries. For example, your code always asks a central place for an implementation:

PricingStrategy strat = PricingStrategyRegistry.getActiveStrategy();

If that registry is set up to return different implementations depending on context or a config, that's a form of injection (the registry injects the chosen implementation). This approach is used in some advanced frameworks, often reading from Custom Metadata (which can specify e.g. "use class X as the PricingStrategy").

Using a DI Framework (like ForceDI) – Not widely required, but there are open-source attempts (like ForceDI by Andy Fawcett) to create a lightweight DI container in Apex. These can allow you to bind interfaces to implementations in a configuration, then request an instance and get the injected graph of objects. For example, ForceDI uses a Custom Metadata to map an interface to a concrete class, and then di_Injector.getInstance(SomeInterface.class) will return an implementation, as we saw in the earlier example. Under the hood, it uses dynamic instantiation and binding rules. While an interesting tool, many Apex teams stick to simpler manual DI because the scale of Apex apps usually isn't so large as to necessitate a full framework.

The key benefits of DI in Apex:

Loose Coupling: Classes do not hard-code their collaborators. This aligns with the Dependency Inversion Principle – high-level modules (logic) do not depend on low-level modules (specific database or API calls); they depend on abstractions (interfaces), and those abstractions are fulfilled by concrete classes injected in.

Testability: If you can inject a mock or stub, you can test a class without relying on the actual implementation of its dependencies. For example, testing OpportunityService's calculateTotal might be done by injecting a dummy PricingStrategy that returns a known value, so you can verify that OpportunityService uses it correctly without worrying about the complex pricing logic in tests.

Configurability: You can swap out behaviors without changing the class code. If a new requirement comes that uses a different implementation, you might just inject a different class in certain contexts. For instance, in a sandbox vs production you might use a different service endpoint; if your code is DI-friendly, you could inject a sandbox service vs a production service easily (or use a factory that checks environment).

Example: DI for an Apex Service Layer

Imagine you have a class AccountReportService that generates some analytics on Accounts. It needs to fetch data about Accounts and also needs to send a summary email. Without DI, the class might instantiate an AccountRepository and an EmailService internally. With DI, you give it those:

```
public class AccountReportService {
  private AccountRepository repo;
  private EmailService emailService;
  public AccountReportService(AccountRepository repo, EmailService emailSvc) {
    this.repo = repo;
    this.emailService = emailSvc;
  }
  public void generateAndEmailReport(Id accountId) {
    Account data = repo.getAccountData(accountId);
    String report = '...build report from data...';
    emailService.sendEmail('report@example.com', 'Account Report', report);
  }
}
```

Now in production, you might call:

AccountReportService svc = new AccountReportService(new AccountRepository(), new EmailService());

svc.generateAndEmailReport(acctId);

In a unit test, you can inject fakes:

AccountRepository fakeRepo = new FakeAccountRepository(); // returns dummy data

EmailService fakeEmail = new FakeEmailService(); // maybe just logs or sets a flag

AccountReportService svc = new AccountReportService(fakeRepo, fakeEmail);

svc.generateAndEmailReport(testAcctId);

// assert that fakeEmail saw the expected content, etc.

This way, you test AccountReportService logic without hitting the database or actually sending an email.

ChatGPT Prompt – DI Scenarios: You can use ChatGPT to explore how to refactor code for DI. For example: *"I have an Apex class that calls [SELECT ...] SOQL in it. How can I refactor it to use dependency injection for easier testing?"* ChatGPT might suggest creating an interface for data access, implementing it with a class that does the SOQL, and changing the original class to use that interface. It may show a step-by-step: create interface, update class to accept it in constructor, update existing calls to pass the real implementation, etc. This is very useful if you're new to DI – it's like having a pair-programming partner guide you.

Also, consider asking: *"Show an Apex example of constructor vs setter injection and discuss when to use each."* The answer might mention that constructor injection is best when the dependency is mandatory, and setter when optional or for some configuration after object creation, and how Apex might use @TestVisible for injecting in tests.

Architect Tip: Embrace DI by designing Apex classes around interfaces and injection points. Define clear interfaces for key services (e.g., PaymentGateway, NotificationService) and have your logic classes rely on those interfaces. Use constructors (or a central factory method) to assemble the required graph of objects. If the team is not used to DI, create a simple example or template (ChatGPT can help produce an internal wiki snippet) that demonstrates the pattern. In the long run, this approach yields a codebase where pieces can be swapped (for example, migrating from one payment provider to another with minimal code changes) and where unit tests can run without heavy setup. As an architect, you might also look into the ForceDI library or similar if you anticipate a lot of dynamic binding needs, but often simpler patterns suffice.

Let's touch on using ChatGPT beyond just code for DI: **voice or conversation** can be effective here too. Sometimes explaining your code design verbally can clarify what needs to be injected. For instance, you might talk through to ChatGPT, "I have class X that does Y and calls Z. How to avoid calling Z directly?" This might yield suggestions to use an interface for Z's functionality and inject it. The conversational aspect can guide you without getting lost in code details prematurely.

In practice, DI in Apex often shows up in tandem with the patterns we've discussed. Strategy pattern is enabled by DI (you inject the strategy to use). The Facade might inject various subsystem handlers it needs. A Command object might be given its receiver via DI. So DI isn't an isolated pattern; it's a facilitator for writing code in a pattern-oriented way.

One challenge in Apex is that we lack a sophisticated runtime wiring mechanism (like Spring's context). So, we often rely on factories or a bit of glue code. For example, if you have multiple implementations of IPricingStrategy, you might have a static PricingStrategy currentStrategy that is set at class load or via some initializer. Or use Custom Metadata to store a mapping and a little factory to read from it (some orgs do, like a Custom Metadata with fields: StrategyName and ClassName, and then a registry that queries this and instantiates accordingly). These are more advanced and can be overkill if not needed. Simpler: often just pass what you need at the point of calling. For triggers, sometimes people pass dependencies through a trigger handler constructor (e.g., a Trigger handler might accept a logging service in its constructor). If nothing is passed, it uses a default. This way tests can pass a mock logger.

Now that we've looked at DI, think about your current Apex code: Are there places where classes instantiate other classes directly? Those are potential spots to apply DI and possibly use a design pattern. The next section will get into writing reusable libraries and utility classes – areas where DI can also play a role, especially in designing them to be flexible.

But before that, consider the role of testers with DI: A **tester** in a Salesforce team (maybe a QA automation engineer writing Apex tests or using API tests) should be aware that if the codebase is DI-heavy, it's actually a boon for testing. Fewer tests need to run as "seeAllData=true" or rely on actual database state, because you can simulate things. Testers can also participate by suggesting where testability is hard and could be improved by more DI. For example, if testers find that a certain class always requires a bunch of records in the database to test because it queries internally, they might suggest introducing an interface for that data retrieval to inject test data.

This collaboration is important: DI might initially seem like extra complexity to some team members, but demonstrating how it simplifies testing and maintenance will win them over.

We've now covered creation, structure, behavior, and injection – a comprehensive set of design best practices for Apex. Next, we will focus on **reusability** from a different angle: not just patterns in the abstract, but concrete approaches to writing Apex code once and using it in many places, through utility classes and libraries.

PROMOTING REUSABILITY WITH UTILITY CLASSES AND LIBRARIES

One hallmark of a mature codebase is the presence of well-defined **utility classes** and possibly shared **libraries** that teams rely on for common functionality. These aren't formal design patterns like the GoF patterns, but they are a crucial aspect of design: how do we organize our Apex code so that we **Don't Repeat Yourself (DRY)** and make common operations easily accessible? In this section, we'll discuss strategies for creating reusable code blocks in Apex – from small utility methods to packaging entire libraries – and how to leverage ChatGPT in building and documenting these.

Utility Classes in Apex

A **utility class** (often called helper class) is typically a class containing a set of static methods (and perhaps constants) that perform generic functions. Examples include:

String or Date manipulation helpers (beyond what String class provides).

Math or financial formula calculators.

Common SOQL query wrappers (like getting user's profile info, org settings, etc. in one call).

Trigger frameworks often have a utility class for things like adding errors to records, or handling recursion.

Utility classes promote reusability by centralizing logic that would otherwise be duplicated. For example, if you have a specific way to format customer names, instead of writing that logic in 5 different places (and possibly inconsistently), you put a method CustomerUtils.formatName(firstName, lastName). Everyone calls that. If the format needs to change, one change in CustomerUtils updates all usages.

When creating utility classes in Apex:

Keep them focused: A utility class should have a coherent purpose. You might have StringUtils, MathUtils, DatabaseUtils, etc., rather than one mega "Utils" class with unrelated methods. This aligns with the single-responsibility principle, even at a class level.

Static or Instance?: Most utility classes in Apex are just a collection of public static methods, meaning you call them without instantiation (like MathUtils.calculateIRR(listOfCashFlows)). This is convenient. The downside is static methods are harder to mock in tests (you can't override them easily). But if a method is deterministic and doesn't call out or depend on state, testing it directly is fine. If you think you might want to inject a utility (like our earlier DI discussion), then design it as an instance class implementing an interface. That might be overkill unless you foresee a need to swap it out (for example, maybe a TaxCalculator utility which could have different versions in different countries – then an interface TaxCalculator with region-specific classes might be better than static methods).

Bulk operations: In Apex, always consider if your utility should handle lists in a bulkified way. For example, EmailUtils.sendSingleEmail(email) vs EmailUtils.sendBulkEmails(list<email>). Often you might provide both variants.

Avoid global state in statics: One caution, if you use static variables in a utility (e.g., a static Map<Id, Account> cache in a util class), remember that persists only within transaction. That can be useful for caching within one trigger context to avoid re-querying data. But also provide a way to reset or clear it if needed (especially in tests, or if there's recursion break conditions).

Salesforce's Apex runtime encourages reusing code because of governor limits. If you've solved a tricky problem of splitting up large data or composing a complex SOSL query, you might encapsulate that in a utility method so that everyone doesn't have to re-code the limit handling.

Enterprise Scale vs SMB context: In an enterprise org (with many developers), having a robust set of utility classes is like having an internal framework. It ensures consistency; for instance, an **Architect** might enforce that all callouts go through a HttpUtils class that wraps Http.send() calls with standardized error handling and logging. In a small org or SMB context, you might not need an elaborate library, but even a two-person team benefits from shared helpers, as it reduces knowledge siloing (the logic is in one place, clear for all to see).

Libraries and Packaging

When we talk about libraries in Salesforce, it often means either:

An organized collection of utility classes (could be just within your org's codebase).

A managed package or unlocked package that provides reusable code (like an open-source library).

Salesforce has some known **open-source libraries** for Apex, for example:

Apex Commons (fflib) – which provides a whole suite of enterprise patterns implementation (UnitOfWork, Domain layer, Selector layer, etc.). If your team uses that, you essentially adopt a library that provides many reusable pieces (like fflib's UnitOfWork to handle DML transactions cleanly). Many of those classes in fflib can be considered advanced utilities/pattern implementations.

FinancialForce Date Toolkit or others – some community projects exist to extend Apex's standard library.

If you find yourself writing code that could be useful beyond your project, you might even consider modularizing it into an unlocked package. For example, if you develop a robust CSV parser in Apex, you could put that in a package and potentially reuse it across multiple orgs (if you work at a consultancy, for instance) or share it with the community. Unlocked packages in Salesforce (especially as of Winter '25, which presumably refined packaging further) allow you to create versioned libraries.

Key considerations for building a reusable library in Apex:

Minimize dependencies: A utility library should ideally not depend on custom objects or specifics of an org. It should be generic (like dealing with collections, strings, callouts).

Use of global: If you package it as a managed package that others install, you'd have to mark classes and methods as global to expose them. But global has downsides (less flexibility to change). In unlocked packages or within one org, you can keep them public.

Documentation: Provide clear docs/comments. If your library is internal, maybe have a Confluence page or an internal site listing the available utilities (ChatGPT can help generate documentation by summarizing methods if you feed it the class code). If it's external, include a README.

Testing: A good library is well-tested. Write thorough unit tests for your utility methods. This not only ensures they work, but also demonstrates usage.

ChatGPT's role: ChatGPT can assist significantly in creating utility code. For example, you might ask:

"Generate an Apex utility class for common string operations like truncating to a certain length with '...' and capitalizing the first letter of each word."

ChatGPT might produce a StringUtils class with static methods truncate(String, Integer) and titleCase(String). You can then fine-tune or incorporate them.

"I need an Apex method to compute the median of a List. Can you help?"

It could produce a snippet for median calculation (sort and pick middle, or the average of two middle values).

"Suggest a design for an Apex Logging utility that logs debug info and can be toggled on/off via Custom Metadata."

Here ChatGPT could outline a Logger class that checks a Custom Metadata (or Custom Setting) flag to decide whether to log, and provides methods like log(String message) or logError(String message, Exception e).

Additionally, for documenting utilities:

"Document the methods of the following utility class:" and paste the code, ChatGPT can produce a Markdown or bullet list summary of each method's purpose. That saves time in writing documentation.

Another advanced use: **image generation** via ChatGPT's tools can be used to create diagrams, but for utility classes, diagrams might not be as useful as they are straightforward. However, if you had a complex library architecture, you could ask for a UML-like diagram of how classes relate. In the future, one could imagine feeding ChatGPT a class diagram request and it using Mermaid or another format to draw it. Right now, describing such to it could result in ASCII art or a pseudo diagram.

Developer Tip: Before writing new code, search your org (or ask senior devs) if a utility already exists for that. It's common to reinvent a JSON parser or a date formatter not realizing there's one already. Investing time in a quick search (or maintaining a catalog of utilities) saves duplication. If nothing exists, build your utility with reusability in mind. Write it as if others will use it – clean interface, handle edge cases, add examples in the comments.

Architect Tip: Enforce a culture of reuse. Perhaps set up an "Apex Commons" repository for your company where common utilities are kept. Architects can periodically review code to find duplicate logic and refactor it into utilities or libraries. For example, if you see many triggers doing similar error handling, factor that out. Over time, this yields a powerful internal library that speeds up new development. ChatGPT can help identify similar code blocks if you feed it chunks, though privacy of code must be considered – maybe use it on open-sourced snippets or a safe subset.

Tester Tip: Utility classes should have dedicated test classes. If you notice a utility method isn't directly tested (perhaps indirectly covered through other tests), consider writing explicit tests for it. This not only improves coverage but ensures the utility truly works in isolation. Also, when utilities involve things like date/time or randomness, testers can add deterministic controls (like an optional parameter for "current date" to test date-sensitive logic). They might request developers add such features for testability. With ChatGPT, a tester could generate some test cases by describing the utility's function: *"What are edge cases for a method that truncates a string and appends '...' if it was longer than max length?"* It would list cases like exactly at length, one over, one under, null string, etc., helping create thorough tests.

One more angle: **Cross-language reuse.** Sometimes you might have logic in Apex that could also be useful in, say, JavaScript (LWC) or in a data pipeline. By having it in Apex, you might reimplement it elsewhere. ChatGPT can translate a piece of logic across languages if needed. For example, if you wrote a complex formula in Apex but now need it in a Node.js script, you could ask ChatGPT to port it, ensuring the logic stays consistent between systems.

Finally, think of reusability beyond just code: your knowledge and design can be reused via ChatGPT. For instance, as a **Product Owner** or manager, you might not code utilities, but you can encourage the team to leverage AI to reduce repetitive work. If a developer is solving a problem that's somewhat standard, prompting them to check with ChatGPT or search online might reveal existing snippets or patterns, saving time and resulting in more reliable code (since it's been used elsewhere). Product owners can thus indirectly push for reuse by asking if a feature can use existing components or libraries, keeping development efficient.

By promoting reusable utilities and possibly packaging them as libraries, a Salesforce team can accelerate development and reduce bugs (since a well-tested utility is used everywhere instead of multiple ad-hoc implementations). This also ties into long-term quality – something we'll emphasize in the summary, as well as the next chapter.

Now, having built up our arsenal of best practices and patterns, it's time to consider how to transition an existing codebase to use them. The final section will address **refactoring Apex code for pattern adoption**, essentially how to retrofit patterns and improve reusability in code that wasn't originally written that way. This is where all the principles we discussed come together in practice.

REFACTORING APEX CODE FOR PATTERN ADOPTION

Adopting design patterns and improving reusability often involves changing existing code – this process is known as **refactoring.** Refactoring is the art of **changing the internal structure of code without altering its external behavior.** The goal is cleaner, more maintainable code that behaves the same (so you don't introduce bugs while improving it). In Apex, where you might have legacy triggers or classes that have grown messy over time, refactoring toward the patterns we've discussed can greatly enhance the longevity and quality of your Salesforce implementation.

This section will explore strategies for refactoring Apex code to introduce design patterns and increase reusability. We'll also highlight how ChatGPT can assist in this non-trivial task and how different roles can contribute to or benefit from refactoring efforts.

Identifying Refactoring Opportunities

First, you need to spot the areas of code that would benefit from a pattern. Common signs:

Massive trigger or class with many responsibilities: e.g., a trigger that does 10 different things in sequence. This is a prime candidate to introduce a *Facade* (to orchestrate tasks) or break into *Commands* (each piece as a command executed in order). It may also indicate need for a *Domain layer* (as per Apex Enterprise Patterns) or at least better separation.

Lots of if/else or switch statements deciding what logic to execute: This often hints at a possible *Strategy pattern*. For example, if you see if(record.Type == 'A') { do X } else if(Type == 'B') { do Y } else if(Type == 'C') { do Z }, consider refactoring so that Type 'A', 'B', 'C' are handled by classes implementing a common interface (Strategy) and the conditional is replaced by selecting the right strategy.

Repeated code blocks in multiple places (violating DRY): This is a chance to extract a utility method or a shared service. Perhaps multiple triggers are sending emails in slightly different ways – create an EmailHelper. Or multiple classes calculate a similar metric – move it to a *Singleton service* that everyone calls.

Tight coupling to specific implementations: For example, class OrderProcessor always instantiates PayPalService. If tomorrow you need to use StripeService, you're stuck. This begs for *Dependency Injection* – refactor OrderProcessor to depend on an interface PaymentService and inject either PayPalService or StripeService. This might involve introducing the interface and perhaps a factory to choose which implementation at runtime (maybe based on org config).

Complex, nested logic that's hard to follow: Maybe it's just begging for a pattern like *Command* to pull out sections of logic into their own execute methods, making the flow more transparent. Or perhaps break a large method into smaller ones that might fit known patterns (like chain of responsibility or others not covered deeply here, but sometimes used).

Tests are hard to write: If you find writing a unit test for a class requires setting up tons of data or context, it indicates the class might be doing too much or tightly coupled. Refactoring using DI or splitting into smaller classes can alleviate that. For instance, if testing a trigger requires 50 records and complex state, maybe the trigger logic could be separated into a domain class or commands which can be tested with simpler inputs.

Step-by-Step Refactoring with Patterns

A safe approach to refactoring is to do it in **small increments:**

Have tests in place (or write them) for the existing behavior before you change anything. This could be just happy-path tests or some critical scenarios. They serve as a safety net.

Apply one refactoring at a time and run tests. For instance, first extract a utility method and replace calls to duplicated code with that method. Verify all tests pass.

Next, introduce an interface and modify a class to use it. Provide a default implementation so behavior stays same. Test again.

Then split a large class: maybe create two classes and move some methods, adjusting calls accordingly. Test again.

Gradually, work your way to the full pattern implementation. It might take multiple deploys if you're cautious (and that's okay, especially in large orgs where you want to minimize risk).

Remove any now-unused code (like old methods that got replaced by new ones) once you're confident in the new structure.

For example, imagine refactoring a monolithic trigger handler into a Strategy pattern:

First, create the interface and one class for one of the strategies, switch one if branch to use that (still possibly called via the old structure).

Later, do for other branches.

Then change the trigger to call a factory that picks the strategy class instead of the if logic.

Ensure triggers still produce the same outcomes with tests.

ChatGPT as a Refactoring Assistant: This is a very promising use of AI:

You can provide ChatGPT with a piece of code and ask for suggestions: *"This Apex class is too long, how can I refactor it using design patterns?"* ChatGPT might identify that it sees multiple concerns and suggest, say, extracting methods or using Strategy if it spots repeated decision logic.

It can even transform code. If you trust it with your code context (ensuring no sensitive info), you could try: "Refactor this code to use the Factory pattern" or "Split this method into smaller ones following single-responsibility principle."

It won't be able to run Apex tests for you, but it can simulate what the code is doing and check logical consistency.

You could also input two versions of code (before and after refactor) and ask, *"Do these behave the same?"* It might catch differences if any logic was inadvertently changed.

Remember not to copy proprietary code into external tools if that violates policy. If using ChatGPT for refactoring, consider abstracting the logic or using a description of the code rather than the exact code, unless you have a secure setup.

Real-world Example

Suppose you have this scenario in Apex (pseudocode for brevity):

```
public class InvoiceProcessor {
  public void processInvoice(Invoice__c inv) {
    //... some code...
    if(inv.Type__c == 'Hardware') {
      // calculate hardware maintenance fee
    } else if(inv.Type__c == 'Software') {
      // different calc
    }
```

```
        //... some more code...
        if(inv.Region__c == 'EU') {
            // apply EU tax rules
        } else if(inv.Region__c == 'US') {
            // apply US tax rules
        }
        //... finalize invoice...
    }
}
```

We spot two dimensions of variation: by Type and by Region. This class could be refactored to use Strategy pattern for calculating fees (hardware vs software strategy) and perhaps another Strategy or a simple utility for tax rules (region-specific tax strategy). Also, it's mixing two concerns (fees vs taxes).

Refactoring plan:

Extract interface IFeeCalculator with method calculateFee(Invoice__c) and classes HardwareFeeCalc, SoftwareFeeCalc. Change code: IFeeCalculator calc = inv.Type__c=='Hardware'? new HardwareFeeCalc() : new SoftwareFeeCalc(); Decimal fee = calc.calculateFee(inv); (and remove the if).

Extract interface ITaxRule for tax calculation, with EUTax and USTax classes. Use similarly.

Now InvoiceProcessor is just coordinating: get fee via strategy, get tax via strategy, etc. Possibly that's fine, or you could even make it a Facade if these calculations were done by separate subsystems.

Run tests to ensure an invoice of each type/region still yields same results.

If this logic was previously scattered, we now also have reusable classes: maybe other parts of the system can reuse EUTax logic.

A subtle refactoring: **Triggers to patterns.** Many older orgs have bulky triggers. A best practice is using a Trigger Handler pattern or the Domain pattern (from Apex Enterprise Patterns) which is effectively a variation of Command and Template patterns. If you refactor triggers:

Ensure one trigger per object remains (that's a best practice).

Move logic into an Apex class (e.g., AccountTriggerHandler.onAfterInsert() etc.). This alone is a refactor (no change in behavior, just moving code).

Then see if within that handler you can apply patterns: Strategy for different record types, or Command for splitting different operations, etc.

Possibly introduce a Unit of Work (if using fflib) to manage DML systematically – that's a bigger refactor but yields future benefits in controlling commits.

Role Perspectives on Refactoring

Architect/Product Owner: Encouraging refactoring in sprint cycles is crucial. Tech debt should be addressed regularly, not just at big bang rewrites. A product owner should allocate some buffer or dedicated stories for refactoring efforts, especially when touching an area of code for new features (the "Boy Scout Rule" – leave the campground cleaner than you found it, i.e., if you modify code, see if you can tidy it up then). Architects can identify high-risk or troublesome code areas (maybe via static analysis tools or incident history) and propose pattern-based refactoring to mitigate future issues.

Developer: When working on a bug fix or enhancement, consider if the surrounding code needs refactoring. It's often easier to refactor *before* adding new functionality than after. Use tools like PMD or CodeScan which might flag "cyclomatic complexity" or "long methods" – those hints often correlate with places needing patterns. And leverage peers or AI (like ChatGPT) to brainstorm a better design. Ensure to communicate with your team – large refactors need code review and possibly coordination (so you don't step on someone else's changes).

Tester: Testers play a key role in verifying that refactoring didn't break anything. If automated tests are in place, they give confidence. But testers might also do exploratory testing around refactored areas. It's wise for testers to have a heads-up: "Hey, we refactored how invoices are processed but outputs should be the same. Can you double-check an invoice creation in EU vs US to confirm tax is still correct?" Testers can compare results in a sandbox (with old code) vs a scratch org (with new code) if needed. The more patterns and DI are used, ironically the easier testing becomes over time – e.g., you can test strategies independently. So testers should actually welcome pattern adoption as it usually means clearer logic and more hooks for testing.

Middle Managers: For those managing multiple teams or concerned with velocity, it might seem refactoring slows things down. But in enterprise thinking, the cost of not refactoring is hidden – it appears later as production bugs or slow onboarding of new devs who can't understand the code. It's cheaper to clean up now than to deal with a crisis later. ChatGPT can even help here: ask for an explanation of code before and after. The after (refactored) should be easier to explain. That's a win for maintainability.

Before concluding, let's mention the use of ChatGPT's "Operator" mode in refactoring: since Operator can perform multi-step tasks like browsing and editing, one could imagine giving it repository access to automatically apply some refactorings (like rename variables, split methods). It's a bit futuristic, but as tools evolve, an AI agent might one day open your Org's unlocked package in a dev environment, apply the steps it described, run tests, and inform you of results. For now, it might help by fetching relevant documentation or similar code examples from the web, which can inspire your refactoring approach.

Refactoring is an ongoing process, not a one-time project. Make pattern-thinking part of your team's DNA. Over time, you'll find new code is written with these patterns upfront (so less refactoring later). And existing code gradually improves.

As we wrap up this chapter, we've covered a lot: from the motivation and theory of design patterns and reusability, through concrete Apex patterns (Creational, Structural, Behavioral), into dependency injection, utilities, and refactoring. In the summary, we'll recap the key points from each section and emphasize how adopting these practices sets the stage for long-term code quality. We'll also highlight how ChatGPT has been and can be a partner in this journey, assisting with pattern identification, code transformation, and ensuring consistency.

By continuously refactoring towards better patterns and higher reuse, your Apex codebase remains **sustainable** — easier to maintain, ready for new requirements, and resilient against attrition of knowledge (because the code itself communicates intent clearly). This paves the way for the next chapter, where we'll focus on long-term Apex code quality practices, building on the foundation of patterns and reusability we established here.

SUMMARY

Design patterns and reusability form the backbone of robust, maintainable Apex solutions. In this chapter, "Apex Design Patterns and Reusability," we explored how applying time-tested design principles in Salesforce Apex can dramatically improve long-term code quality and scalability. Let's recap the key ideas from each section:

Introduction – We began by defining design patterns as reusable, proven solutions to common software design problems. We saw why they matter for Apex: patterns and reusable code lead to more organized, scalable applications that can evolve over time without degrading. The introduction highlighted how different stakeholders benefit: architects and managers get more maintainable systems, developers get clearer guidance, testers get more testable code, and product owners see more reliable feature delivery. We also set the expectation that ChatGPT can assist at every step, from explaining concepts to helping implement them (even via voice input or by analyzing uploaded code).

Introduction to Apex Design Patterns – This section laid the groundwork by categorizing patterns (Creational, Structural, Behavioral) and relating them to Apex. We noted Apex's unique environment (governor limits, multi-tenancy) which patterns must accommodate. Key takeaways were that patterns encourage separation of concerns and consistency. We also introduced how ChatGPT can help identify where patterns might apply in Apex code. For example, describing a code scenario to ChatGPT can yield suggestions like using Strategy instead of lengthy if-else logic. The section included an Architect Tip on using ChatGPT to document pattern usage guidelines, reinforcing an enterprise approach to design.

Creational Patterns: Factory and Singleton – We examined two creational patterns:

Factory Pattern: Provides a flexible way to create objects. In Apex, factories are useful for dynamically choosing which class or record type to instantiate based on context (e.g., deciding which helper class to use for a given SObject type). The section gave use cases like dynamic record creation and integration adapters, and how to implement a simple factory method. We saw how ChatGPT can generate Apex factory examples on the fly, and a Developer Tip reminded us to centralize object creation logic for consistency.

Singleton Pattern: Ensures one instance of a class per transaction and provides global access to it. We discussed common uses in Apex such as caching settings or limiting expensive operations to once per execution. Importantly, we noted that Apex singletons are per-transaction (e.g., one per trigger context) due to statelessness between transactions. The section provided a code template for an Apex singleton and cited examples like a SettingsManager. A Tester Tip here pointed out considerations for static state in tests and the need for proper resetting to avoid cross-test interference.

These patterns improve resource use and organization. A subtle theme was that they can work together (e.g., a Factory might produce a Singleton or vice versa in certain frameworks).

Structural Patterns: Facade and Adapter – Structural patterns help manage complexity:

Facade Pattern: Simplifies interactions with complex subsystems by providing a unified interface. We saw how an Apex facade (like OrderFacade) can wrap multiple service calls (order creation, payment, shipment) into one method call. Advantages include easier usage, loose coupling between subsystems and callers, and centralized control for processes (like a business transaction). The section also gave a Product Owner Tip highlighting that facades make large processes easier to adjust in the future since the complexity is isolated.

Adapter Pattern: Allows classes with incompatible interfaces to collaborate by translating calls. In Apex, adapters shine when integrating external APIs or reconciling differences between old and new implementations. We covered an example of adapting a new third-party API to an existing Apex interface. The section emphasized that adapters often work hand-in-hand with Apex integration (e.g., wrapping an HTTP response into a domain object). An Architect Tip here suggested using adapters for any external integration points, making future swaps or upgrades much easier.

Throughout, we saw how ChatGPT can help design these patterns—like producing skeletons for facade methods or suggesting how to implement an adapter for a given API. Structural patterns ultimately improve the composition of your code, making it more modular and easier to reason about or change.

Behavioral Patterns: Strategy and Command – We dove into patterns that influence program flow:

Strategy Pattern: Encapsulates interchangeable algorithms so you can select one at runtime. We demonstrated this with examples like lead scoring or validation strategies, replacing complex conditional logic with polymorphism. This pattern keeps code open for extension (new strategies) without modifying the context class, aligning with Open/Closed Principle. It's especially relevant to Salesforce where business rules often change—new strategy classes can be added with minimal impact. A Developer Tip in this section encouraged refactoring conditional logic to strategies for clarity and easy testing (each strategy has its own small class to unit test).

Command Pattern: Turns requests into standalone objects containing all information needed for the request. We saw that commands in Apex can be useful for queuing actions (async jobs), implementing undo operations, or simply structuring complex logic into smaller actions (like "commands" for each step of a multi-step process). We related Command to Apex features: for example, a Queueable class could take a

Command and execute it, decoupling the enqueueing from the actual work.

Behavioral patterns focus on flexibility in how actions are carried out. A highlight was how testers benefit: with Strategy, they can test each variation in isolation; with Command, they can simulate sequences of actions or test undo logic systematically. We also noted ChatGPT can assist in brainstorming strategies or commands (even generating pseudo-code for how to parameterize and queue them).

Implementing Dependency Injection in Apex – This section addressed a fundamental principle for decoupling code. Dependency Injection (DI) allows an Apex class to receive its dependencies from the outside, rather than hard-coding them. We covered:

Constructor injection and setter injection in Apex, with examples of injecting a PricingStrategy or EmailService into a service class instead of instantiating them inside. This makes the code more flexible and **testable**, because you can swap in mock implementations during testing.

The use of Apex interfaces to facilitate DI (classes depend on an interface, and at runtime you provide an instance of some class implementing that interface). We referenced how this aligns with the Dependency Inversion Principle and noted Apex Enterprise Patterns' take on DI, including mention of the ForceDI library for advanced scenarios (where bindings can be configured via Custom Metadata).

Concrete examples, like an AccountReportService that is given an AccountRepository and EmailService, showed how DI leads to easily switchable components (e.g., use a FakeEmailService in tests to avoid sending emails).

DI is not a single pattern but an approach that enables patterns (e.g., injecting different Strategy implementations). The section wrapped up with an Architect Tip about institutionalizing DI for critical boundaries (like external services) and a note that ChatGPT can help refactor existing code to use DI (for instance, by suggesting where to introduce interfaces or how to redesign constructors).

Promoting Reusability with Utility Classes and Libraries – Here, we shifted focus to practical reusability tactics:

We explained how to create and manage utility classes (e.g., StringUtils, DateUtils) that collect static helper methods to avoid code duplication. A key best practice highlighted was to keep utility classes focused and generic, and consider bulk-friendly designs for Apex (since operating on lists is common).

We discussed building libraries – which could be simply well-organized classes in your org or packaged sets of functionality (unlocked packages) for broader reuse. Examples like the fflib Apex Commons were mentioned as existing libraries that many enterprise projects leverage. We touched on the process of packaging reusable code for distribution (with considerations about using global and maintaining backward compatibility).

Throughout, the importance of documentation and consistency was stressed. Reusable code is only useful if others know about it and how to use it. ChatGPT was noted as a tool to auto-generate documentation or usage examples, or even to convert a utility method from one language to another for consistency across a multi-platform project.

We included a Developer Tip urging devs to search for existing utilities before writing new code (prevent reinventing the wheel) and a Tester Tip to ensure utilities are thoroughly unit-tested (since so much depends on them). Overall, this section underscored that encouraging reuse (via libraries or common classes) is a hallmark of a high-performing team, benefiting both enterprise and smaller orgs by reducing redundant work and errors.

Refactoring Apex Code for Pattern Adoption – In the final section, we confronted the challenge of evolving existing codebases. We summarized refactoring as improving internal code structure without changing behavior , and offered guidance on how to gradually introduce patterns into legacy Apex:

We gave clues for spotting refactoring opportunities (e.g., long if-else chains signal Strategy, duplicated code signals need for a utility or Singleton, tightly coupled classes signal need for DI).

The section detailed a methodical approach: small incremental changes with tests at each step, rather than giant overhauls. For example, extract a method, then an interface, then swap out logic—checking that tests (or behavior in a sandbox) remain correct at each stage.

A real-world-like example illustrated refactoring a monolithic Invoice-Processor with multiple conditionals into using Strategy for fee calculation and tax rules, splitting the concerns into separate classes. This showed concretely how a messy method becomes a set of cleaner, more maintainable classes.

We emphasized that ChatGPT can act as a refactoring assistant, suggesting patterns to apply or even rewriting a snippet in a cleaner way. The human developer, of course, validates and tests those suggestions.

We also looked at roles: managers and product owners should allocate time for refactoring (it's an investment in future velocity and stability), developers should cultivate the skill and habit of continual refactoring, and testers ensure that refactoring hasn't changed expected outcomes. We stressed that patterns make code more testable, so refactoring with patterns often results in more reliable code – aligning everyone's interests.

By refactoring, teams can gradually transform a brittle codebase into a robust one. Each pattern adopted solves a class of problems and positions the codebase to handle growth. We ended the section by tying this to the upcoming theme of long-term quality: patterns and reusability are preventive care for code, which will be expanded on in the next chapter on sustaining Apex code quality.

In closing, **design patterns and reusability are foundational to sustainable Apex solutions.** They enable teams to handle complexity with elegance: a well-implemented pattern can make a complex task look simple from the outside (facade), or allow adding new functionality with minimal fuss (strategy), or let you change pieces of your system in isolation (DI and adapter). Reusability through utilities and singletons ensures that you write (and debug) logic once and leverage it everywhere needed, greatly reducing inconsistencies and bugs.

Crucially, we highlighted how **ChatGPT is a valuable ally** in this domain:

It can help identify patterns that suit a given problem, acting like an on-demand consultant for design discussions.

It can accelerate refactoring by proposing code transformations or providing templates (e.g., "here's how a Singleton would look for this use case") – saving developer time and avoiding common mistakes.

It aids in enforcing consistency – from generating documentation for your utilities to ensuring you don't forget a step in implementing a pattern (like making a constructor private for Singleton, which ChatGPT would remind you).

Through methods beyond plain typing (like voice explanations, or possibly using Operator to search codebases), ChatGPT lowers the friction to use best practices. For instance, a developer stuck on how to implement a factory can quickly ask ChatGPT and get unblocked, maintaining development momentum.

By leveraging both the wisdom of classic patterns and the power of modern AI assistance, Salesforce teams can craft Apex code that not only meets current requirements but is resilient to future changes. This future-aware approach is essential as Salesforce orgs live for many years and undergo constant evolution with new business needs and platform updates (e.g., anticipating what Winter '25 or Spring '26 features might bring and designing with flexibility to incorporate them).

As we move on to the next chapter on long-term Apex code quality practices, remember that the patterns and techniques covered here are not one-time fixes but part of a continuous discipline. Write code with patterns in mind, review code for adherence to these principles, and refactor relentlessly when something can be improved. With these practices ingrained, your Apex codebase will remain clean and adaptable year after year.

Salesforce Enterprise Thinking means building solutions that stand the test of time and scale. By mastering design patterns and reusability in Apex – and using tools like ChatGPT to support us – we ensure our

Salesforce implementations are robust foundations on which the business can continuously innovate.

In the next chapter, we will look at **long-term Apex code quality practices**, where we'll see how patterns and reusability interplay with other aspects like testing strategies, code reviews, static analysis, and performance tuning to uphold excellence in Apex development over the lifetime of an application. The journey towards ever-better Apex code continues, building on the solid pattern-based approach we've established here.

13

MAINTAINING APEX CODE QUALITY OVER TIME

Maintaining high code quality in a Salesforce Apex codebase is an ongoing journey, not a one-time task. As business requirements evolve and Salesforce releases new features and updates, Apex code that was written a year or even a month ago may need revisiting. Code quality is the foundation for reliability, performance, and security in your applications – it ensures your customizations continue to run smoothly with each Salesforce seasonal release (Spring, Summer, Winter) and as your org grows. In this chapter, we will explore how to preserve and improve Apex code quality over time, using both proven engineering practices and modern AI assistance. You'll learn how disciplined processes combined with tools like ChatGPT can keep your code robust and ready for the future.

Apex code quality isn't just about writing code that works – it's about writing code that **stays** working and maintainable as changes happen. This means adopting habits that catch bugs and design flaws early, and adapting code to new needs proactively. Neglecting code quality can lead to accumulation of technical debt: hacks and shortcuts that make the code harder to change later. By contrast, investing effort regularly in quality (through refactoring, reviews, and testing) pays off by reducing outages and speeding up future development. In the Salesforce context, where orgs can live for many years and serve thousands of users, maintaining high-quality Apex is especially critical to ensure trust and success release after release.

Fortunately, modern development practices and tools make it easier than ever to manage code quality. Time-tested techniques such as code reviews, refactoring, thorough documentation, and continuous integration (CI) help teams spot issues and improve code continuously. Newer innovations, including AI-powered assistants like ChatGPT and Salesforce's own Einstein GPT for Developers, add another layer of support. They can automate parts of the process, provide guidance on best practices, and even generate code or documentation drafts. In 2025, a Salesforce development team has an expanded toolkit: you can have automated scanners flagging potential bugs, and ChatGPT answering questions or reviewing code suggestions 24/7. Embracing these tools doesn't replace human judgment, but it augments your team's capabilities and reduces routine burdens.

Throughout this chapter, we will highlight how you can leverage ChatGPT in various ways to sustain and enhance Apex code quality. Each section jumps into a key practice – from regular code reviews to leveraging AI for improvement – and offers example ChatGPT prompts you can try. You'll see not only how to **type** requests to ChatGPT for help, but also alternative ways to interact with it, such as using voice input to ask questions hands-free or uploading code files for analysis (with appropriate privacy precautions). For instance, a developer might upload an Apex class file and ask ChatGPT for a critique, or even use a voice query on the ChatGPT mobile app to get a quick explanation of a tricky Apex concept. We'll also remind you how to protect sensitive code and data when using these AI tools, ensuring your company's intellectual property remains secure.

This chapter is organized into seven in-depth sections, each focusing on an aspect of long-term code quality maintenance: **Regular and Effective Apex Code Reviews**, **Proactive and Structured Apex Refactoring**, **Maintaining and Updating Apex Documentation**, **Implementing Continuous Integration and Deployment Practices**, **Building Effective Apex Code Metrics and KPIs**, **Cultivating a Collaborative Apex Development Culture**, and **Leveraging Automation and AI for Continuous Quality Improvement**. In each section, we'll discuss state-of-the-art best practices as of 2025, including any new trends or tools on the horizon. Expect discussions on how Salesforce teams are adopting things like Einstein GPT, GitHub Copilot, and the Salesforce Code Analyzer CLI in their workflows, and how these might evolve. We'll clearly note what's established best practice and what's an educated guess about the future, so you can plan accordingly.

Whether you are a Salesforce architect setting standards, a developer writing and reviewing Apex code, a QA engineer or tester ensuring everything works, or a product owner coordinating the team's efforts, this chapter offers insights for you. Throughout the sections, look for callout tips specifically for architects, developers, testers, and product owners – these highlight how each role can contribute to and benefit from maintaining code quality. The tone here is forward-looking and practical: we aim to equip enterprises and small teams alike with methods that improve code quality in a sustainable, collaborative way. High quality Apex isn't just "nice to have" – it aligns development with business agility and trust, which is vital for organizations of all sizes.

Maintaining Apex code quality over time requires commitment, the right processes, and smart use of tools. As we proceed, we'll start with one of the most powerful practices for keeping quality high: regular code reviews. By continuously reviewing each other's code, teams create a first line of defense against bugs and a mechanism for spreading knowledge. Let's begin by examining how to conduct Apex code reviews effectively in the era of AI-assisted development, and how this practice sets the stage for all other quality initiatives to come.

REGULAR AND EFFECTIVE APEX CODE REVIEWS

Code reviews are one of the fundamental practices for sustaining quality in any codebase, and Apex is no exception. In a regular Apex code review, developers inspect each other's code changes (for example, via pull requests in a source control system) to catch issues and suggest improvements before the code is merged and deployed. The benefits of consistent code reviews are well-proven: they help detect errors early and ensure coding standards are followed, and they facilitate knowledge sharing among the team. In other words, a code review is not only about finding bugs – it's also about spreading best practices and domain knowledge. Every review is an opportunity for the team to learn and to maintain a collective understanding of the codebase. Over time, this leads to more uniform, maintainable code because everyone is aligning to common standards and learning from each other.

To make Apex code reviews effective, it's important to establish a structured process and a supportive culture. *Frequency:* Reviews should happen regularly, ideally for every significant change. Many teams adopt a policy that no code goes into the main branch without at least one peer review. This might be formal (e.g., requiring an approved pull request) or informal (pair programming or over-the-shoulder review), depending on your team size and governance needs. *Scope:* Review not just for functionality, but for adherence to best practices and potential long-term issues. In Salesforce Apex, reviewers should pay attention to Apex-specific pitfalls such as SOQL or DML inside loops, proper bulkification of code to handle large data volumes, and respecting governor limits. They should also check for security patterns (like enforcing user permissions, using "with sharing" on classes where appropriate) and ensure the code follows the project's architecture patterns (for example, using a trigger handler framework if one is in place, or proper layer separation between controller, service, and helper classes).

Using a **code review checklist** can help ensure consistency. Many Salesforce teams maintain a checklist of items to verify during reviews. This typically includes code style and naming conventions, correct usage of Apex best practices, test coverage verification, and more. For example, a reviewer might go through a mental (or documented) list: *Are variables and methods named clearly? Is the code bulkified to handle many records? Does it avoid hard-coded IDs and use Custom Settings/Metadata for config? Are exceptions handled and logged properly? Is there any repeated logic that should be refactored into a utility or kept DRY (Don't Repeat Yourself)?* Such checklists often derive from Salesforce's own recommendations and frameworks. If your team doesn't have one yet, you can ask ChatGPT to help draft a code review checklist. For instance, you might prompt: *"ChatGPT, list the top 10 Apex code review checklist items focusing on performance and security."* This could yield a starting point that you tailor to your org's needs. Here are some example prompts to guide code reviews with ChatGPT:

"Generate a code review checklist for Apex classes and triggers, including bulkification and security checks."

"Review the following Apex method for any anti-patterns or governor limit issues: [paste code]."

"As a Salesforce code reviewer, what questions should I ask when reviewing a trigger implementation?"

"Explain why having a SOQL query inside a for-loop is problematic in Apex."

"What are some Apex code smells I should look out for during code reviews?"

By using such prompts, you can quickly gather insights or verify your own review comments. Some developers even use voice input to speed up this process – for example, dictating *"Explain the best practices for Apex error handling"* into a voice-enabled ChatGPT interface and listening to the guidance while reviewing code. This hands-free approach can make the review process more interactive, especially if you like to talk through code changes.

In practice, **ChatGPT can act as an assistant reviewer** when used wisely. Imagine you're reviewing a colleague's Apex class and something looks off in the way a collection is handled or a regex is used. You could copy a simplified version of that code (nothing proprietary or sensitive, of course) into ChatGPT and ask, *"Do you see any potential issues with this code?"* The AI might point out a governor limit risk or a logic bug you missed. This doesn't replace the human judgment – you still decide what feedback to give – but it provides a safety net and learning aid. Always double-check AI suggestions, because ChatGPT might occasionally misunderstand the code or suggest irrelevant changes. Still, as of 2025, these AI models are very knowledgeable about Apex patterns and common mistakes, so they can be quite helpful in flagging things like unbulkified code or missing null checks.

When incorporating ChatGPT into code review workflows, **privacy and security** are paramount. If you choose to upload a chunk of code or an entire file for ChatGPT to analyze, be sure your company's policies allow it. OpenAI and similar services may retain data for a period, so avoid pasting any sensitive business logic or secrets unless you're using an instance of ChatGPT that offers data privacy assurances (for example, ChatGPT Enterprise with data controls). You can also abstract the code when asking for help – describe the scenario instead of pasting raw code. For instance, rather than sharing the exact class, you might ask: *"In an Apex trigger that updates contact addresses when an account address changes, what edge cases should I consider?"* This way, you get guidance without revealing proprietary code. As a reminder, if you do use file upload features (say, attaching a.cls file for ChatGPT to examine), double-check that your ChatGPT account settings or enterprise OpenAI settings are configured to not use that data for training and to delete it as needed.

A healthy code review practice also involves the **right mindset and roles.** It's important that reviews are seen as collaborative and positive, not as personal critiques. A few tips for various roles in the review process:

Tip for Developers: Before submitting your code, run a self-review (or even ask ChatGPT to do a quick pass) to catch obvious issues. This not only makes the reviewer's job easier but also helps you learn. For example, you can prompt ChatGPT with *"Review my Apex method for any improvements: [code]"* and see if it catches something you overlooked (like a missing else or an inefficient query). This way, you come to the human reviewer with cleaner code, and you might even mention "I checked that there are no SOQL-in-loop issues" which shows due diligence.

Tip for Reviewers/Architects: Establish a clear code review guideline document for your team. This might include specific Apex patterns your org uses (like "we use the fflib Service layer pattern for all trigger logic – ensure any new trigger follows it") and links to Salesforce best practice documentation. Use ChatGPT to help explain the reasoning behind certain guidelines. For instance, if junior developers often ask "Why do we only allow one trigger per object?", you could have ChatGPT generate an explanation or even an analogy that you include in your internal docs. As an architect, you might also use ChatGPT to stay updated on new Apex features that could influence your code standards (for example, *"What's new in Apex in the Summer '25 release that affects code patterns?"*).

Tip for Testers: While testers might not review code for syntax, they can be involved in reviewing **Apex test classes.** Ensuring that test methods truly assert business requirements and cover various scenarios is part of quality. If you're a QA engineer with some coding knowledge, consider reviewing the team's test code or using ChatGPT to generate additional test cases. You could ask, *"Given this Apex method logic, what are some edge cases I should test?"* This helps maintain not just code quality but quality of testing over time.

Tip for Product Owners: You may not dive into code, but you can foster a culture that values code review by building it into the development process (for example, making it a non-optional step in Definition of Done for user stories). If you're ever presented with technical feedback from a code review (say, a developer says "We need to refactor X because the review found a design flaw"), ChatGPT can help translate that into business terms. You might ask, *"Why is it important to refactor code found in a review, in terms of business risk?"* The answer could help you explain to stakeholders that a small delay now prevents larger issues later, such as system downtime or costly fixes.

Regular code reviews, done in a supportive way, create a robust feedback loop that keeps code quality on track. They ensure that no piece of code is solely in one person's head and that common standards are upheld. Over time, as team members internalize the review feedback, the overall quality of new code improves – meaning fewer issues to catch in review in the first place. It's a virtuous cycle. And with tools like static analyzers and AI assistants supplementing human effort (we'll talk more about those in later sections), reviews can focus more on design and clarity rather than trivial style issues. In fact, many teams use **Salesforce Code Analyzer** or PMD to automatically scan for style and basic errors before the human review, so the reviewer spends time on higher-level concerns. (For example, Code Analyzer might flag a missing null check or a SOQL in loop, so the developer fixes those, and the reviewer can concentrate on whether the algorithm makes sense or if all business cases are handled.)

By implementing regular, effective code reviews, you set the stage for long-term quality. But finding issues is only half the battle – we also need to fix and improve code as we go. Next, we'll look at how to proactively refactor and improve Apex codebases over time, turning those review findings and technical debt items into action. This structured approach to **refactoring** will ensure your codebase stays clean, optimized, and adaptable to change.

PROACTIVE AND STRUCTURED APEX REFACTORING

Even with rigorous code reviews and good initial design, any long-lived Apex codebase will accumulate some *cruft* and technical debt over time. Business logic evolves, Salesforce platform capabilities change, and what was once a great solution might now be outdated or inefficient. This is where **refactoring** comes in – the process of improving the internal structure of code without changing its external behavior. Proactive, structured refactoring is essential to maintain Apex code quality in the long run. It means regularly scheduling time to rewrite or clean up code **before** it becomes a serious problem, rather than only refactoring as a last resort when something breaks. In this section, we'll discuss strategies for refactoring Apex code systematically, and how ChatGPT can help identify refactoring opportunities and even assist in the refactoring process.

First, let's clarify *why* refactoring is so important in a Salesforce context. Salesforce orgs are often maintained for many years, and requirements can change frequently with each new sales, service, or compliance need. Code that was written to quickly meet a deadline might not follow best practices and could be slowing down the org or making further changes risky. For example, perhaps an older trigger was written without using a trigger framework or without considering bulk operations; it might work for small data volumes but starts hitting governor limits as the company grows. Or maybe you have a utility class with a method that has grown too large and is doing too many things, making it hard to debug. Proactive refactoring addresses these issues incrementally. Rather than waiting until a major incident or a complete system overhaul, the team continuously improves the code in small pieces – maybe dedicating part of each sprint to paying down technical debt or scheduling a "refactoring sprint" every few months. This prevents the *"big ball of mud"* scenario where code becomes so entangled that adding new features is slow and perilous (in fact, a lack of ongoing refactoring and design can directly lead to bad technical debt in Salesforce orgs).

A **structured refactoring approach** starts with identifying what to refactor and planning how to do it safely. Identification can come from various sources: code review feedback (e.g., "This module is getting hard to maintain, maybe split it up"), static code analysis reports highlighting high complexity or duplicate code, performance issues observed in logs, or team members simply noticing "this code is confusing." It's helpful to keep a log of such observations – some teams maintain a *technical debt backlog* or mark Jira tickets with a "refactor needed" tag. That way, you have visibility into areas of the code that need love. Prioritize these by impact and risk: e.g., code that is run frequently in production or core to the business should be kept especially clean and efficient. Also, consider Salesforce's updates: if the platform introduces a better way to do something (say a new Lightning API or a flow that could replace some Apex), you might refactor to use the new approach and simplify your code.

Once you decide to refactor a certain piece, follow best practices to do it safely. Write or update Apex tests *before* you change the code (or use existing tests to ensure you don't break functionality). Then, make small, incremental changes rather than a giant rewrite whenever possible. In Apex, a common pattern is to introduce a new class or method and switch usage to it gradually. For example, if you're refactoring a monolithic trigger into a Trigger Handler pattern, you can create the new handler class and have the trigger invoke it, then move pieces of logic into the handler one by one, testing as you go. Each small refactoring step should keep all tests green. This way, if something does break, you know exactly which change caused it.

ChatGPT can be a surprisingly useful companion during refactoring. Here are several ways you might leverage it:

Brainstorming improvements: If you know a section of code is problematic but aren't sure how to improve it, you could describe the situation to ChatGPT and ask for suggestions. For instance: *"I have an Apex method that's 300 lines long and very hard to read. It does multiple things: queries data, applies business logic, then sends emails. How can I refactor this?"* ChatGPT might respond with ideas like splitting the method into smaller helper methods, implementing the Strategy pattern for different business rules, or moving the email logic to a separate service class. These are ideas you as a developer can then evaluate and implement as appropriate.

Identifying code smells: You can feed portions of your code to ChatGPT and ask if it sees any "code smells" or potential refactoring opportunities. For example: *"Here is an Apex class [paste partial class]. What improvements or refactors would you suggest for better maintainability?"* The AI could notice things like duplicated code, overly long methods, usage of outdated patterns, etc. Maybe it points out that you have multiple if-else branches that could be simplified with a switch or polymorphism, or that you're doing a repetitive task in several methods which could be unified. This doesn't absolve you from thinking critically, but it's like having a pair programmer who never gets tired of analyzing code structure.

Step-by-step refactoring help: For more straightforward refactors, you can even ask ChatGPT to rewrite a snippet in a cleaner way. For instance: *"Refactor this Apex code to use a map for efficient lookup instead of multiple loops: [code]."* The model might return a rewritten version of the code using a Map<Id, SObject> to avoid repeated queries or loops. Of course, you'll need to adapt and test any such output, but it can accelerate the grunt work. Developers have successfully used ChatGPT to convert code that uses one pattern into another (say, replacing nested loops with collection operations, or breaking apart a giant method into structured pieces). Always verify the logic after such transformations – the AI might introduce a subtle bug if some context is missed – but it often does a decent job on mechanical refactoring tasks.

Documentation and justification: After refactoring, it's good practice to document what was changed and why (we'll talk more about documentation in the next section). ChatGPT can help articulate the rationale. For example, *"Explain the benefits of using a Trigger Handler pattern over a raw trigger in Salesforce."* The answer might highlight maintainability, reusability, and order-of-execution control. You can use that explanation in your code comments or design docs to help future maintainers understand the purpose of the refactor. This turns a potentially abstract "clean-up" task into a learning artifact for the team.

One powerful way to use ChatGPT beyond just typing text is by leveraging *file uploads* during refactoring sessions. Suppose you have a large Apex class that you want to refactor. Instead of copying snippet by snippet into the chat, you might upload the file (if the interface supports it) and ask ChatGPT to analyze the whole thing. *"Here is our Opportunity-Processor class (attached). What design improvements would you suggest? Please consider splitting responsibilities if appropriate."* The AI can then read the entire file and provide a holistic analysis. Always remember to remove or obfuscate any sensitive data in the code before sharing and ensure you're comfortable with the level of exposure – ideally using an enterprise-grade version of ChatGPT that guarantees privacy. As a **safety tip**, confirm that your OpenAI settings don't share your code with their training dataset (this is usually an opt-out in the user settings or a feature of enterprise accounts). By taking these precautions, you can safely get a second opinion on a large piece of code in one go, which is immensely helpful for complex refactors.

Refactoring should also be **collaborative and continuous.** Encourage team members to flag code that "doesn't smell right" and create an environment where improving code is valued, not seen as wasted time. Often, product owners worry that refactoring is merely technical churn with no user-visible benefit. It can help to tie refactoring to clear outcomes like performance improvements, increased stability, or easier feature additions. For example, instead of a vague "refactor the XYZ class," frame it as "improve the XYZ class to reduce page load times" or "to enable adding new product types more easily next quarter." This way, everyone understands the business value of the effort.

Tip for Developers: Don't hesitate to refactor in small ways as you touch code. A principle from the Agile world, the "Boy Scout Rule," says to leave the campground cleaner than you found it. If you're adding a new feature in a class and you notice a bit of messy code nearby (maybe a poorly named variable or a chance to remove a redundant check), do that small refactor as part of your change. Just be sure to test and mention it in your pull request notes. Small, incremental improvements add up significantly over time.

Tip for Architects: Set aside time for refactoring in your project plans. For instance, allocate a certain percentage of each sprint for technical debt work, or schedule a refactoring sprint every few releases dedicated to cleaning up critical areas. This sends a clear message that quality is a priority. You can use ChatGPT to help identify high-debt areas by asking for potential issues in certain patterns. Some architects use static analysis tools (like Code Analyzer or SonarQube) to get metrics and then have ChatGPT suggest which types of issues to tackle first. An example prompt could be, *"We have 50 instances of 'method complexity too high' and 20 of 'duplicate code' in our static analysis report – which should we address first and why?"* The answer can help prioritize refactoring efforts.

Tip for Testers: Whenever a refactor is done, it's an opportunity to augment or improve tests. As a QA engineer, you can use ChatGPT to generate ideas for additional test cases that might not have been covered before. For example, if a complex method was broken into smaller ones, consider if there were any implicit assumptions in the old code that now need explicit testing. Ask ChatGPT, *"What edge cases should be tested after refactoring an Apex method that calculates discounts?"* – it might suggest boundary values or different data combinations you hadn't considered. This ensures that refactoring truly maintains behavior and that any changes are caught.

Tip for Product Owners: When planning refactoring tasks, try to quantify the expected benefit. For instance, "refactoring the order allocation code will reduce maintenance effort when we launch in new countries" or "will cut down org CPU time by 20% on large deals." ChatGPT can assist you in translating tech speak into business outcomes. You might ask, *"How would you explain the importance of refactoring duplicate code to a non-technical stakeholder?"* The response can give you a succinct analogy (like eliminating redundant work to reduce errors and speed up future changes). Including such rationales in your backlog items can make it easier to justify them to leadership and ensure they don't get constantly deprioritized in favor of only new features.

By engaging in structured refactoring, you essentially **renew** your Apex codebase continually. Think of it like regular maintenance on a car – you wouldn't skip oil changes and tire rotations and expect the car to perform well for years. Similarly, regularly refactoring keeps the code healthy, improves performance, and lowers the risk of critical failures. It also keeps the code aligned with current business needs; sometimes a refactor involves deleting code that's no longer needed, which simplifies the system (few things are higher quality than code that no longer exists at all, because it can't fail!).

ChatGPT and AI tools can significantly reduce the effort to refactor by providing suggestions, doing some of the heavy lifting in rewriting code, and educating the team on better designs. As these AI tools evolve, we can imagine even more advanced capabilities – perhaps future Salesforce development environments will have an "AI Refactor" button that automatically applies common optimizations or updates code to new Salesforce API versions. Already, we see early signs of this: GitHub Copilot and Einstein for Developers can suggest new code as you type; it's not a stretch to think that they might also one day scan your existing code and propose refactoring changes proactively. For now, the combination of human insight and AI assistance is powerful. Use the AI as a partner – bounce ideas off it, let it draft some code, but guide the final outcome based on your understanding of the requirements and context.

After refactoring, it's crucial to update the related documentation and ensure that the whole team is aware of the changes. In the next section, we will dive into best practices for **maintaining and updating Apex documentation**. Code changes over time, and so must the documentation that explains how it works and why. We'll see how to keep documentation in sync with the code (a challenge for many teams) and how tools like ChatGPT can generate and refine documentation effortlessly, thus closing the loop of continuous improvement.

MAINTAINING AND UPDATING APEX DOCUMENTATION

Writing and maintaining documentation is an often overlooked aspect of code quality, but it is absolutely vital for long-term maintainability. As your Apex codebase evolves through new features and refactoring, the documentation should evolve with it. Up-to-date documentation ensures that future developers (or your future self) can understand the intent behind the code, the design decisions made, and how to use the classes and methods without having to reverse-engineer them from scratch. In this section, we'll explore strategies for creating useful Apex documentation and keeping it current over time. We'll also look at how ChatGPT can help generate documentation drafts, summarize code, and even create diagrams or visual aids to complement your text documentation. By treating documentation as a first-class part of your code quality efforts, you reduce onboarding time for new team members, decrease the risk of misusing code, and improve overall team efficiency.

Types of Apex documentation: Documentation can take many forms, and it's important to have the right mix. Inline code comments in Apex classes and triggers are the most immediate form – these include comments explaining complex logic, TODO notes for future improvements, and ApexDoc-style comments (similar to Javadoc) for describing classes and methods (parameters, return values, etc.). A step above the code-level docs, you might maintain external documentation: for example, a design document for a major feature that includes the architecture of how different classes and Salesforce components (like Apex, flows, triggers, etc.) interact. Many teams use wikis or Confluence pages for such documentation, covering things like "Architecture of the Lead Conversion process" or "Data model and Apex interactions for Project X". There are also **runbooks** or support docs – if an Apex job fails or an error occurs, documentation can guide someone on how to troubleshoot or fix the issue. Lastly, user-focused documentation (like how to use a custom Salesforce app) might be owned by product or training teams, but developers often contribute by explaining what the code does from a user's perspective. All these types of documentation benefit from being kept up-to-date as the code changes.

One challenge is ensuring that when code is modified, the related docs are also updated. A good practice is to include documentation updates as part of the definition of done for a task. For example, if a developer adds a new method to a service class, they should also update the class header comment and any relevant external docs. Code reviews can enforce this: a reviewer might say, "Please update the README or design doc to reflect this new parameter you added." It's similar to how we treat tests: just as every new feature should come with tests, it should also come with necessary documentation.

Using ChatGPT to generate and improve documentation: This is a huge time-saver. Developers sometimes find writing documentation tedious or struggle with wording. ChatGPT, however, excels at turning technical information into coherent explanations. Here are some ways to use it:

Summarizing code: You can literally feed a chunk of Apex code to ChatGPT and ask for a summary. For example: *"Explain what the following Apex class does and why it's useful: [paste class code]."* The AI will analyze the code and produce a human-readable description. You might get a response like, *"This class OpportunityProcessor implements a batch job that recalculates Opportunity roll-up summaries. It queries related Line Items, computes total values, and updates each Opportunity. It's useful for ensuring the Opportunity's total reflects the latest Line Items in scenarios where automatic roll-ups are not available or have been turned off."* Such a summary can be tweaked and then placed at the top of your class as a comment, or put into a wiki page. It's much faster than writing it from scratch, and it provides a baseline that you can refine for accuracy.

Generating documentation drafts: Suppose you maintain an Apex API document for your org – basically a guide for internal developers on how to use various classes or services. If you create a new class, you can ask ChatGPT to generate a usage example. *"Provide an example of how to use the new InvoiceService Apex class to generate an invoice PDF."* If you've provided the context (maybe the class code or method signature), ChatGPT can output a short example code snippet with explanatory text. You can include that in your documentation so other developers see how to call the methods properly. This is like having an assistant technical writer who can work from your code to produce helpful guides.

Explaining complex logic: Some Apex code involves intricate algorithms or calculations (think of a complex discounting algorithm or a custom sharing rule computation). For such cases, beyond just commenting the code, you might want a separate document or section explaining the approach in plain language, possibly with a flowchart. You can use ChatGPT to break down the logic step by step. *"Explain the algorithm used in method X which calculates insurance premium, in simple terms."* The result can be used to create a documentation paragraph that says, "First, the method filters out ineligible records, then it calculates base premium using formula Y, and finally applies adjustments for each risk factor A, B, C." Having this explanation means that even someone not deeply familiar with the code (or not fluent in Apex) can understand what's happening. This is great for testers or business analysts who might read the docs.

Multilingual or simplified explanations: If your organization is global or if you need to communicate with non-developers, you can ask ChatGPT to rephrase documentation for different audiences. For instance, *"Rephrase this technical explanation in non-technical terms for a business stakeholder."* Or even, *"Translate this code comment to Spanish."* This can help create documentation that's accessible to all necessary audiences without requiring the developer to be a language expert. (Always have a fluent speaker review AI translations for critical docs, of course, but it's a good starting point.)

Another interesting way to document systems is through **diagrams and visuals.** Salesforce architecture can be complex (data model diagrams, flow charts of how a trigger hands off to a batch job, etc.). ChatGPT can assist in creating textual descriptions that you can turn into diagrams. For example, you might prompt: *"Describe the flow of data when an Account is created, covering triggers, classes, and any automation."* The description you get can be the basis for a sequence diagram or a flow-chart. There are tools and plugins that can take structured text (like a list of steps) and render a diagram. You could even ask ChatGPT to output something in Mermaid or PlantUML syntax (common text-based diagramming languages). For example: *"Based on this Apex trigger logic, provide a Mermaid sequence diagram of the process."* With a bit of tweaking, you can get a diagram script that you then render to an image. This is a bit advanced, but it shows how you can go beyond just written documentation – you can actually have AI help create **visual documentation.** Some folks also use image-generating AI to illustrate concepts (like generating an infographic-style image of code architecture), though that's more experimental. If you do create diagrams from AI output, always verify that they accurately represent the system.

Now, **keeping documentation updated** is the hard part. One approach is to periodically review and audit your documentation. For example, each quarter or release, assign someone to read through key documentation pages and compare with the current state of the code. If something is outdated (maybe the code changed a parameter but the doc still shows the old one), update it. You can again utilize ChatGPT here: *"The code now does X instead of Y; suggest an update to this documentation paragraph."* This yields a quick diff in narrative that you can accept or refine. Also, encourage a culture where team members *use* the documentation and thus spot inaccuracies. If developers routinely check the design docs before coding, they will notice when the doc doesn't match reality and can correct it.

Tip for Developers: When writing code, especially complex code, write the documentation *as you develop*. If you have a particularly tricky bit of logic, drop a detailed comment in the code right there. You can always ask ChatGPT to polish that comment later. For example, while coding you might write a rough comment like "// calc premium: filter ineligible, base = age*rate, apply discount if any". After finishing the coding, take that and ask ChatGPT, *"Turn this comment into a full sentence explaining the calculation."* It might produce: "// Calculates the insurance premium by first filtering out ineligible applicants, then setting a base rate (age multiplied by a predefined rate), and finally applying any discounts eligible to the applicant." Now your code has a clear explanation in it. Also, if you have an external doc, don't wait till the end to update it – open it while coding and jot down notes, then refine later. This habit ensures documentation isn't an afterthought.

Tip for Architects: Standardize documentation practices. For instance, use a consistent format for ApexDoc comments (descriptions for classes and methods). Perhaps adopt a tool that generates an **Apex documentation website** from your code comments (there are community tools to generate HTML docs from ApexDoc comments). This can act like an internal reference manual. Encourage architects or senior devs to review not just code but also the documentation for new features. Additionally, maintain high-level diagrams of system architecture (integrations, major processes). You can use ChatGPT to narrate changes to those diagrams when architecture evolves. If you integrate new systems or flows, update the visuals. Some architects include documentation updates as a deliverable for any significant project – treating it with the same importance as delivering code.

Tip for Testers: Documentation isn't only for developers. As a tester or QA engineer, you should also contribute. For example, if you discover through testing that the system behaves a certain way not clearly documented, you can update the docs or at least flag it. Test cases themselves can be a form of documentation – they show usage examples and expected outcomes. Consider writing a short summary of what a group of test methods in a class is verifying, and include that as a comment or in a test plan document. You might ask ChatGPT to generate a summary of a test class: *"Summarize what scenarios are covered by this test class."* The result could serve as a quick guide to what's being validated (e.g., "This test class covers the scenario of valid data (expected success), missing required field (expected failure), and boundary conditions for the discount calculation.").

Tip for Product Owners/Managers: While you may not write Apex docs, you should ensure that user-facing change logs or release notes are updated. Often developers will provide a technical summary, and you need to translate it for end users or stakeholders. ChatGPT is excellent for this translation. For instance, if a dev writes in a release note: "Refactored the opportunity trigger to use a new handler class," you might want to phrase it as "Improved the stability and maintainability of the Opportunity update process (internal change)" for an executive audience. You can run the dev note through ChatGPT with a prompt like *"Rephrase this technical note for a non-technical audience."* This ensures the value is communicated without jargon. Also, ensure that any attachments (like files or data) you provide to ChatGPT for this purpose are free of sensitive info, and use it in a logged, approved way.

As we approach 2025, there are also **emerging tools for automated documentation.** For example, GitHub Copilot has a Labs feature that can generate documentation or explanations for code. Salesforce might integrate some of these capabilities in the future – we can anticipate a world where upon saving an Apex class, an AI might auto-generate or update its ApexDoc block comment for you. There's already research into AI-generated code comments that match the code's behavior. These are not perfect yet, but the trend is clear: AI will play a bigger role in documentation. Some teams might even experiment with a chatbot trained on their org's documentation and code, so that any team member can ask "How does feature X work?" and get an instant answer drawn from the docs. If you have Salesforce's Einstein GPT or other AI hooked into your knowledge base, this could become a reality, boosting team productivity by making documentation queryable in natural language.

In summary, documentation is the guardian of long-term code quality – it ensures knowledge isn't lost and that changes are made with full understanding. By keeping documentation up-to-date and leveraging AI tools like ChatGPT to ease the burden, you make sure that your Apex code's **intent** and **usage** remain clear over time. This reduces miscommunication and errors when maintaining or extending the system. With your code clean (through refactoring) and well-documented, the next piece of the puzzle is to integrate these practices into your release process. In the following section, we'll discuss how to implement **Continuous Integration and Deployment (CI/CD)** for Salesforce projects. CI/CD ties everything together by automatically enforcing quality checks (like tests and static analysis) on every change and making deployments repeatable and reliable – a key step in sustaining Apex code quality in a fast-paced development environment.

IMPLEMENTING CONTINUOUS INTEGRATION AND DEPLOYMENT PRACTICES

Thus far, we've focused on practices that developers and teams perform somewhat manually – reviews, refactoring, documentation. Now, we turn to the automated backbone of modern software development: **Continuous Integration (CI) and Continuous Deployment/Delivery (CD).** In the Salesforce world, CI/CD has historically been a bit challenging due to the metadata-driven nature of deployments and org complexities. However, in recent years Salesforce and the community have provided tools and approaches to bring Apex development into the CI/CD paradigm successfully. Implementing CI/CD for your Apex code means that every change is automatically built and tested, and that deployments to various orgs (dev, test, UAT, production) are done in a consistent, scripted way. This automation is crucial for maintaining code quality over time because it ensures that quality checks are uniformly applied and reduces human error in the release process. It also enables faster, more frequent releases – interestingly, companies that achieve fast, continuous release cycles tend to see higher ROI from Salesforce , likely because they can deliver value quickly and respond to issues promptly.

Continuous Integration (CI): At its core, CI means that whenever a developer commits code to the repository, an automated process kicks off to validate that change. In a typical Apex context, this includes: compiling the code (for Apex, saving to an org is the "compile"), running all Apex tests, and perhaps running static code analysis (like Code Analyzer or PMD rules) and lint checks. If any test fails or any critical static analysis issue is found, the CI process flags the build as failed, and the team can address it immediately. This way, problems are caught early – ideally before the code even gets to a higher environment or causes a bug in production.

Setting up CI for Salesforce often involves using scratch orgs or sandbox environments where the code can be deployed and tested in isolation. For example, using Salesforce DX, a CI server (like Jenkins, GitHub Actions, Azure DevOps, etc.) can spin up a scratch org, push the source code into it, and run sfdx force:apex:test:run to execute tests. The results then inform whether the integration is successful. If your project uses unlocked packages, the CI might build the package and run tests in a test org with that package installed. Tools like **Salesforce CLI** (sfdx) and the newer sf commands provide the automation hooks to do all this via scripts.

Continuous Deployment/Delivery (CD): CD picks up where CI leaves off. After code is integrated and validated, how does it move to various environments and ultimately to production? With CD, you automate these deployments. For example, you might have a pipeline such that once changes are merged into a main branch, the CI runs tests and then automatically deploys to a staging or UAT org. After user acceptance testing is done, a release manager can trigger a production deployment, which is also scripted. Some teams even go for true continuous **deployment**, meaning every change that passes tests goes straight to production in a fully automated way – though in Salesforce this is less common due to governance and the risk of deploying without human oversight. More often it's continuous **delivery**, where every build is deployable and automation assists the deployment, but a human or schedule triggers the final step.

Implementing CI/CD for Apex code has several best practices:

Source Control First: Store all your code and metadata (as much as possible) in a version control system (git is the most popular). This is the single source of truth. Developers commit changes to git, not directly in production. This allows the CI process to run off the code repository. Salesforce's move to source-tracking with DX has made this easier. Ensure your team is disciplined about branching strategies (feature branches, and then merging into integration branches, etc.). This way, CI can be configured to run on certain branches (e.g., every pull request to main triggers the full test run).

Automate Tests: Make running tests a non-negotiable part of CI. Fortunately, Salesforce requires 75% code coverage to deploy to prod, but that's a floor, not a ceiling. Aim higher (e.g., 90%+) and, importantly, ensure tests are asserting correct behavior, not just covering lines. A good CI practice is to enforce a quality gate: for instance, the build fails if code coverage drops below a threshold or if any test fails. You can also incorporate tools like Apex PMD/Salesforce Code Analyzer in the CI: for example, fail the build if any "blocker" severity issues are reported by static analysis. The new Salesforce Code Analyzer 5, for instance, supports generating results in SARIF format which can integrate with GitHub to annotate pull requests with issues. This means your CI can add comments like "This line has a SOSL injection risk" automatically on a PR.

Include Metadata Deploys: Apex rarely stands alone – you likely have accompanying metadata like Custom Objects, Validation Rules, Flows, etc. Your CI/CD should also handle deploying these, either through source format conversion and the metadata API or unlocked packages. Ensuring the declarative components and code stay in sync is part of quality. If a Flow or Validation Rule change isn't deployed along with the Apex that depends on it, you get problems. So treat metadata as code too. There are CI tools (like Gearset, Copado, Flosum, etc. which are Salesforce-specific solutions) that can help manage this in a more Salesforce-tailored way if you prefer not to build everything with generic tools.

Use scratch orgs or isolated test orgs: One issue in Salesforce CI is tests that depend on data. Ideally, Apex tests create their own data and don't rely on existing org data (see Isolation of Tests in Salesforce docs). Your CI org (scratch org) will be essentially empty except what you deploy. So design tests to run in isolation. If you have required setup (like certain custom metadata or records), your CI process can load those as part of setup (maybe via a JSON data load or so). This ensures consistent test results. If a test only passes in prod because prod has some record present, that's a fragile test – better to fix the test to create what it needs.

Parallel and Faster CI: Running an entire Salesforce test suite can be time-consuming if you have thousands of tests. Salesforce has introduced ways to run tests in parallel and improved test runtime over the years. Still, you might optimize by running only tests affected by a change (some CI setups use git diff to run a subset, though with Apex this can be complex to determine). Another strategy is to use Test Levels: Quick tests (no @IsTest(SeeAllData=true) tests) for quick feedback on commits, and full tests before a release. A forward-looking trend is using AI to predict which tests are likely impacted by a change (some companies do this for large systems); perhaps in the near future, Einstein or Copilot could analyze code changes and suggest which test classes should be run, optimizing CI time. This is speculative but an interesting possibility.

Where does ChatGPT and AI come into play for CI/CD? A few areas:

Setting up CI/CD pipelines: If you're new to CI, writing the pipeline scripts (YAML for GitHub Actions or pipelines for Jenkins) can be daunting. ChatGPT can help bootstrap this. For example: *"ChatGPT, give me a sample GitHub Actions workflow for Salesforce that checks out the repo, authenticates to a Dev Hub, creates a scratch org, deploys metadata, runs tests, and reports coverage."* You might get a pretty good YAML snippet that uses the sfdx CLI to do these steps. You'd still need to tweak it (like adding your auth details securely via secrets), but it's a big time-saver. ChatGPT might even know about specific actions like sfdx-actions (a community GitHub Action for Salesforce CLI) and incorporate those.

Troubleshooting build errors: When something fails in CI, you often get a log or an error message. ChatGPT can be like a second pair of eyes to diagnose it. If a deployment failed with a cryptic message about a dependency or a test failure that's hard to replicate, you can paste the error (ensuring no sensitive info) and ask *"What might be causing this deployment error in Salesforce CI?"* Many times, it can identify "Ah, this might be due to a missing permission or an unmigrated picklist value" from the text. It's similar to how we use Stack Exchange or Google, but ChatGPT can synthesize an answer from the error and known patterns.

Optimizing pipelines: Perhaps your deployment to a sandbox is slow. You could describe your process to ChatGPT and ask for optimization tips. *"My Jenkins pipeline for Salesforce takes 1 hour to run tests. How can I speed it up?"* It might suggest splitting tests, using the recent "test run" feature to run in parallel, or excluding certain tests, etc. Or if a certain step is flaky (like CLI authentication), it might suggest better ways (like JWT auth or using the sf CLI which has some improvements).

Policy enforcement: Some teams might use ChatGPT in a proactive way by integrating it via API to comment on pull requests. For example, an automated bot could use an LLM to analyze a PR's code diffs and provide suggestions or highlight potential issues (like "I notice you updated a validation rule but not the related Apex that uses that field, is that intentional?"). This is cutting-edge and not common yet, but as an idea it shows how AI might assist CI beyond just failing a build – it could give semantically rich feedback. GitHub is working on things like Copilot for Pull Requests which will do some of this (e.g., suggest better descriptions, or point out possible bugs in the changes). By 2025, we expect such features to be more available.

While setting up CI/CD requires an initial investment, it pays huge dividends. It standardizes the deployment process, reducing those late-night production hotfixes because "someone missed moving a component". It also builds confidence that your Apex code is always in a releasable state. Teams with good CI/CD can release more frequently with less risk, which ironically means higher quality because you ship smaller changes more often (so if something goes wrong, it's easier to pinpoint and fix). It also enforces discipline: if a quick-and-dirty code change can't pass the automated tests or lint, then it doesn't get in – protecting the codebase from regression.

Salesforce has been investing in making CI/CD easier: the introduction of **Salesforce DevOps Center** (a Salesforce-provided UI-based change and release management tool) is a sign of that. DevOps Center, GA since 2022, provides a more declarative interface for moving changes but under the hood it uses source control and pipelines. By 2025, many orgs (especially those with less technical admins) might be using DevOps Center as a stepping stone to full CI/CD. It's worth noting that whether you use Salesforce's tools or your own scripts, the principles remain: automate what you can, and ensure quality checks (tests, static analysis) are part of the process.

Tip for Developers: Embrace the CI process – don't see it as a hurdle, but as a safety net. If the CI finds an issue, it's much better than that issue slipping to production. When you get a CI failure, use it as a learning moment. For example, if a static code analysis rule flags something on your commit, take time to understand why that rule exists. Maybe you got flagged for "Avoid logic in triggers" or "CRUD/FLS check missing". Ask ChatGPT, *"Why is this rule important and how do I fix it?"* – this could reinforce best practices. Also, try running the same CI steps locally if possible (like there's a command sfdx scanner:run for Code Analyzer, or you can run all tests in your dev org) to catch things before pushing.

Tip for DevOps Engineers/Architects: If you're designing the pipeline, make sure to include not just happy-path deployment but also rollback or backout plans. Salesforce doesn't have a native rollback, but your strategy could be to keep the last known good metadata to redeploy if needed. Include monitoring of the CI: e.g., have it output code coverage numbers, and perhaps trend them over time (some use tools to track coverage history or static analysis issue count trends as a KPI – more on that in the next section). Also, consider using **feature flags** for risky changes. That's not directly a CI/CD thing, but it complements it: deploy the code turned "off" via a feature flag, test in production silently, then enable it once confident. Tools like LaunchDarkly can be integrated with Salesforce (e.g., via Platform Events) to do runtime feature flags. This reduces the need to do emergency deployments, since you can turn off a problematic feature without code changes.

Tip for Testers: Integrate your testing work with CI. If you have automated UI tests (Selenium, Cypress, etc.) for Salesforce, consider having them run after the deployment stage of CI (maybe in a test org with test data). Some teams hook up post-deployment tests – for instance, after a successful UAT org deploy, run a smoke test suite to ensure basic flows still work. As a tester, you can also monitor the CI results. If you see a particular test failing intermittently in CI, investigate it; it could be a flaky test or a sign of a timing issue in code. Use ChatGPT to discuss flakiness: *"Our test sometimes fails with error X, how can we make it more robust?"* – it might suggest adding a wait, or adjusting test data setup. Additionally, contribute to the CI by writing good, reliable tests that can run unattended.

Tip for Product Owners: While you might not deal with pipelines, you'll feel the impact of CI/CD in the team's delivery. If the team sets up a solid CI/CD, releases become more predictable and less scary. Encourage the team to demo not just features, but also their improved process. For instance, they could showcase in a sprint review: "Look, when we merge code, automatically an org is built and tested – this gives us confidence." This might seem technical, but it's tied to business agility. In fact, as mentioned, companies with continuous release capabilities see better ROI , something you could mention up the chain to justify the time spent on infrastructure. If an executive asks, "Why are we investing time in pipeline automation?", you can answer that it leads to faster time-to-market and fewer production issues. ChatGPT can help phrase that: *"Explain the business value of CI/CD in Salesforce in one paragraph."* Use the answer in your pitch to stakeholders.

With CI/CD in place, your team has a robust mechanism to catch issues early and ensure consistent quality with each deployment. The code is reviewed, refactored, documented, and now continuously tested and delivered. You've established a strong engineering rhythm. But how do you measure the effectiveness of all this? That's where **metrics and KPIs** for code quality come into play. In the next section, we'll discuss how to build and use Apex code quality metrics – from code coverage percentages to bug trends – to quantitatively track your code's health over time. These metrics will help you identify areas for improvement and demonstrate the value of your quality initiatives to the broader organization.

BUILDING EFFECTIVE APEX CODE METRICS AND KPIS

As the saying goes, "you can't improve what you don't measure." To maintain and improve Apex code quality over time, it's important to establish metrics and Key Performance Indicators (KPIs) that give you insight into the health of your code and development process. Metrics can help answer questions like: Are we improving code coverage? Is our code becoming easier to maintain, or is complexity growing? How often are we introducing bugs, and how quickly do we fix them? By tracking such information, teams can make data-driven decisions on where to focus their quality efforts. In this section, we'll outline various code quality metrics relevant to Apex, how to gather them, and how to use them effectively. We'll also look at how tools and AI can assist in analyzing these metrics. The goal is to create a feedback loop where metrics inform action, and actions (like more refactoring or training) improve the metrics over time, leading to continuously better code.

Key code quality metrics for Apex (and Salesforce development) might include:

Test Coverage: The percentage of Apex code covered by unit tests. Salesforce enforces a minimum of 75% org-wide coverage for deployment, but internally you might track the coverage of each project or even each class. If one module has only 80% coverage while others are at 95%, that might indicate an area of risk. Over time, you want to see coverage stay high or improve. However, beware of chasing coverage for its own sake – ensure tests are meaningful. A good complementary metric is *test effectiveness*, sometimes measured by counting production bugs that could have been caught by tests.

Static Code Analysis Results: If you use a tool like Salesforce Code Analyzer, PMD, or SonarQube, you'll get counts of rule violations categorized by severity (critical, major, minor) and type (security, performance, style, etc.). Tracking the number of open issues over time is insightful. Ideally, the trend should be downward (as you fix issues faster than you introduce new ones). For example, you might measure "number of critical security findings in Apex code" each month. A spike might occur after adding a lot of new code – that's a sign to pause and address those findings. Many teams set a KPI like "No critical Code Analyzer issues in codebase" or "All new code has 0 critical/major static analysis issues". With Code Analyzer 5, new features like duplicate code detection (CPD) can produce metrics – e.g., "duplication percentage" in code. High duplication might lead you to refactor into shared methods.

Complexity Metrics: Cyclomatic complexity or cognitive complexity per method or class is a measure of how complicated the code is (essentially how many distinct paths or how hard to understand). Tools like SonarQube or PMD can compute this for Apex. If a method's cyclomatic complexity is, say, 25 (lots of if/else and loops), that's quite complex for Apex. You might have a policy that "no method should exceed 10 complexity" and track compliance. Average complexity of methods across the codebase can be a KPI to watch: is it going down (code is getting simpler) or creeping up (perhaps a sign that new code is being written in a convoluted way or too much logic per method)?

Code Churn and Size: These measure how much code is changing. Churn can be captured by version control stats – e.g., lines of code added/modified in a release. If a particular area has extremely high churn, maybe it's unstable or someone is struggling with it (churn isn't inherently bad if it's deliberate refactoring, but unexpected churn can indicate an issue). Code base size (lines of Apex code) is also interesting: if it's rapidly growing, ensure you're not reinventing the wheel or adding overly complex solutions where smaller ones exist. In the Salesforce context, also watch the sheer number of components (Apex classes, triggers, flows) – an explosion in count might signal a need for consolidation or better design.

Defect Metrics: Track bugs related to Apex code. For example, how many production incidents or support tickets were caused by Apex logic issues in a given quarter? What's the average time to resolve them (mean time to repair, MTTR)? Also, perhaps categorize them: e.g., of bugs found, how many were regression (introduced by recent changes) vs long-standing? A reduction in regressions over time might show that your testing and CI is catching more issues before release. If certain areas of code produce many bugs, that area likely needs more testing or a refactor.

Performance Metrics: Since Apex runs on a multi-tenant platform, we have governor limits rather than raw CPU/hardware metrics. But you can still measure performance-related indicators. For instance, monitor the average CPU time or heap size of key Apex executions (Salesforce's Apex logging and Event Monitoring can provide such data). If you have a scheduled job or a trigger that frequently runs, track if it's getting closer to governor limits as data grows. A KPI could be "No single transaction exceeds 50% of CPU time limit" or "Batch jobs complete within X minutes". Salesforce's logging can be aggregated (some use tools like New Relic or Splunk with Salesforce log data). If you notice trends like a batch that used 2 sec CPU last month now uses 5 sec, it could signal a need to optimize as data volume increased.

Deployment/CI Metrics: These aren't code quality per se, but they relate. E.g., percentage of successful builds, average deployment frequency, lead time from code commit to production. These DevOps metrics (akin to DORA metrics) tell you how smoothly your process is running. If your build failure rate is high, maybe code quality is an issue (e.g., brittle tests). If you can deploy often with low failure, it means quality is under control enough to allow that.

Now, collecting these metrics often requires tools:

Salesforce Code Analyzer (CLI): can output a report of static analysis issues and complexity metrics. You might run sf code:analyze as part of CI and then push the results to a dashboard or file.

SonarQube or Code Climate: These are platforms where you ingest your code (via an export or CI plugin) and they provide a dashboard of many metrics (coverage, complexity, duplications, rule violations, etc.). SonarQube, for example, gives an overall quality gate status and a maintainability rating. It also tracks "technical debt" in terms of estimated time to fix all issues.

Custom Apex Scripts: Some teams write Apex scripts to compute metrics like "count of triggers per object" (should generally be 1) or "which classes have no test class".

Bug tracking system: Make sure to label or tag bugs with components or root cause (if it was a code bug) so you can query "bugs by module" easily.

Once you have metrics, **using them effectively** is key. Don't drown in data – pick a few KPIs that matter most and set targets or thresholds. For example:

"We will keep code coverage above 90% at all times."

"We will reduce the number of critical static analysis issues to zero by end of Q2."

"No single Apex class should have more than 500 lines; large classes will be split (max 20% of classes can exceed this, and decreasing)."

"Our aim is less than 5% of stories result in a production defect within a month of release (a measure of feature quality)."

Review these regularly, say in a monthly or sprint-end retrospective. If something is off track (e.g., code coverage dipped because a lot of new code was added quickly without enough tests), then you decide actions (like write more tests in the next sprint). Metrics should prompt questions, not blind adherence. If a metric is consistently green, maybe you can focus elsewhere – or maybe the metric is too easy and you should tighten the standard.

One caution: **avoid gaming the metrics.** If you set a rigid goal like 95% coverage, someone might write trivial tests just to hit the number (tests that assert 2+2=4 to cover lines without truly testing logic). To counter this, combine metrics. For instance, high coverage but also track *assertion* *density* (how many assertions per test or per lines of code). Or do code reviews of tests to ensure they are meaningful. Another example: an "issue count" going down is good unless people are just lowering the severity of issues to remove them from count. Encourage a culture of genuinely improving the code, not just the numbers. The numbers are indicators, not the end goal themselves.

Now, how can ChatGPT help with metrics and KPIs?

Understanding metrics: If you're not familiar with a metric, ask ChatGPT. "What is cyclomatic complexity and what is a good value for Apex methods?" You'll get an explanation. Or "How does SonarQube define technical debt and maintainability for Apex?" This helps you interpret dashboards.

Choosing metrics: You could consult ChatGPT on what metrics make sense. *"What code quality metrics should a Salesforce development team track?"* Perhaps it will list similar ones to above, and you might pick from that.

Analyzing trends: Suppose you have some data, like code coverage over 6 months or number of bugs. You can present that to ChatGPT (maybe as a small table) and ask for insights. *"Our code coverage went 92%, 90%, 87%, 88%, 91% in the last 5 releases. What might be causing the drop in the middle and how to address it?"* It might reason, "the drop to 87-88% could be due to a large addition of code without tests, perhaps during a big feature push. To address it, identify those new areas and add tests, and enforce coverage in CI to prevent future drops."

Brainstorming improvement ideas: If a metric is not improving, ChatGPT might help. For example, *"Our static analysis issues are not decreasing despite efforts. What strategies can reduce code violations in Apex?"* The AI might suggest things like training devs on common issues, integrating checks earlier, or even using automated fixes for some issues.

Communicating metrics: Sometimes, you need to report these to management or a broader audience. ChatGPT can help frame the narrative. *"Help me summarize our code quality metrics for the quarter: we improved coverage from 88% to 93%, fixed 15 security issues, but our bug count went up from 2 to 5 due to a new integration.*" It can generate a nice little report or talking points, which you can then adjust.

Another aspect is using **visualization.** Metrics are best understood with charts. Export your data to a spreadsheet or use a tool (SonarQube has its UI, but you might want a custom dashboard). If you have the data in CSV, you could even use ChatGPT (with its Advanced Data Analysis capabilities, formerly Code Interpreter) to plot graphs. For example, feed it a CSV of "Month, Coverage%, Bugs" and ask for a chart. It could output a line chart image of coverage and a bar chart of bugs. Embedding that in your internal wiki or slides helps convey the story. In fact, many teams include quality metrics in their Definition of Done or project status reports to ensure it's as much a priority as delivering features.

Tip for Architects/Team Leads: Select a small set of **team KPIs** related to code quality and make them visible. For instance, display the current code coverage and static analysis issue count on a team dashboard or a monitor in the workspace (if in-office) or in a Slack channel via a bot. This keeps quality front-of-mind. Periodically rotate focus: maybe this quarter emphasize security fixes (reduce security issue count), next quarter focus on reducing code complexity. Over time, you cover all bases. Use these metrics in performance reviews or retrospectives to celebrate improvements (e.g., "We achieved our goal of 0 PMD critical violations – kudos to the team!"). Positive reinforcement around quality helps the culture.

Tip for Developers: Pay attention to the metrics that tie to your work. If you see a report that your new feature decreased overall coverage, take initiative to add more tests. If complexity for a method you wrote is high, consider refactoring that method into simpler units. Developers often have the most direct ability to change these numbers through their daily work. Also, personal metrics can be insightful (for self-improvement, not as a managerial tool): e.g., track how many code review comments your PRs typically get about quality issues – over time, aim to reduce that by learning from past feedback. Or how many defects were attributed to code you wrote – use that as motivation to add more negative tests or think more deeply in design. Remember, the goal isn't to assign blame but to improve your craft.

Tip for Testers: Some metrics like code coverage and defects involve you directly. Ensure that the team isn't just focusing on unit test coverage but also integration test results. You might maintain a metric like "percentage of test cases passed in UAT" or "number of critical bugs found in UAT vs production." If you find that many bugs slip to production, that suggests enhancing either the development testing or the UAT process. Communicate that with data: "We found 10 serious bugs in production last year, 7 of which could have been caught by better unit tests." That can justify more investment in testing. Also, collaborate in maintaining a traceability matrix (mapping requirements to tests) – not exactly code metrics, but quality metrics of coverage in terms of requirements tested. If using AI, you could ask ChatGPT to help identify gaps: *"We have tests for A, B, D – did we forget C?"* It might or might not directly know, but you can prompt it with requirement descriptions to suggest test scenarios.

Tip for Product Owners/Managers: Quality metrics can be part of your project health reports. It's worth translating them to business terms when needed. For example, instead of saying "static analysis issues went down," say "our codebase is now conforming better to best practices, which reduces the risk of runtime errors and security issues." If deployment frequency increased after CI/CD, highlight that as "we can deliver features to users faster now." Essentially, tie the technical metrics to business value: reliability, security, agility, cost savings (less time firefighting issues). ChatGPT can assist in crafting those messages: *"Explain the benefit of having 95% code coverage to a CFO who is not technical."* Perhaps it will respond with something about risk reduction and confidence in releases, which you can use.

Armed with metrics and KPIs, a team can quantitatively monitor their journey of maintaining Apex quality. But one must not forget: behind all these practices and numbers are the **people** writing the code. The best tools and processes will falter without a supportive team culture. In the next section, we'll focus on the human aspect: **cultivating a collaborative Apex development culture** that prioritizes quality. We'll discuss how team norms, communication, and learning can ensure that all the practices we've covered truly take root and continue to thrive. After all, quality is ultimately a team responsibility and an ongoing mindset, not just a checklist.

CULTIVATING A COLLABORATIVE APEX DEVELOPMENT CULTURE

Technology and tools aside, the long-term quality of a Salesforce codebase is heavily influenced by the culture of the team maintaining it. An **enterprise-oriented, collaborative culture** ensures that best practices are not just one-off efforts but a sustained way of working. Culture is about how team members communicate, how they approach problems, how they value quality, and how they learn and improve together. In this section, we'll explore ways to build and nurture a team culture that supports continuous Apex code quality. This includes encouraging knowledge sharing, continuous learning, openness to feedback, and a sense of collective code ownership. We will also highlight how ChatGPT and similar tools can be used in group settings or team workflows to enhance collaboration – for example, by serving as a knowledge repository or by breaking down communication barriers. Both large enterprise teams and small Salesforce dev teams can benefit from these cultural principles.

Knowledge sharing and mentorship: A collaborative culture means developers regularly share what they know with each other. This prevents silos where only one person understands a piece of code (which is risky for maintenance). Tactics include pair programming, peer code reviews (which we discussed – they also serve to spread knowledge), and formal or informal "lunch and learn" sessions. For example, if one developer is an expert in Apex performance tuning, they might host a short session to teach others how to bulkify or avoid common mistakes. Or after a major refactor, the person who did it can present the new design to the team, explaining why it's better. Encouraging questions is key – no one should feel afraid to ask "Hey, I don't understand how this part of the trigger framework works." When a question arises, sometimes ChatGPT can be brought in as a quick assistant. A developer might say, "I'm not sure why we need to use with sharing here," and another might quickly consult ChatGPT (perhaps via a Slack integration or just on the side) to get an explanation of Apex sharing rules to help answer that question.

Some teams integrate ChatGPT (or an internal knowledge bot) into their team chat (like Slack or Microsoft Teams). For instance, a developer could tag a bot with a question: "@CodeBot what does this error FIELD-_CUSTOM_VALIDATION_EXCEPTION mean?" and the bot (powered by something like ChatGPT) could reply with an explanation and maybe a link to relevant Salesforce docs. This way, even shy team members can ask "dumb questions" to the bot and get help without fear of judgement, or the whole team benefits from the answer if it's shared in chat. Of course, one must ensure the bot has proper access and respects privacy (no posting sensitive code to external services without checks).

Blameless Post-mortems and Continuous Improvement: When issues occur – say a production bug or an outage caused by a faulty Apex logic – a collaborative culture responds by focusing on learning, not blaming. The team can do a post-mortem meeting: analyze what happened, why the process failed to catch it (maybe tests missed a case, or requirements were unclear), and identify action items to improve. For example, if a bug slipped because code review missed something, perhaps action is "update the code review checklist to include that scenario." The key culturally is that it's "we as a team" that owns the code, so "we" collectively learn and fix the process, rather than pointing fingers at the individual who wrote the bug. This encourages people to be open about mistakes or concerns early (instead of hiding them). ChatGPT can help draft a post-mortem report by organizing the findings or even role-playing a scenario ("How to explain to stakeholders that we had an outage due to a code bug, emphasizing what we learned?"). It might seem odd to use AI for something so human, but it can provide a neutral perspective or suggestions on phrasing that keep things constructive.

Encouraging innovation and staying updated: A forward-looking culture doesn't just stick to old ways; it keeps an eye on new Salesforce features, tools, and methodologies. For example, if Salesforce introduces a new Apex annotation or a feature like Lightning Web Components (LWC) that can offload what was previously done in Apex, the team should discuss and adapt. Encouraging team members to try out new tools like Einstein for Developers or GitHub Copilot (as they become available) can be culturally enabled by giving time for experimentation. Maybe a hackathon day to let devs play with AI tools on some nonproduction code, then share what they learned. A supportive culture ensures people don't feel their jobs threatened by these tools, but rather see them as aids to make their work more interesting (less grunt work, more creative). Leadership and architects can lead by example by openly trying these tools and discussing their pros/cons.

Cross-functional collaboration: In many Salesforce projects, developers, admins, testers, and business analysts all work together. Code quality isn't just a dev concern – it affects how easily an admin can configure something or how reliably features can be delivered to users. A collaborative culture breaks down the walls between coders and "clickers" (a friendly term for those doing declarative config). They should be partners. For example, if an admin config change (like a new validation rule) keeps breaking an Apex trigger, the admin and developer should sit together and find a solution (maybe the code needs to handle that validation scenario). Regular meetings or at least channels where admins and devs communicate helps a lot. Product owners and testers also should be in the loop. One way to foster this is to have *joint design sessions* where a new feature is discussed by everyone: the architect might propose an Apex design, the admin might suggest "hey we can do part of this with a Flow instead," the tester says "I will need a way to inject test data here," etc. These discussions ensure the solution is holistic and not just thrown over the wall from one role to another.

Using ChatGPT beyond just coding – for team communication: Teams have even started using AI to help with things like writing user stories or acceptance criteria (by asking for templates or examples), drafting meeting agendas, or summarizing long Slack threads. For instance, if there was a heated Slack discussion about a technical issue overnight, the morning stand-up could benefit from a quick summary. ChatGPT could be fed the chat transcript and asked for a concise summary of decisions and open questions. This can save time and ensure everyone is up to speed without wading through 100 messages. (Again, do this carefully with respect to privacy – if using an external AI, make sure not to leak internal chat contents that are sensitive. There are also on-prem solutions for this if needed.)

Another cultural element is **collective code ownership.** This means no single person "owns" a piece of code; the team does. Anyone should feel empowered to fix a bug or improve a module, not thinking "Oh, that's Alice's code, I shouldn't touch it." Achieving this requires trust and consistency in coding style (so one dev's code isn't too hard for another to pick up – using consistent patterns helps). Code reviews help because multiple eyes have seen the code. Pair programming can rotate who works on what. If someone goes on vacation, another can step in without terror because the documentation is there and the team has overall knowledge. Encourage rotation: e.g., if one dev always does integrations and another always does CPQ, have them swap tasks occasionally to broaden knowledge. ChatGPT can assist someone stepping into unfamiliar code: *"I'm new to the Subscription billing code module, can you explain how it works?"* The team might have documentation, but an AI summary of the code can reinforce understanding quickly.

Psychological safety and feedback: Team members should feel safe to voice concerns about code quality. If a junior developer thinks a piece of code by a senior is confusing or potentially problematic, they should be encouraged to speak up (respectfully). Leaders can model this by being open to feedback and not pulling rank in code discussions. If someone's suggestion is wrong, correct it kindly and use it as a learning moment, not a ridicule. ChatGPT can even be a practice ground for juniors: if they are unsure about commenting on a senior's code design, they might first ask ChatGPT *"Is using a lot of static variables in Apex bad practice and why?"* to gain confidence in their point, then raise it. It's like getting a second opinion to bolster one's courage in discussions.

Tip for Team Leads/Managers: Build time for team activities that strengthen culture. For example, have a monthly "quality spotlight" where one team member presents a cool tip (could be as simple as a new VS Code shortcut, or how they used ChatGPT to solve a tricky problem). Acknowledge and reward not just features delivered, but also refactors done, complex bugs solved, excellent test coverage, helpful code reviews given, etc. This signals that quality work is valued. If you have new AI tools available, make sure everyone gets training or at least a demo, so no one is left behind or feeling threatened. Perhaps pair a less experienced dev with a more experienced one to explore Copilot together, leveling up both on AI usage and coding.

Tip for Developers: Embrace the mindset that "it's our code" rather than "my code." When giving feedback, be kind and specific – remember there's a person behind the code. When receiving feedback, try not to take it personally; it's about the code in that moment, and even great developers have things to learn. Use tools like ChatGPT to clarify things you're uncomfortable asking others yet, but do eventually ask the team – chances are someone else has the same question. Participate in team knowledge sharing, and consider writing short internal blog posts or emails about things you learn. For example, if you discover a nifty approach to mock callouts in tests, share it. If writing isn't your forte, draft something and then let ChatGPT polish the language. It can turn your bullet points into a coherent mini-article for the team.

Tip for Testers: You are a critical part of the dev team, not separate. Get involved early in design discussions. A culture of quality means testers' voices are heard when they say, "This scenario is hard to test" or "We need more time to create automated tests for this." Advocate for quality on behalf of end-users. If you find recurring issues, suggest process improvements, not just bug fixes. For instance, if many bugs are due to null reference errors, maybe the dev team should adopt more null handling patterns or use static analysis rules – raise that in retrospectives. Use ChatGPT to explore technical topics to better understand the code you're testing. If the devs talk about "Selectors" or "Unit of Work pattern" and you're not familiar, ask the AI for an explanation so you can follow along and contribute.

Tip for Product Owners/Analysts: Quality culture needs support from the product side too. For example, if developers ask for time to do a refactor or write more tests, understand why and back them up. Don't measure velocity purely in features delivered; include quality metrics or at least qualitatively discuss quality improvements as achievements. When planning work, include acceptance criteria around quality (e.g., "code must follow security best practices, no high-severity lint issues"). If you use ChatGPT in your role (maybe to analyze user feedback or generate report summaries), share those productivity gains with the team – it encourages an overall culture of working smart. Also, translate between technical and business often: explain to stakeholders why the team spends time on "invisible" quality tasks, and conversely, ensure the dev team understands the business impact of quality issues (like "that last outage due to a bug caused us to lose some customer trust"). This holistic understanding binds the team with a common purpose.

A truly collaborative Apex development culture is one where everyone feels responsible for the health of the code and supports each other in maintaining it. In such an environment, the practices of reviews, refactoring, documentation, CI, and metrics we discussed become much easier to implement because people see their value and collectively uphold them. The introduction of AI tools like ChatGPT is seamlessly integrated as just another way team members get help and share knowledge – it doesn't replace human interaction but supplements it.

With a strong culture in place, teams are well-positioned to take advantage of the latest advancements in tooling. In the final section, we will look at **leveraging automation and AI for continuous quality improvement.** We'll examine current and upcoming AI-driven tools (including Salesforce's own Einstein for Developers and GitHub Copilot X) that can further automate code quality processes, and how a forward-looking team can integrate these into their workflow. This is the cutting edge of maintaining code quality: using intelligent tools to work smarter, while the team focuses on the creative and complex aspects of development. Let's explore that future-forward topic next.

LEVERAGING AUTOMATION AND AI FOR CONTINUOUS QUALITY IMPROVEMENT

In the not-so-distant past, maintaining code quality relied entirely on human effort and traditional tools, but today we stand on the frontier of an AI-augmented development era. Automation and Artificial Intelligence (AI) are increasingly woven into the software development lifecycle, and Salesforce Apex development is benefiting from this trend as well. In this final section, we'll discuss how you can leverage automation and AI to continuously improve Apex code quality. Some of these approaches are already accessible, while others are emerging trends that we anticipate will play a bigger role going forward. We will touch on Salesforce's own AI offerings like Einstein GPT for Developers, third-party tools like GitHub Copilot, and various automation techniques that reduce manual toil. Throughout, we'll consider how these tools fit into an enterprise development workflow and how to use them responsibly (AI is powerful but not infallible). The tone here is forward-looking – a glimpse into how Apex development might evolve in 2025 and beyond, and how you can position your team to ride that wave.

One of the most exciting developments is **Salesforce Einstein for Developers**, which became available in beta in late 2023. Einstein for Developers is a generative AI coding assistant built into the Salesforce Platform, specifically trained for Salesforce languages (Apex, LWC, etc.). It can turn natural language prompts into Apex code, suggest code completions as you type, and even scan your code for vulnerabilities in real-time in the IDE. Because it's integrated with your org (with appropriate permissions and the Salesforce AI Trust Layer), it can make context-aware suggestions – for example, it might know your naming conventions or use of certain frameworks and tailor its output to that. By enabling Einstein for Developers in your org and VS Code (via the extension), developers can dramatically speed up routine coding tasks. Instead of writing boilerplate from scratch, you might type a comment like "// TODO: Query contacts by specific criteria and update fields" and Einstein can generate the Apex code to do that. It's using Salesforce's own large language model (CodeGen) under the hood.

Example of using Einstein GPT for Developers: The VS Code Einstein Sidebar (top) shows a natural language prompt requesting an Apex class *ContactSum* that counts contact records linked to an Account, and the AI-generated Apex code (bottom) fulfilling that request. Such AI tools accelerate development by producing boilerplate code and following best practices (note the *with sharing* keyword for security). Developers should still review and test the code, but this demonstrates how generative AI can assist in writing Apex.

As shown in the example image above, you can ask Einstein to create not only classes but also accompanying test classes. This means a lot of the grunt work – creating test data, writing assert statements, etc. – can be handled by AI suggestions. It's important to emphasize that while these suggestions are impressively accurate much of the time, they are not a replacement for human review. You might treat them as starting points or accelerators. The developer should review the generated code for correctness, adherence to business logic (the AI doesn't know your exact requirements unless you specify them very clearly in the prompt), and alignment with your team's style. In practice, though, if Einstein suggests a block of code, it likely saves you from a few syntax mistakes or forgetting a null check because it tends to include common best practices. Some teams have reported that such tools help maintain quality by **reducing human error** in repetitive code – for instance, the AI won't typo a variable name or skip a necessary field, whereas a human might on a tired day.

GitHub Copilot, another AI pair-programmer tool, has been around a bit longer and many Salesforce developers use it too. It was trained on public code, including Apex code on GitHub, so it is surprisingly adept at generating Apex snippets and even entire classes. Copilot supports Apex, Visualforce, LWC, etc., with users noting it works very well with Apex specifically. If you comment your code or write a function signature, Copilot can often suggest the full implementation. It's great for mundane tasks like writing a loop to process records or constructing a JSON serialize/deserialize routine – things that have predictable patterns. GitHub is also advancing Copilot with features like **Copilot X**, which will include automatic test generation, integrated chat for explaining code, and AI-assisted pull requests (like generating PR descriptions or even suggesting changes). Imagine opening a pull request and an AI already highlights a potential bug or suggests adding a test for a certain scenario – that could be reality soon. In fact, by 2025, we might see integration where an AI can act as a "virtual reviewer" on your pull request, leaving comments just like a human reviewer would. This doesn't replace your colleagues' reviews but supplements them by catching obvious issues or giving a neutral analysis of the changes.

Beyond coding assistants, we have **automation in quality processes:**

Automated Code Scanning: We touched on static analysis tools in metrics. Automation here means you can set up these tools to run not just in CI, but continuously or on a schedule, and perhaps automatically create tickets for issues. For example, a nightly job could run Salesforce Code Analyzer across your codebase and if it finds any new critical issues, automatically log a work item in your agile tool assigned to the module owner. Or integrate with GitHub's code scanning alerts (since

Code Analyzer now supports SARIF format, you can feed results to GitHub security alerts). This way, nothing falls through cracks; issues are tracked like any bug.

Automated Testing and Healing: Creating tests is still largely manual, but as mentioned, Copilot X and Einstein can assist in test code generation. There are also tools for test automation on the UI side (like Provar for Salesforce UI tests) that can adapt to changes – some use AI to locate fields or buttons if labels change, making UI tests more resilient. While not Apex-specific, this contributes to overall quality by reducing false test failures when the UI changes slightly.

Intelligent Monitoring: Salesforce provides event monitoring logs and error logs. There's an emerging area of AI Ops where machine learning looks at your error logs and identifies patterns or anomalies. For instance, if a certain Apex error (maybe an unhandled exception) suddenly starts happening more frequently, an AI could alert you saying "This error is trending up in the last week; it may indicate a new issue introduced in the latest release." Previously, you'd rely on manual monitoring or user reports to catch such patterns. Now AI can sift through the noise of logs and surface relevant info. Salesforce's platform might integrate such capabilities (they already have some anomaly detection in Einstein Analytics that could be repurposed for dev ops).

ChatGPT for DevOps: We already used ChatGPT for writing CI scripts and troubleshooting. We could take it further: some forward-thinking teams create custom ChatGPT plugins or internal tools so that team members can interact with the CI/CD pipeline via natural language. For example, a Slack bot that you ask "Deploy the latest build to UAT" and it triggers the pipeline (after some authentication). Or asking "What is the code coverage difference between this release and last release?" and it fetches that info from your dashboards. This isn't standard yet, but not far-fetched – it's about making dev tools more conversational and accessible, which is what ChatGPT is great at.

A noteworthy addition on Salesforce's roadmap is the evolution of Code Analyzer (the static analysis tool). The version 5 (Beta) introduced things like scanning Flows and detecting copy-paste code , which shows a trend: Salesforce is aiming to cover more than just Apex code in its analysis – it's going towards holistic org analysis. We can anticipate future versions might include AI-powered rule suggestions or an AI that scans your org configuration alongside code to spot inconsistencies (for example, "you have a validation rule that conflicts with this Apex logic, causing errors"). Einstein for Developers already scans code for vulnerabilities; that's a specialized AI analyzing code semantics, arguably an AI-driven static analysis. As these tools mature, some tasks like code reviews for certain patterns might become almost fully automated (the AI flags them as you write, you fix them before even committing).

One question often asked is: will AI replace developers? The stance in 2025, and likely for a long time, is **no – but it will change the developer's role.** AI can churn out lines of Apex or find issues, but it doesn't truly understand the business context or the user needs. Developers and architects still have to decide *what* to build, ensure it fits requirements, and have the creativity to design solutions. What AI does is take over a lot of the repetitive and detailed work, allowing developers to focus on higher-level problems. It's similar to how calculators didn't replace mathematicians; they just freed them from doing arithmetic by hand. That said, developers should upskill to work effectively with AI: knowing how to prompt it well, how to verify its outputs, and how to integrate multiple tools together. A developer in 2025 might spend as much time reviewing and tweaking AI-suggested code as writing original code – but the output is higher quality and produced faster than before.

It's also worth mentioning **limitations and precautions**:

AI suggestions might occasionally be outdated (training data might not include the latest API changes). For example, if Salesforce adds a new method in Summer '25, Copilot might not know about it yet. Always verify new or strange code suggestions with the official documentation.

AI can introduce subtle bugs if you trust it blindly. Maybe it writes a SOQL that doesn't cover a certain condition properly. Treat AI code like you would code from a junior developer: helpful, but needs review and testing.

Ensure use of AI tools complies with data governance. Some companies may restrict usage of cloud AI for code due to intellectual property concerns. Salesforce's Einstein has an advantage here – it's within the Salesforce trust boundary and they emphasize data isn't used to train external models. GitHub Copilot has settings to not use your code for training. Make sure those settings are enabled if required by your legal team. And when using ChatGPT or others, avoid putting proprietary logic verbatim if that's a concern; instead, abstract the problem.

Keep humans in the loop. For example, if an AI flags a security issue, have a security expert confirm it. AI can have false positives or negatives.

To integrate AI smoothly, consider adding usage guidelines to your team's best practices. For instance: "Use Einstein/CoPilot to generate baseline code, but all AI-generated code must be reviewed in code review as if it were written by a team member." Or "We use ChatGPT for research and explanations, but final decisions must be based on Salesforce documentation or proven best practice." These guardrails ensure AI remains a helpful servant, not a misguided master.

Tip for Architects: Stay abreast of new AI tooling in the Salesforce ecosystem. Salesforce is likely to continue embedding AI in the platform (e.g., AI-assisted code refactorings in the future, or smarter performance analysis). Be ready to pilot these tools in a controlled way. For example, enable Einstein for Developers in a sandbox first and let developers try it on non-critical code to see how it works for your team. Also, consider the **ethical use** of AI: make sure it's not inadvertently introducing any compliance issues. If your org is in a regulated industry, ensure AI outputs (especially if they can affect data or security) are vetted. On the flip side, push for AI adoption where it clearly reduces risk – like using automated security scanning and letting AI help remediate those (some security tools can auto-fix common vulnerabilities, which you then review). In architecture decisions, factor in that "some maintenance can be offloaded to AI" but also "some new design might confuse AI assistance if it's very novel," which is an interesting consideration.

Tip for Developers: Embrace the new tools. If you have access to Copilot or Einstein GPT, use them to the fullest to save time, but never stop learning the fundamentals. Think of AI suggestions as code snippets from StackExchange – they can be great, but you need to understand them. If Copilot writes an Apex method for you, walk through it and make sure you understand each line. This way, you learn from it as well. If you find the AI often gives a suboptimal solution, that's an opportunity: you might feed back better examples (Copilot improves as you accept/reject suggestions) or you can even suggest improvements to Salesforce (they likely gather feedback from Einstein usage). Also, help train your team in these tools: if you figure out a particularly useful prompt (like how to get ChatGPT to generate bulkified code properly), share that knowledge. In code reviews, don't discriminate against AI-written code vs human – hold it to the same standards. Sometimes, you might spot an "AI-ism" (like an odd variable name that clearly came from a different context); clean those up.

Tip for Testers: AI can assist you too – generating test cases or even generating test data. You might use a tool to create realistic test records (there are AI data generators that produce names, addresses, etc. that look real without using actual PII). Also, keep an eye on how AI is used in code so you know where potential weak spots are. For instance, if a piece of code was largely AI-generated, maybe pay extra attention in testing that area until the team builds trust in the AI outputs. Conversely, AI might help you by analyzing logs. If a test fails and produces a log, you could paste it into ChatGPT and ask "what likely went wrong here?". It might identify the null pointer or logic branch that failed, saving you time in root cause analysis.

Tip for Product Owners: From a product perspective, faster and better quality delivery is the payoff of using AI and automation. Monitor the impact: if the team adopts Copilot, do you see features being delivered faster or with fewer bugs? Those are talking points to justify continued investment in these tools (they often require licenses, e.g., Copilot has a monthly fee). Also, consider AI as an assistant in your realm – for writing user stories, summarizing user feedback, etc. If you use it and save time, that gives you more bandwidth to focus on quality-related aspects like refining requirements (many bugs stem from unclear requirements). There's also an aspect of managing expectations: higher-ups might think "Oh, you have AI now, you can do 2x work with half the team!" Be realistic and use data from the team to set expectations. Yes, AI can improve productivity, but it's not magic; it's part of a broader strategy including all the practices we discussed. Finally, encourage an innovation mindset in the team: maybe allocate some budget or time to experiment with new automation ideas (like a hackathon to connect ChatGPT to your internal knowledge base, etc.). Such experiments could yield big quality or productivity wins.

We've journeyed through a wide array of topics in this chapter – from human-centric practices like code reviews and culture to cutting-edge AI tools transforming how we code. The common thread is **continuous improvement**. Maintaining Apex code quality over time isn't a single task, it's a cycle of improvement, feedback, and adaptation. We started with the basics and ended with the future, but all pieces work together. As you implement these practices, remember that each team is unique; you'll need to tailor these recommendations to fit your context, whether you're a small startup or a large enterprise Salesforce shop.

In summary, automation and AI act as force multipliers for your existing quality efforts. They can watch your back 24/7 (through automated checks), handle tedious tasks (like writing boilerplate or tests), and provide insights that might not be obvious at first glance (through intelligent analysis of code and logs). By leveraging these, you free up the creative and analytical capacity of your team to focus on delivering value and designing better systems. The role of a Salesforce developer is evolving – it's an exciting time where you get to collaborate not just with humans but with AI assistants. Those who learn to make the most of this collaboration will lead the way in productivity and quality.

As we conclude this chapter and this book, let's recap the key takeaways and look ahead to how you can continue the journey of excellence in Apex development. We'll wrap up with a summary and a thank-you, as well as an invitation to keep in touch as fellow Trailblazers in the Salesforce ecosystem.

SUMMARY

In this chapter, we explored **Maintaining Apex Code Quality Over Time** from every angle – process, people, tools, and trends. We began by underscoring that code quality is a continuous journey, requiring regular attention and adaptation. Through seven sections, we looked at actionable best practices and forward-looking insights:

Regular and Effective Apex Code Reviews: We learned that frequent peer reviews catch issues early and spread knowledge within the team. By using checklists tailored to Salesforce (ensuring bulkification, security, proper design patterns, etc.) and leveraging ChatGPT for quick advice or code explanations, teams can make reviews thorough yet efficient. We emphasized creating a positive review culture where feedback is welcomed and aided by AI when appropriate, like asking ChatGPT to double-check logic or clarify a Salesforce feature during the review process.

Proactive and Structured Apex Refactoring: We discussed treating refactoring as an ongoing activity to combat technical debt before it becomes a crisis. Approaches include scheduling refactoring tasks and using static analysis to identify code smells. ChatGPT can suggest refactoring ideas or even generate refactored code snippets, acting as a catalyst for improvements. The importance of safe refactoring (with good tests and small steps) was highlighted, as well as involving product owners so they understand the long-term value of this work.

Maintaining and Updating Apex Documentation: Documentation was reinforced as a first-class citizen for code quality. We covered maintaining inline comments, ApexDoc, design docs, and even diagrams, ensuring they evolve with the code. AI shines here – we saw how ChatGPT can summarize code, generate user-friendly explanations, and even help create diagram descriptions. By integrating documentation updates into the development workflow (and using AI to reduce the effort), teams keep tribal knowledge to a minimum and make onboarding or context-switching easier.

Implementing Continuous Integration and Deployment Practices: We examined how a robust CI/CD pipeline automates quality checks and enforces consistency. With every commit tested and analyzed, teams catch problems early and reduce deployment risk. We noted that companies with fast, continuous release cycles see higher ROI , underlining the business impact of CI/CD. Using ChatGPT to assist in writing pipeline scripts or troubleshooting build errors can accelerate setting up these practices. The result is more frequent, reliable releases – quality delivered to users continuously, not in big uncertain jumps.

Building Effective Apex Code Metrics and KPIs: We identified key metrics like test coverage, static analysis issue counts, complexity scores, and bug rates that help quantify code quality. Setting targets for these and tracking them over time provides objective feedback on whether quality is improving or slipping. We discussed using tools (SonarQube, Code Analyzer) to gather data and how ChatGPT can help interpret results or communicate them. The takeaway is to use metrics as a guide, not a crutch – to prompt meaningful conversations and decisions around code health.

Cultivating a Collaborative Apex Development Culture: Perhaps most importantly, we highlighted the human factors. A culture of collaboration, learning, and shared ownership is the soil in which all these practices flourish. By encouraging knowledge sharing (pair programming, internal workshops), maintaining psychological safety, and aligning everyone (devs, testers, admins, product managers) around quality goals, teams create an environment where doing the right thing for code quality is the path of least resistance. We saw how ChatGPT and similar tools can be embedded in team communication to answer questions and spread knowledge, but ultimately it's the team's values and habits that drive long-term success.

Leveraging Automation and AI for Continuous Quality Improvement: Finally, we peered into the state-of-the-art tools and what's on the horizon. AI assistants like Einstein GPT for Developers and GitHub Copilot are game-changers, providing on-the-fly code generation and suggestions that adhere to best practices. Automation extends to intelligent code scanning and test generation, further easing the maintenance burden. We advised using these tools as accelerators – they amplify human talent when used wisely. With proper oversight, AI and automation can handle repetitive tasks and surface insights, allowing the development team to focus on creative problem-solving and delivering business value.

Throughout each section, we included **ChatGPT prompt examples** to illustrate how readers can engage with AI to learn and adapt the advice to their specific needs. We also mentioned alternative ways to use ChatGPT beyond typing: from voice input for hands-free assistance, to uploading files like code or logs for analysis, to even generating images or diagrams for visual aid. Additionally, we offered tips for various roles – architects, developers, testers, product owners – recognizing that maintaining code quality is a cross-functional effort. Everyone has a part to play, and when each role is empowered with the right practices and tools (AI included), the synergy leads to higher quality outcomes.

As of 2025, the Salesforce development landscape is rapidly evolving. Best practices today may further mature tomorrow, especially as AI becomes more ingrained. We've noted where some advice is speculative – for instance, anticipating more AI-driven features in tools like Salesforce Code Analyzer or Copilot's upcoming capabilities. The key for readers is to remain adaptable and continuously curious. The principles in this chapter (and book) form a resilient foundation: focus on clarity, correctness, collaboration, and continuous improvement. With those in place, adopting new tools or adjusting to platform changes becomes much easier, because you have a strong culture and process to build on.

To conclude, maintaining Apex code quality over time is a marathon, not a sprint. It's about setting up an ecosystem – of people, process, and technology – that naturally produces and sustains high-quality code. By applying the practices outlined in this chapter, you can avoid the slow decay that often plagues long-lived codebases and instead keep your Salesforce org's custom code nimble, reliable, and secure. This ensures that as your business grows and changes, your Apex code is an enabler, not an obstacle.

Thank you for joining us on this journey through Apex best practices with ChatGPT. We hope you found valuable insights and practical tips that you can implement in your projects. Quality software is a team endeavor, and we're excited to see how you and your teams will take these ideas forward and perhaps even improve upon them.

As part of the **Salesforce Enterprise Thinking** series, our goal is to foster a community of professionals who strive for excellence and are eager to share and learn from one another. We, the authors, are continuously learning too – especially in this fast-moving era of AI and cloud innovations. We invite you to connect with us on LinkedIn to continue the conversation, share your experiences, and stay in touch. Your feedback and insights are most welcome, and who knows – they might shape the next edition of this work!

Thank you, and happy coding. Let's keep pushing the boundaries of what we can achieve with Apex, ChatGPT, and a collaborative spirit in the Salesforce ecosystem. Here's to building high-quality solutions that stand the test of time and drive success for our organizations for years to come.

APPENDIX A: APEX CODING STANDARDS HANDBOOK

This handbook defines the official coding standards for Apex development in Salesforce. It is written for experienced Apex developers and aims to ensure that all code in the organization is consistent, reliable, and maintainable. By adhering to strict coding standards, teams can improve code readability, reduce errors, and simplify code reviews and maintenance.

Purpose of Coding Standards

The purpose of Apex coding standards is to establish a common language and set of expectations for all developers. These standards eliminate guesswork by providing clear rules on how code should be written. Consistency in code style makes it easier for any developer to understand and modify code written by others. Moreover, following standards can prevent common mistakes. For example, consistently using braces even for single-line if statements avoids ambiguity and bugs:

```
// Correct: Always use braces, even for one-line if statements

if (isActive) {

   executeAction();
```

```
}
// Incorrect: No braces (can lead to errors if expanded later)
if (isActive)
    executeAction();
```

In the correct example above, braces make it explicit what code is controlled by the if condition, enhancing clarity. The incorrect example without braces could introduce errors if another line is added unintentionally inside the if. **In summary, coding standards like this improve code safety and clarity.**

Scope and Applicability

This handbook focuses on **enforceable technical coding standards** for Apex code. It covers naming, formatting, commenting, language-specific rules, testing, security, resource usage, asynchronous processes, integrations, use of metadata/annotations, source control, and governance of standards. We **avoid purely theoretical best practices or design patterns** – every rule here is meant to be concrete and checkable (either via code reviews or automated tools). For instance, "avoid SOQL inside loops" is included as a standard because it can be automatically detected and has a clear impact on governor limits, whereas a general design guideline like "use a Service layer" is not included because it's more of an architectural choice than a strict rule.

These standards apply to all Apex classes, triggers, controllers, batch classes, test classes, and other Apex constructs in the codebase. They should be followed in all development, whether writing new code or modifying existing code (legacy code should be brought up to standards when touched, if feasible). By making these rules universal, we ensure that the entire codebase follows the same conventions.

Using This Handbook

Each chapter of this handbook addresses a specific area of Apex coding standards. Within each chapter, you will find multiple sections that describe individual standards. Each section provides:

A Heading describing the standard.

Purpose: A brief statement of why the standard exists or what it aims to achieve.

Explanation: A technical explanation and details of the rule.

Code Example: At least one Apex code example demonstrating the standard (with a correct implementation, and sometimes a contrasting incorrect example).

Readers can use the handbook as a reference during development and code reviews. For example, if you are naming a new Apex class, refer to **Chapter 2: Naming Standards** to ensure the class name meets our conventions. If writing a trigger, consult both naming standards for triggers and **Chapter 8: Governor Limits** to ensure the trigger is bulkified.

The rules herein are intended to be followed strictly. Automated tools (see **Chapter 13**) and peer reviews will be used to enforce compliance. Over time, adherence to these standards will become second nature, resulting in a cleaner and more reliable codebase.

Summary (Introduction): In this chapter, we established why coding standards are critical in Apex development. We outlined the scope of this handbook – focusing on clear, enforceable rules rather than generic advice. We also explained how to navigate and use the handbook. With this foundation, the following chapters will dive into specific standards in areas like naming, formatting, documentation, and more, each with concrete examples to follow.

NAMING STANDARDS

Introduction: Naming conventions are fundamental to readable and maintainable code. In Apex (as in Java), well-chosen names make code self-documenting. This chapter defines strict naming standards for all identifiers: classes, triggers, methods, variables, constants, and other Apex elements. Consistent naming helps developers understand the role of a class or variable at a glance and avoids confusion. We follow Java-style naming conventions as recommended by Salesforce , with some Salesforce-specific rules for things like trigger naming.

Class Names

Purpose: Standardize how classes are named for clarity and consistency across the codebase. A class name should clearly indicate its purpose or entity, using a consistent style.

Standard: Class names must be written in PascalCase (also known as UpperCamelCase), starting with an uppercase letter and capitalizing each subsequent word. Class names are typically nouns or noun phrases. They should be descriptive but not excessively long. For example, a class representing an account trigger handler might be named AccountTriggerHandler.

If the class is a test class, it should end with Test (or Tests) to clearly distinguish it as a test. Example: AccountServiceTest for tests covering AccountService class.

Example:

```
// Correct

public class InvoiceProcessor {

  // class implementation

}

// Correct (test class)

@isTest

private class InvoiceProcessorTest {

   static testmethod void testProcessing() {

      // test code

   }

}

// Incorrect: not using PascalCase, or ambiguous name

public class invoiceprocessor { /*... */ }

public class ProcessInv { /*... */ }
```

In the incorrect examples, invoiceprocessor is not capitalized and ProcessInv is abbreviated to the point of obscurity. The correct examples follow PascalCase and clearly communicate their intent. Remember that the Apex file name must match the class name exactly (case-sensitive), and there should be only one top-level class per file (the platform enforces this).

Trigger Names

Purpose: Establish a clear naming scheme for trigger metadata, differentiating triggers from classes and indicating the object they operate on.

Standard: Use one trigger per object, and name the trigger after the object. The trigger name should be ObjectNameTrigger. For example, if you have a trigger on the Account object, name it AccountTrigger. Because we allow only one trigger per SObject (to maintain a unified entry point for logic), no further description (like event or order) is needed in the name. The trigger's logic for different events will be managed inside or delegated to handler classes.

Example:

```
// Correct: one trigger for Account, clearly named

trigger AccountTrigger on Account (before insert, before update, after update) {

    AccountTriggerHandler.handleTrigger(Trigger.new, Trigger.oldMap);

}

// Incorrect: multiple triggers on the same object or unclear naming

trigger account_before_insert on Account (before insert) { /*... */ }

trigger AccountUpdate on Account (after update) { /*... */ }
```

In the incorrect example, having separate triggers account_before_insert and AccountUpdate on the Account object is against the standard. We enforce a single AccountTrigger to handle all relevant events, which simplifies control of execution flow and avoids conflicts. The correct example shows a single trigger named AccountTrigger handling multiple events and delegating logic to a handler class (the handler class naming would follow class naming standards, e.g., AccountTrigger-Handler).

Method Names

Purpose: Ensure method names clearly convey the action or computation they perform, following a consistent style that differentiates them from class names.

Standard: Method names must be written in camelCase, starting with a lowercase letter, and typically use a verb or verb phrase. A method performs an action, so its name should usually include a verb (e.g., calculateTotal, sendEmailNotification). Use descriptive names that make the method's purpose obvious. Do not use underscores in method names. Getter and setter methods (if used instead of Apex properties) should be named according to their value (e.g., getTotalAmount). Conversion of acronyms: if the method name contains acronyms or abbreviations, treat them as words (e.g., use parseXmlResponse() not parseXMLResponse() for readability).

Example:

```
public class OpportunityService {

  // Correct: descriptive verb-named methods

   public static void calculateCommission(List<Opportunity> opps) { /*... */ }

  public Integer findHighestYearToDateSales() { /*... */ return 0; }

   // Incorrect: Method names should not start with uppercase or have underscores

   public void Calculatecommission() { /*... */ } // Starts with uppercase - not allowed

   public void send_email() { /*... */ }          // Contains underscore - not allowed

}
```

In the correct examples, calculateCommission and findHighestYearToDateSales clearly describe what the methods do. They follow camelCase (initial lowercase, subsequent words capitalized). In contrast, the incorrect examples violate the naming rules: Calculatecommission starts with an uppercase letter (and is harder to read due to missing camelCase on "commission"), and send_email uses an underscore (Apex methods should use camelCase instead). Consistent method naming makes it easier to distinguish methods from classes at a glance and understand code intent.

Variable and Property Names

Purpose: Define how to name variables (including object properties, local variables, and class fields) so their purpose is clear and they are distinguishable by case conventions.

Standard: Variables should be named in camelCase, starting with a lowercase letter, much like method names. Choose meaningful names that reflect the stored value or purpose. For example, totalAmount, accountList, isActive. Avoid single-letter or cryptic names except in very short scopes (like a loop index i). Do not use Hungarian notation or prefixes to denote type; the type is evident from the declaration. Instance and static variables of a class should also follow camelCase (e.g., private static Integer maxRetryCount;). Booleans should be prefixed in a way that implies a true/false question, for example, isComplete, hasAccess – this enhances readability (e.g., if (isComplete) {...} reads naturally).

If using Apex properties (with { get; set; }), the property name should also follow camelCase (often properties represent data fields, so treat them like variable names).

Example:

// Correct variable naming

Integer recordCount = 0;

String accountName = 'Acme Inc.';

Boolean isEligible = false;

```
List<Contact> contactList = new List<Contact>();

Account currentAccount; // instance of an Account object

// Incorrect variable naming

Integer RecordCount = 0;   // Starts with capital letter - not camelCase

String acct_name = 'Acme'; // Uses underscore - not allowed

Boolean flag = false;    // Non-descriptive name

List<Contact> Contacts = new List<Contact>(); // Capitalized variable name
```

In the correct examples, each name clearly indicates content or purpose (recordCount, accountName, isEligible, contactList). The boolean isEligible is prefixed with "is", implying a condition. The incorrect examples show what to avoid: capitalized RecordCount (should be lowercased first letter), acct_name (underscore used), a vague name flag, and a capitalized Contacts list. By following these naming rules, code becomes self-explanatory. For instance, contactList immediately signals a collection of contacts, and isEligible signals a boolean condition.

Constant Names (Final Static Variables)

Purpose: Provide a clear convention for naming constants so they stand out and indicate immutability.

Standard: Constants (such as public static final fields) should be named in ALL_CAPS with words separated by underscores. This convention makes them easily distinguishable from other variables. Only use this for true constants – values that are final and do not change after initialization. For example: public static final Integer MAX_RETRY_ATTEMPTS = 3;. Non-constant static variables (that can change) should *not* use all-caps, they should follow normal variable naming.

Example:

```
public class MathUtil {

  public static final Decimal PI = 3.14159;
```

```
  private static final Integer MAX_RETRIES = 5;

  public static Decimal circleArea(Decimal radius) {

    // Use of constant in calculation

    return PI * radius * radius;

  }

}

// Incorrect usage of constants naming

public class Foo {

  public static final String defaultName = 'ABC'; // Should be DEFAULT-
_NAME in caps

    public static final Integer MaxCount = 10;        // Should be
MAX_COUNT

}
```

In the MathUtil example, PI and MAX_RETRIES are constants: they're declared static final and their names are in all caps with underscores. This immediately signals that these are constant values. In the incorrect Foo class, defaultName and MaxCount are declared final but not following the ALL_CAPS convention, which reduces clarity. By using the all-caps convention, we ensure constants are easily recognizable throughout the code.

Naming Conventions for Tests and Apex Components

Purpose: Outline naming schemes for test methods and other Salesforce-specific constructs to maintain consistency.

Standard: Test classes should be named after the class they test, with a suffix Test or Tests. Test method names should describe what they are testing or the scenario, often starting with test (though the @isTest annotation is what makes them test methods, a naming pattern is for readability). For example, in AccountServiceTest class, a method could be static void testCalculateAnnualRevenue() to indicate it tests the calculation of annual revenue.

Triggers were covered earlier (one per object with the object name). If you have batch classes, you might suffix them with Batch (e.g., OrderProcessingBatch) to indicate their purpose. Schedulable classes might end with Scheduler or Job. Similarly, Apex Aura or LWC controller classes (Apex controllers for Lightning) might end with Controller if appropriate. These are not strictly enforced by the platform but are helpful conventions.

Visualforce component controllers and extensions, if any in the project, should be named with suffixes like Controller or Extension for clarity (e.g., OpportunityTableController).

Example:

```
@isTest

private class AccountServiceTest {

  static testMethod void testCalculateAnnualRevenue() {

    //...

  }

      static testMethod void testCalculateAnnualRevenue_noOpportunities() {

    //...

  }

}

// Suppose we have a batch class

public class OrderProcessingBatch implements Database.Batchable<sObject> {

  //...

}

// A schedulable class
```

```
public class DataCleanupScheduler implements Schedulable {
  //...
}
```

In the above examples, naming communicates role: AccountServiceTest clearly relates to AccountService. The test method names like testCalculateAnnualRevenue tell the reader exactly what scenario is being verified (optionally including a condition like _noOpportunities for a specific case). OrderProcessingBatch and DataCleanupScheduler immediately indicate their function. Adhering to these naming suffix conventions, while not a platform requirement, makes it easier to identify the purpose of classes across the codebase.

Summary (Naming Standards): In this chapter, we established consistent naming rules for Apex code elements. Classes and triggers use PascalCase (with triggers specifically named as ObjectTrigger and only one per object). Methods and variables use camelCase, with methods often starting with a verb and booleans prefixed with *is/has*. Constants are in ALL_CAPS with underscores. We also discussed naming for test classes, batch/scheduled classes, and other components. Following these naming standards ensures that anyone reading the code can infer meaning and intent simply from the names, leading to a more understandable and maintainable codebase.

FORMATTING AND LAYOUT STANDARDS

Introduction: Formatting and layout standards ensure that Apex code is presented in a clean, consistent manner, making it easy to read and navigate. While formatting does not change code execution, it profoundly affects maintainability and collaboration – poorly formatted code can hide errors and slow down reviews. This chapter specifies standards for indentation, braces, line length, spacing, and overall file layout. Adhering to these rules means any developer can open an Apex class and find it structured in a familiar way.

3.1 Indentation and Whitespace

Purpose: Maintain consistent indentation and spacing so that the structure of code blocks is visually clear. Proper indentation delineates code blocks (such as loops, if/else, try/catch) and makes nesting levels obvious.

Standard: Use 4 spaces per indentation level (do not use tabs). Every time you open a new block (after a { brace), increase indent by 4 spaces; when the block ends (}), decrease indent by 4. Never mix tabs and spaces – use spaces only. Also, do not leave trailing whitespace at the end of lines. Keep whitespace consistent: one space between keywords and parentheses (e.g., if (), and around operators (a + b, not a+b), for readability.

Example:

```
public class OrderService {
    public void processOrders(List<Order__c> orders) {
        for (Order__c ord : orders) {
            if (ord.Status__c == 'New') {
                // Indented inside if block
                ord.Processed__c = true;
            } else {
                ord.Processed__c = false;
            }
        }
    }
}
```

In this example, each block level (class, method, loop, if/else) is indented +4 spaces inside its parent block. The use of spaces (not tabs) ensures consistency across editors. The structure is easy to scan: you can match opening and closing braces by their indent level. The spacing around ==

and after if makes the code less dense and more readable. If the code were mis-indented or had inconsistent spacing, it would be much harder to follow logic flow.

Braces and Curly Bracket Placement

Purpose: Define a uniform style for placing { and } braces so that code blocks are easily recognizable and no one has to guess or hunt for matching braces.

Standard: Use K&R style (Kernighan & Ritchie) for braces placement. This means: the opening brace { goes at the end of the preceding line (not on a new line) for class definitions, method definitions, and control structures. The closing brace } goes on its own line, aligned with the beginning of the construct it closes. Always use braces for if/else, for, while, and for... else blocks, even if the block is a single statement. This standard prevents errors due to missing braces and improves readability.

Example:

```
// Correct K&R style and braces usage:
if (status == 'READY') {
   startProcess();
} else if (status == 'PENDING') {
   queueProcess();
} else {
   handleError();
}
// Incorrect:
if (status == 'READY')
{
   startProcess();
}
```

else

handleError();

In the correct example, the opening brace for the if is on the same line as the if statement, and each closing brace aligns under the if or else if. All conditional blocks use braces, even the else with one statement. In the incorrect example, the braces are on separate lines (which is a different style not used in our standards) and the else block has no braces which could cause confusion or errors if another statement is later added. By using a consistent brace style and always including braces, the code's structure remains clear and maintainable.

One Statement Per Line

Purpose: Improve readability by not cramming multiple actions into one line. Each statement deserves its own line to clearly separate logic steps.

Standard: Write only one statement per line. Terminate each statement with a semicolon and a line break thereafter. Do not write multiple variable declarations or method calls in a single line separated by semicolons. Also, do not chain method calls on the same line if it causes the line to become too long or hard to read – consider line wrapping (covered later) if needed.

Example:

// Correct:

Account acct = new Account(Name='Acme');

insert acct;

System.debug('Account inserted: ' + acct.Id);

// Incorrect (multiple statements on one line):

Account acct2 = new Account(Name='Acme'); insert acct2; System.debug('Account inserted: ' + acct2.Id);

The correct example clearly separates the creation of an account, the insert operation, and the debug logging, each on its own line. This makes it easy to comment out or modify one action independently. The incorrect example places three statements on one line, which is hard to read and prone to errors during code reviews (one might overlook a subsequent statement). One-per-line rule ensures each operation is visually distinct.

Line Length and Wrapping

Purpose: Avoid lines that are too long to read easily or that require horizontal scrolling. This enhances readability, especially in side-by-side code reviews or standard editors.

Standard: Limit line length to 100 characters (120 maximum). If a line of code would exceed this length, break it into multiple lines in a logical way. Common scenarios for wrapping lines include: method declarations with many parameters, method call chaining, or SOQL queries with many fields. When wrapping, indent the continuation lines by at least +4 spaces from the original line indent. Break lines at sensible breakpoints – after commas, operators, or before a high-level element. Also, ensure that continued lines align in a readable manner (often continuation lines are indented once or twice to clearly separate them from the start of the statement).

Example:

```
// A long SOQL query broken into multiple lines for readability:

List<Account> recentBigAccounts = [

  SELECT Id, Name, AnnualRevenue, Industry, Type,

     BillingCountry, ShippingCountry

  FROM Account

  WHERE IsActive__c = TRUE

  AND AnnualRevenue > 1000000

  ORDER BY LastModifiedDate DESC

];
```

```
// Method call with multiple parameters, wrapped:
IntegrationService.callExternalSystem(
    order.Id,
    order.Status__c,
    order.Customer__r.Name,
    order.Items__c
);
```

In the SOQL example, the query is wrapped after the SELECT and then some fields are listed on the next line, indented. The WHERE and subsequent clauses each start on new lines, making the query much more readable. The line breaks come at logical points (after commas or keywords like FROM, WHERE). In the method call example, instead of a single very long line, the parameters are each on a new line, indented under the parenthesis of the call, which makes it easy to see each argument separately. This formatting ensures that no line exceeds the recommended length and that the code's structure is apparent.

Blank Lines and Spacing

Purpose: Use vertical space (blank lines) to separate logical sections of code and horizontal space to improve readability, without excess that could cause fragmentation.

Standard: Use blank lines to group related code, but avoid excessive blank lines. For example, put a blank line between method definitions in a class, and between logical sections inside a long method (such as before a new block of logic or a comment section explaining something). Do *not* put multiple blank lines in a row – one blank line is sufficient for separation. At the top and bottom of a class definition, it's common to avoid blank lines after the opening brace or before the closing brace (keep braces tight to class content). Within methods, use blank lines sparingly to break up code into steps (e.g., after declarations, before return statements if it improves clarity).

For horizontal spacing: put a space after commas in arguments and list initializers, after // in a comment (to start the comment text), around assignment and arithmetic operators, and after : in ternary operations. However, do not put spaces just inside parentheses or brackets (e.g., use if (x == 5) not if(x == 5)). Also, no space between a method name and the ((e.g., doSomething(param) not doSomething (param)).

Example:

```
public class PaymentService {

  // Class variables

  private Decimal exchangeRate;

  private Decimal feePercentage;

  // Constructor

    public PaymentService(Decimal exchangeRate, Decimal feePer-
centage) {

    this.exchangeRate = exchangeRate;

    this.feePercentage = feePercentage;

  }

  // Method to calculate total payment amount

  public Decimal calculateTotalAmount(Decimal baseAmount) {

    Decimal converted = baseAmount * exchangeRate;

    Decimal fee = converted * feePercentage;

    // Separate the return for clarity

    return converted + fee;

  }

    public Decimal applyDiscount(Decimal amount, Decimal
discountRate) {

    // Simple one-liner, no blank line needed after this comment
```

```
    return amount * (1 - discountRate);
  }
}
```

In this example, blank lines are used to separate class variables, constructor, and methods, making the structure of the class clear. Inside calculateTotalAmount, a blank line before the return is used to visually separate calculation from the return statement, which can improve readability. Notice spaces around operators: baseAmount * exchangeRate has spaces around *, and similarly for other operations. Commas in parameter lists have a space after them (Decimal amount, Decimal discountRate). These spacing choices make the code less cramped and more legible. There are no trailing spaces at line ends, and no unnecessary blank lines (e.g., no blank line immediately after the method signature line). The result is a cleanly formatted class.

File Structure and Organization

Purpose: Define a standard order of contents in an Apex class file for consistency, so that developers know where to find certain elements (e.g., member variables, constructors, methods).

Standard: Organize class contents in a consistent order:

Class documentation comment (if any, see commenting chapter).

Class declaration (public class ClassName {...}).

Static constants and static variables (grouped together at top).

Instance member variables.

Constructors.

Methods (public methods before private methods is a common convention).

Inner classes or enums (if present, usually at bottom).

Also, within each category, order should be logical (e.g., perhaps alphabetically, or grouping related methods). All fields should be declared before methods. Avoid mixing fields and methods randomly. If you have multiple constructors, group them together. If the class implements an interface or overrides methods, it can be helpful to group override methods together or mark them clearly.

Example Outline:

```
/**
* @description Utility class for managing invoices.
*/
public class InvoiceManager {
  // 1. Static constants
  public static final Integer MAX_INVOICES = 500;
  // 2. Static variables
  private static Integer activeInvoicesCount = 0;
  // 3. Instance variables
  private List<Invoice__c> invoices;
  // 4. Constructors
  public InvoiceManager(List<Invoice__c> invoices) {
    this.invoices = invoices;
  }
  // 5. Public methods
  public void processAll() { /*... */ }
  public Decimal calculateTotalAmount() { /*... */ }
  // 6. Private methods
  private Decimal convertCurrency(Decimal amount) { /*... */ }
  // 7. Inner classes or enums (if any)
  private class InvoiceHelper { /*... */ }
```

```
}
```

In this structure, anyone opening InvoiceManager.cls will find constants at the top, then static and instance fields, then the constructor, then public methods, then private helper methods, etc. This predictability saves time when navigating code. For example, you don't have to hunt through the file to find a field declaration; by convention it's near the top. The formatting (indentation and spacing) should of course also follow earlier rules. By maintaining a consistent file structure, merging changes and code reviewing become more straightforward.

Summary (Formatting and Layout): This chapter set the standards for how Apex code should look. Key points include using 4 spaces for indentation (no tabs), K&R style braces with every control structure using braces, one statement per line, keeping lines around 100 characters max and wrapping longer lines thoughtfully, using blank lines and spaces to make code readable but not sparse, and organizing class content in a logical order. Adhering to these formatting rules results in code that is neat and uniform, so developers can focus on the logic rather than deciphering formatting. Consistent formatting can also be automatically enforced by tools (like Prettier or PMD formatting rules) to ensure compliance across the team.

COMMENTING AND DOCUMENTATION STANDARDS

Introduction: Clear code documentation is essential in large codebases. This chapter describes how to properly comment Apex code and use documentation comments (ApexDoc) so that the purpose and behavior of code are well-understood. The standards focus on writing comments that are useful and maintaining a consistent style, including header comments for classes/triggers and explanatory comments for methods and complex logic. Good commenting practices help new developers grasp the code faster and aid in generating external documentation if needed.

Class and Trigger Header Comments

Purpose: Provide high-level context for each class or trigger by documenting its purpose, authorship, and revision history if applicable. This helps anyone opening the file to immediately know what it's about.

Standard: Every class and trigger should begin with a header comment block that describes its role. The header should include at least a brief description of the class/trigger's functionality. Optionally, it can include metadata like author, date of creation, and a change log, depending on team practices. Use the ApexDoc style format for these comments: a block starting with /** and containing tags like @description, @author, etc., so tools can pick them up if needed.

Example:

/**

* @description Controller class for the Invoice page. Handles user actions and data loading for invoices.

* @author Jane Doe

* @date 2025-05-01

* @group Billing

* @group-content Invoice Management functions

~

* This class provides methods to create, update, and validate invoices from the UI.

*/

public with sharing class InvoicePageController {

//...

}

For triggers, a similar header comment should be included:

```
/**

* @description Trigger on Account that propagates changes to related Contact records.

* @author    John Admin

* @date      2025-04-15

*/

trigger AccountTrigger on Account (before update, after update) {

  // trigger logic...

}
```

These comments use a consistent format. The @description gives a one-line summary, followed by possibly more detail in plain text. Including author and date is optional but can be useful in an enterprise setting for traceability. If a change log is maintained, it can be appended or included in the comment (though too much detail can clutter; often version control history is sufficient). The key is that any developer can read the header and understand the overall purpose of the class or trigger without reading the entire code.

Method Documentation Comments

Purpose: Document what each significant method does, its inputs and outputs, so that other developers (and automated documentation tools) can understand how to use the method without digging into implementation details.

Standard: Public and protected methods should be preceded by an ApexDoc comment (using /**... */) that describes the method's purpose, parameters, and return value. Use @param tags to describe each parameter and @return to describe the return value. For @isTest methods, documentation is less critical but still helpful if the test's intent isn't obvious. Private methods can be documented with regular comments if needed (if they have complex logic), but the emphasis is on documenting the interface (public/protected API of your classes). Keep the method comment concise but informative about any important behaviors (e.g., side effects, exceptions it might throw, assumptions).

Example:

```
/**

* @description Calculates the commission for a list of opportunities.

* @param opportunities List of Opportunity records to calculate commission for.

* @param commissionRate Decimal commission rate (e.g., 0.1 for 10%).

* @return Map<Id, Decimal> mapping Opportunity Id to calculated commission amount.

*/

public static Map<Id, Decimal> calculateCommissions(List<Opportunity> opportunities, Decimal commissionRate) {

  Map<Id, Decimal> commissionsByOpp = new Map<Id, Decimal>();

  //... calculation logic...

  return commissionsByOpp;

}
```

This comment clearly explains what calculateCommissions does, what inputs it expects (opportunities list and a commissionRate), and what it returns (a map of results). Such comments are invaluable for others who want to call this method – they don't need to read the code to know how to use it properly. Additionally, if you run an ApexDoc tool, these comments will be pulled into generated documentation.

All parameters listed in the method signature are explained with @param, and the return is explained with @return. If the method could throw an exception that callers should handle, mention that as well (there's no ApexDoc tag for exceptions specifically, but it can be noted in the description).

It's also a standard that **methods have an explicit access modifier** (e.g., public, private), because ApexDoc will only recognize methods with declared scope. So ensure methods are not using the old testMethod keyword or default access without annotation if you want them documented – instead use @isTest and an access modifier.

Inline Comments for Logic

Purpose: To clarify complex or non-obvious code segments within methods using inline comments, without cluttering trivial lines. This helps maintainers follow the algorithm or rationale, especially when something is done for a specific reason that might not be immediately apparent.

Standard: Use single-line comments (//) to explain non-obvious code logic or important caveats in code. Place the comment above the line or at the end of the line it explains, but ensure it remains concise. Do not simply restate what the code does (avoid redundant comments). Instead, explain why something is done, or any important detail about how it's done. For multi-line explanations, you can still use // on each line, or a block comment /*... */ for a paragraph of text, but long explanations in code should be rare – if needed, it might indicate the code could be refactored for clarity.

Also, if a piece of code has a workaround or a temporary solution, clearly comment it and possibly tag with TODO: or FIXME: along with an explanation, so it can be easily found later.

Example:

```
Decimal totalAmount = baseAmount * exchangeRate;

// We apply exchange rate first as per requirements (commission is on converted amount)

Decimal commission = totalAmount * commissionRate;

// TODO: Remove hardcoded default commissionRate once config is in place

if (commission > MAX_COMMISSION) {

    commission = MAX_COMMISSION; // Cap the commission to avoid excessive payout
```

```
}
```

In this example, the first comment explains *why* we multiply by exchange rate first – perhaps the business rule is that commission is calculated on the converted currency. Without the comment, another developer might wonder why the order isn't reversed. The second comment is a TODO noting a future improvement (using configuration instead of a hardcoded commissionRate) – this signals to developers that this is intentional and needs attention later. The third comment, inline after capping commission, clarifies the reason for that line (to avoid excessive payout).

Such comments should be kept up-to-date as code changes. An outdated comment can be more dangerous than none. Therefore, if code is modified, any associated comment should be reviewed for accuracy.

Comment Style and Tone

Purpose: Ensure comments are written in a consistent style and professional tone, as they are part of the code that others will read. This includes avoiding certain pitfalls like commented-out code remnants or inappropriate content.

Standard: Write comments in clear, complete sentences when giving explanations. Use proper spelling and grammar to maintain professionalism. If referring to system names or variables, use the same casing as in code. Avoid using comments for jokes or personal messages – keep them technical and relevant. Do not include sensitive information in comments (e.g., passwords, secrets, or confidential logic).

Remove any commented-out code before committing. If a piece of code is not needed, it should be removed rather than kept in comments. The source control history can retrieve old code if needed; leaving it commented only clutters the file. An exception is if you are temporarily toggling something during development, but such commented code should not remain in the final committed version.

Use /*... */ for large comment blocks sparingly – for example, a large explanation at the top of a class (like a file header or a long rationale for an approach) might use a block comment. In general, // is preferred for regular inline use because it's visually clear and easy to add/remove.

Example:

```
// Correct style:
if (!userHasAccess) {
  // User doesn't have access to perform this action, log a warning and exit.
  System.debug('Warning: User attempted action without access');
  return;
}
/* Incorrect style examples:
//This code multiplies X and Y
result = x*y; // (Comment just restating the obvious operation)
/* Hard-coded ID, fix later */
String recordId = '001xx000003DGb7'; // Sensitive: should use config, not a direct ID
```

In the correct style example, the comment is a concise sentence explaining a block of code (why we are logging and returning). It's clear and to the point. In the incorrect examples: The first is a comment that simply repeats what the code does (x*y multiplication) – this is unnecessary and cluttering. The second shows a comment that indicates a problem (hard-coded ID) but is written in a terse way and actually includes the sensitive ID in code – the standard would be to avoid this situation entirely by not hardcoding (as per later standards). Also, that comment is inside a block that might have been used to disable code (notice the /* starting a block but not properly closed in the snippet). Such misuse of comments can lead to confusion and even syntax errors. Always close comment blocks properly and prefer deleting unused code.

Summary (Commenting and Documentation): In this chapter, we established that every Apex class and trigger should have a descriptive header comment, and key methods should be documented with purpose, params, and return information. We also emphasized writing meaningful inline comments for complex logic and avoiding noise comments or leftover commented code. Following these standards leads to a well-documented codebase where the *why* and *how* of the code are clearly communicated. This reduces onboarding time for new developers and aids in knowledge transfer. Remember: code tells *how* things are done; comments should usually tell *why*.

APEX LANGUAGE AND SYNTAX STANDARDS

Introduction: Beyond naming and formatting, there are specific ways Apex code should be written to make best use of the language's features and avoid common pitfalls. This chapter covers standards for idiomatic Apex usage: using appropriate language features (such as access modifiers, properties, and collections), structuring conditionals and loops properly, and handling Apex-specific nuances. These standards ensure that we write Apex code that is not only consistent but also leverages the language effectively and safely.

Class and Method Modifiers

Purpose: Make the intended accessibility of classes and methods explicit and to use modifiers in a way that aligns with Apex best practices (e.g., avoiding global unless necessary).

Standard: Always explicitly specify the access modifier for classes and methods (public, private, protected, global, or webservice where appropriate). Do not rely on default access. In Apex, the default for inner classes and methods is private, but it's clearer to state it. Use public for classes and methods that are used by other classes within your namespace. Use protected for methods intended to be overridden by subclasses in the same namespace. **Avoid using global** unless absolutely required (for example, classes in a managed package that need to be accessible to subscriber orgs). Global opens your API to outside namespaces and cannot be reduced in scope later. For triggers, the access modifier is not applicable (they are implicitly system context but controlled by sharing as per class with handler). For AuraEnabled or web service methods, use public or global thoughtfully and annotate with the appropriate annotation (we'll cover in Chapter 11).

Additionally, mark methods or classes as virtual (or use the newer abstract/override approach) if they are intended to be extended. If a method overrides an inherited method, use the override keyword for clarity. Mark utility classes that should not be instantiated with a private constructor.

Example:

```
public virtual class PaymentProcessor {
   public PaymentProcessor() { /*... */ }    // explicit public constructor
   public virtual Decimal calculateFee(Decimal amount) { /*... */ }
}
public class CustomPaymentProcessor extends PaymentProcessor {
   // Using override to show intent
   public override Decimal calculateFee(Decimal amount) { /*... */ }
}
// Utility class example:
public class MathUtils {
   private MathUtils() {} // prevent instantiation
   public static Integer add(Integer a, Integer b) { return a + b; }
}
```

In this example, PaymentProcessor is declared public virtual explicitly, and its method calculateFee is also public virtual. The subclass uses public override to override the parent method. This explicitness is clear to anyone reading and to the compiler. The MathUtils class has a private constructor to indicate it's not meant to be instantiated (it only contains static methods). By following this, we avoid ambiguities and control the usage of classes. If a class is meant to be used only internally, keep it private or at most package public. If it's for an API, global might be used but then it must adhere to managed package API versioning rules. Being deliberate with modifiers prevents accidentally exposing things or restricting things incorrectly.

5.2 Use of Properties vs. Getters/Setters

Purpose: Simplify code and adhere to Apex idioms by using built-in property syntax instead of verbose getter and setter methods when no extra logic is required.

Standard: Use Apex properties ({ get; set; }) for simple data encapsulation instead of writing explicit getter and setter methods. Apex allows you to declare properties which automatically handle storage and retrieval. This leads to cleaner and more readable classes. Only write a custom getter or setter method if you need to perform additional logic during get or set. In those cases, use the getSomething/setSomething naming convention or property with custom accessor bodies.

Example:

```
public class Order {

  // Using property for automatic getter/setter

  public Decimal totalAmount { get; private set; }

  public String status { get; set; }

  public Order() {

    this.totalAmount = 0;

    this.status = 'New';

  }

  // Custom setter for status if needed

  public void setStatus(String newStatus) {

    if (newStatus == 'Cancelled') {

      this.totalAmount = 0; // reset total if cancelled

    }
```

```
    this.status = newStatus;
  }
}
```

In this Order class, totalAmount and status are declared as properties. totalAmount has a private setter, meaning outside code can read it but not modify it directly (enforcing some encapsulation), whereas status has public read/write. The code is succinct. If we needed custom behavior on setting status, we could implement a setStatus method (or make status a property with a custom setter block). By default, just use { get; set; } instead of writing trivial methods:

```
// Avoid doing this for no reason:
public Decimal getTotalAmount() { return totalAmount; }
public void setTotalAmount(Decimal val) { totalAmount = val; }
```

The above is unnecessary because Apex properties handle it. Using properties makes the class definition shorter and clearer.

Conditional and Loop Structures

Purpose: Encourage clear and safe conditional logic and looping constructs in Apex, avoiding mistakes like missing braces or overly complex expressions that reduce clarity.

Standard: Always use braces for if/else and loop bodies (as stated in formatting standards) – this is reiterated here because it is both a style and a safety rule. For if-else if-else ladders, structure them with one if, any number of else if, and an optional else at the end; avoid separate disconnected ifs when one is logically the "otherwise" of another. Use switch statements (available in newer Apex versions) for multiple discrete values instead of long if-else chains when it improves clarity; always include a default branch in switch to handle unexpected values.

For loops: use the appropriate loop type. Use for (Type var : collection) for iterating over lists or sets, which is concise. Use traditional for (init; condition; increment) loops when you need an index or more control.

Avoid while loops that could be infinite; ensure loop conditions will eventually break.

Keep conditional expressions simple; if a condition is complex, consider breaking parts into boolean helper variables with meaningful names. For example, instead of:

```
if ((status == 'Approved' && isActive && userRole == 'Manager') || (status == 'Pending' && isAdmin)) {... }
```

This can be made clearer by computing sub-conditions:

```
Boolean approvedAndActiveManager = (status == 'Approved' && isActive && userRole == 'Manager');

Boolean pendingAndAdmin = (status == 'Pending' && isAdmin);

if (approvedAndActiveManager || pendingAndAdmin) {... }
```

This isn't a strict rule but a suggestion to improve readability of complex logic – the standard is to strive for clarity.

Example:

```
// Using a switch for clarity:

String stage = opp.StageName;

Decimal discount = 0;

switch on stage {

  when 'Prospecting' {

    discount = 0;

  }

  when 'Negotiation' {

    discount = 0.05;

  }

  when 'Closed Won' {

    discount = 0.1;
```

```
  }
  when else {
    discount = 0;
  }
}
// For loop examples:
List<Integer> scores = new List<Integer>{5, 7, 9};
Integer total = 0;
for (Integer score : scores) {
  total += score;
}
// Traditional for when index is needed
for (Integer i = 0; i < scores.size(); i++) {
  System.debug('Score ' + i + ': ' + scores[i]);
}
```

The switch on stage example demonstrates using a switch statement instead of multiple if/else. It covers known stages and has a when else as a default. This is more readable and less error-prone for such scenarios. The loop examples show the use of the enhanced for-each loop for summing scores, and a traditional indexed loop when we needed the index for debugging output. Both loops use braces and clear loop bounds.

By consistently structuring conditionals and loops in these ways, we reduce logic errors and make the flow of code easier to follow.

Exception Handling

Purpose: Ensure that exceptions are properly handled or propagated in

Apex code, and that we don't accidentally swallow errors or throw generic exceptions without context.

Standard: Use try-catch blocks around code that can throw exceptions, especially when performing DML, callouts, or other operations that commonly throw. Catch only exceptions you can meaningfully handle; otherwise, let them propagate. When catching exceptions, at minimum log the error (e.g., System.debug or better, use a centralized logging mechanism) so that the failure can be traced. Do not catch generic Exception just to suppress errors – this would hide problems. If you catch an exception and cannot resolve it, consider rethrowing a custom exception or at least wrapping it in something that adds context.

Never leave an empty catch block. An empty catch block is a silent failure which is almost always against standards. If intentionally suppressing an expected exception, at least comment why (though better is to avoid throwing it in the first place by checking conditions).

Also, use specific exception types in catch where possible (e.g., DmlException, NullPointerException, CalloutException) to handle different cases separately if needed.

When throwing exceptions in your code (like validating business rules), throw specific and descriptive exceptions. Prefer creating a custom exception class (extend Exception) if you need to signal something specific (e.g., new OrderValidationException('Order amount cannot be negative')). This helps the callers catch that specific exception vs every exception.

Example:

```
public void updateRecords(List<SObject> records) {

  try {

     update records;

  } catch (DmlException e) {

      // Log the error and rethrow as a custom exception for higher-level
handling

     System.debug('Error updating records: ' + e.getMessage());
```

```
        throw new UpdateFailedException('Failed to update records in updateRecords: ' + e.getMessage());
    }
}

// Using multiple catch blocks for specific exceptions
public Account fetchAccount(Id accountId) {
    try {
        Account acct = [SELECT Id, Name FROM Account WHERE Id = :accountId];
        return acct;
    } catch (QueryException e) {
        System.debug('No account found for Id ' + accountId);
        return null; // or throw a custom NotFoundException
    } catch (Exception e) {
        System.debug('Unexpected error: ' + e);
        throw e; // rethrow unexpected exceptions
    }
}
```

In updateRecords, we catch DmlException specifically, log it, and then throw a custom UpdateFailedException (which would be a class extending Exception defined elsewhere) with a message that includes context. This way, the method doesn't just fail silently or with a generic error; it transforms the low-level exception into a higher-level one that calling code could catch if it wanted to specifically handle update failures.

In fetchAccount, we handle the case where the query returns nothing by catching QueryException (which happens if the query doesn't find a record or if more than one when using a single fetch). We then decide to return null to indicate not found (alternatively, some might throw a

NotFound type exception). We also have a generic catch for any other unexpected Exception, which we log and rethrow – this ensures truly unexpected issues aren't suppressed.

The key points: don't ignore exceptions, use meaningful context in errors, and keep exception handling logic clear and purposeful.

Best Use of Apex Collections and Types

Purpose: Leverage Apex's collection types and other type features for clearer and more efficient code, while avoiding outdated or less safe patterns.

Standard: Use Apex collections (List, Set, Map) appropriately:

Use List<Type> for ordered collections and when duplicates are allowed.

Use Set<Type> for unique collections, to easily check membership or remove duplicates.

Use Map<Key, Value> to store mappings from one key to a related value (e.g., Id to SObject, or Id to some aggregate result).

Prefer using these collection classes over older array syntax. For example, instead of Account[] accounts = new Account[0];, use List<Account> accounts = new List<Account>(); – the latter is more idiomatic and flexible.

When iterating and needing to check membership, using a Set or Map can drastically simplify code (e.g., use a Set of Ids for quick contains checks instead of looping through a list each time).

When performing bulk operations, use maps to key records by Id for efficient lookup (like retrieving related records and mapping them by parent Id to link with children, etc.). This is partly a performance and governor limit consideration (avoid nested loops by using map lookups, covered more in Chapter 8), but it's also a coding style: using the correct data structure leads to clearer intent.

Use type-specific methods when available: e.g., String.isBlank() or isNotBlank() to check emptiness instead of manual checks, List.isEmpty() instead of checking size, etc., for clarity.

Example:

```
// Using Set for quick membership check

Set<String> premiumCustomerEmails = new Set<String>{ 'vip1@example.com', 'vip2@example.com' };

for (Contact con : contacts) {

  if (premiumCustomerEmails.contains(con.Email)) {

    con.Status__c = 'Premium';

  }

}

// Using Map to avoid nested loops when matching related records

List<Opportunity> opps = [SELECT Id, AccountId, Name FROM Opportunity WHERE...];

List<Account> accts = [SELECT Id, Rating FROM Account WHERE Id IN :new Set<Id>(opps.stream().map(o -> o.AccountId))];

// (Above assumes a utility to map opps to account Id set; if not available, loop to collect Ids)

Map<Id, Account> acctById = new Map<Id, Account>(accts);

for (Opportunity opp : opps) {

  Account parent = acctById.get(opp.AccountId);

  if (parent != null && parent.Rating == 'Hot') {

    // do something with hot account's opportunity

  }

}
```

In the first snippet, a Set is used to mark premium customers by email. This avoids a nested loop or repeated conditional logic to check if a contact's email is one of a few special addresses. It's clear and efficient: contains on a set is O(1) average.

In the second snippet, we create a Map of Accounts by Id so that when iterating opportunities, we can directly get the related Account without looping through accounts each time. The code intent is clear: acctById.get(opp.AccountId) directly fetches the matching account. (We used a pseudo-code opps.stream().map(...) to get account Ids; in actual Apex, since we don't have lambda, you would loop through opps to build the set of accountIds first. The key is using that set in the SOQL and then mapping results.)

By using collections and maps, code becomes both more performant and easier to understand. It expresses the developer's intent (e.g., "I need to check membership" or "I need to map these by Id") directly.

SOQL and DML Usage in Code

Purpose: Standardize how we incorporate database operations (queries and data modifications) in Apex code in terms of style and safety, and to align with language capabilities.

Standard: Use static SOQL queries whenever possible instead of dynamic queries. Static SOQL (the bracket syntax) is checked at compile time and is easier to read. Use dynamic SOQL only when you truly need to construct queries at runtime (and even then, prefer bound variables and safe constructions to avoid injection issues – more on that in security chapter). For static SOQL inside Apex, format them as per formatting standards (multiline for readability if they have many fields or conditions).

When performing DML, if you have options, use the method form (Database.insert, etc.) with partial success handling only if you need that behavior, otherwise the insert, update keywords are fine and simpler. But be aware of the difference (Database methods don't throw exceptions on partial failures, instead return results). If using Database methods, always check the Database.SaveResult or similar to handle errors.

Always check the results of DML operations (either by catching exceptions for the keyword form or examining the SaveResult for database method form). Do not assume success – e.g., after an insert inside try-catch, the catch can log failure.

Do not hard-code IDs in SOQL or code – as stated later in Metadata chapter, but we echo it here: always query or pass IDs as parameters; never embed an Id like 'a012345...' in an Apex string. Hard-coded IDs break when moving between orgs (different IDs).

Use query parameters (binding variables) in SOQL to include Apex variable values, instead of string concatenation, to prevent SOQL injection and to keep code safe and clear.

Example:

```
// Static SOQL with binding

public List<Contact> getContactsByState(String state) {

  List<Contact> results = [SELECT Id, Name, Email

                FROM Contact

                WHERE MailingState = :state

                ORDER BY LastName];

  return results;

}

// Handling DML results using Database method

public void updateAccounts(List<Account> accountsToUpdate) {

   Database.SaveResult[] results = Database.update(accountsToUpdate,
false);

  for (Integer i = 0; i < results.size(); i++) {

    if (!results[i].isSuccess()) {

      Database.Error err = results[i].getErrors()[0];

          System.debug('Failed to update Account ' + accountsToUp-
date[i].Id + ': ' + err.getMessage());

        // (Consider throwing exception or collecting errors for further
handling)

    }
```

```
    }
}
```

In getContactsByState, the query uses a bound variable :state rather than building a string. It's static SOQL, easy to read, and the query is checked at compile time. The formatting is multiline for clarity. There's no risk of injection as the platform handles the binding safely.

In updateAccounts, we use Database.update with the second parameter allOrNone set to false, meaning it will do partial updates. This returns a SaveResult list. We then loop through results to check for errors and log them. This is the correct way to use the Database class for partial success – you must handle the errors or they'll be lost. If we didn't care about partial success and wanted to stop at the first error, we'd simply do update accountsToUpdate; inside a try-catch for DmlException. Both patterns are acceptable as long as you handle errors appropriately and choose the right one for the scenario.

Additionally, as a style point: do not perform DML or SOQL in constructors or static initializers if avoidable – it's better to have explicit methods for operations, so that they can be controlled and tested separately. This is more of a design guideline but worth noting: doing heavy logic in a constructor can make unit testing harder (since tests that instantiate the class will trigger that logic implicitly).

Summary (Apex Language and Syntax): In this chapter, we set rules for writing idiomatic and robust Apex code. We covered explicit use of access modifiers (and avoiding global), leveraging properties for simple accessors, structuring conditionals and loops with clarity and always with braces, proper exception handling (never silently swallowing exceptions), usage of collections and maps for clarity and performance, and safe usage of SOQL/DML (static queries, checking DML results, no hardcoded IDs). By following these standards, our Apex code will be more predictable, less error-prone, and easier for others to follow, which is vital in an enterprise environment.

APEX TESTING STANDARDS

Introduction: Testing is a first-class citizen in Apex development. Salesforce requires 75% code coverage for deployment, but beyond that, good tests ensure code quality and prevent regressions. This chapter lays out standards for writing Apex test classes and methods. These standards ensure that tests are reliable, maintainable, and truly validate that the code meets requirements. We focus on enforceable aspects like naming of test methods, not using certain test annotations or patterns, making sure tests have assertions, and managing test data properly.

Test Class Structure and Naming

Purpose: Keep test classes organized and easily identifiable, ensuring they only contain test code and related utility code.

Standard: All test classes must be annotated with @isTest and named with a suffix "Test" (or "Tests"). For example, if testing OrderService, the test class should be OrderServiceTest. The class should be private or public as needed (private is fine for tests unless you need cross-namespace testing). Do not include non-test logic in a test class. You may include helper methods within the test class (marked @testSetup or just private static methods to create data), but they should relate to testing (e.g., test data factory methods).

Within a test class, test methods should either be annotated with @isTest or use the testMethod keyword (the annotation is preferred, and the old testMethod syntax is deprecated). Each test method should ideally target a specific piece of functionality or scenario, and the method name should reflect that scenario (e.g., testCalculateFee_withValidInput() or testCalculateFee_exceedsLimit()).

Ensure test classes are self-contained – they set up their own data (using test setup methods or within each test) and don't depend on data from other tests or the org (see next section on data). Also, test classes should be in the same package/namespace as the code they test (except in special cases of package extension, which is advanced).

Example:

```
@isTest
private class OrderServiceTest {
  @testSetup
  static void setupTestData() {
    // create common test records used by multiple tests
    Account acct = new Account(Name='Test Account');
    insert acct;
    // perhaps create some products, orders, etc.
  }
  @isTest
  static void testCalculateTotalOrderValue_basic() {
    // Setup unique data for this test
      Order__c order = new Order__c(Name='Test Order', Total_Amoun-
t__c=100);
    insert order;
    // Execute functionality
    Decimal total = OrderService.calculateTotalOrderValue(order.Id);
    // Verify outcome
     System.assertEquals(100, total, 'Order total should be 100 for initial
amount 100');
  }
  @isTest
  static void testCalculateTotalOrderValue_withLineItems() {
    //...another scenario...
  }
}
```

In this OrderServiceTest class, we use @isTest at the class level (which makes all methods default to test context). We included a @testSetup method to create an Account that might be used by all tests (to attach orders to, for example). The test methods themselves are annotated with @isTest. The naming testCalculateTotalOrderValue_basic clearly indicates what's being tested (the calculateTotalOrderValue method under normal conditions). The second test might handle a scenario with line items affecting total, etc.

The test class is marked private (which is fine; tests can be private as the testing framework still runs them). The use of @isTest on methods is current best practice (the code comment highlights Salesforce's recommendation to use @isTest annotation rather than testMethod keyword).

Test Data and Isolation

Purpose: Ensure tests do not rely on org data or other tests, and that they cleanly manage their own data for consistent results.

Standard: Every test must create the data it needs or use @testSetup to share setup data among tests in the class. Do not query or depend on real data in the org (use of SeeAllData=true in tests is highly discouraged and should be avoided). By default, tests run with no access to org records (aside from a few special objects like PriceBook or Users), so you should insert any records required for the test within the test context.

Use @testSetup methods to create common data that can be used by all tests in the class, which saves each test from needing to repeat those inserts (but keep in mind all test methods see the data created in testSetup as if it's fresh for them).

Each test method should ideally commit or rollback its own DML independent of others – the test framework isolates tests by rolling back after each method, so there's usually no interference. However, be cautious not to write tests that depend on running order. Tests in one class can run in any order, and tests in different classes run in parallel by default (in Salesforce DX environments), so never have an interdependency.

Also, **cleanup is automatic** (all data is rolled back), so you don't need to delete records at the end of tests.

Example:

```
@isTest
static void testOpportunityProcessing() {
  // Don't rely on existing Opportunities; create what we need
  Account acct = new Account(Name='TestCo');
  insert acct;
  Opportunity opp = new Opportunity(Name='Test Opportunity', StageName='Prospecting', CloseDate=Date.today(), AccountId=acct.Id);
  insert opp;
  // Perform some operation that processes the opportunity
  OpportunityProcessor.process([opp]); // hypothetical method
  // Now query the opportunity to see if fields were updated
  Opportunity updatedOpp = [SELECT Id, Status__c FROM Opportunity WHERE Id = :opp.Id];
  System.assertEquals('Processed', updatedOpp.Status__c, 'Opportunity status should be updated to Processed');
}
```

In this test, we explicitly insert an Account and an Opportunity to use. We do not, for example, assume an Account exists with a particular name in the org. Even something seemingly harmless like querying [SELECT Id FROM Account LIMIT 1] to get "some" account is not allowed unless SeeAllData=true, and that breaks test isolation. So we avoid that and always create the account.

By following this, tests can run on any org (dev, QA, prod) the same way, since they bring their own data. It makes tests more reliable and easier to debug.

Assertions and Test Verification

Purpose: Ensure that tests actually verify outcomes, not just execute code. A test without assertions may run without error but not actually confirm correctness.

Standard: Each test method must include one or more assertions (System.assert / assertEquals / assertNotEquals, etc.) to validate behavior. It is not enough to call code and not check results. Use System.assertEquals(expected, actual, message) to compare outcomes, or System.assert(condition, message) for boolean conditions. Ideally, follow the pattern: *setup data, execute logic, verify results via asserts*.

There is a PMD rule and general best practice that every test should have at least one assert. We enforce that standard. The only exception might be tests that are expected to throw an exception (negative tests); in those, you use System.assertThrows (in newer Apex) or a try-catch in the test to assert that an exception was thrown. But that still counts as an assertion of behavior (i.e., asserting an exception occurs).

Additionally, use informative messages in assertions. The third parameter of System.assertEquals allows you to describe what's being tested, which is very helpful if the assertion fails and you need to read logs.

Example:

```
@isTest

static void testCalculateCommission() {

  // Setup data

      Opportunity opp = new Opportunity(Name='Test Opp', StageName='Closed Won', CloseDate=Date.today(), Amount=10000);

  insert opp;

  // Execute

      Decimal commission = CommissionService.calculateCommission(opp.Id);

  // Verify
```

System.assertNotEqual(0, commission, 'Commission should be calculated and not be zero');

System.assert(commission > 0 && commission < opp.Amount, 'Commission should be a positive value less than the opportunity amount');

}

Here we have two assertions: one to ensure commission is not zero, and another to ensure it's within an expected range relative to opp.Amount. If CommissionService.calculateCommission had logic based on stage or amount, we'd assert the exact expected value if we know it (e.g., if commission rate is 10%, we'd expect 1000 for 10000 amount, so System.assertEquals(1000, commission, 'Commission should be 10% of 10000');). The key is, we **assert.** If this method changed unintentionally and returned 0 or something, a failing assert will catch it.

Without asserts, the test might pass (no exceptions) but we wouldn't know the functionality broke. So we require asserts in every test.

Also, assert the important outcomes: state changes (field values), returned values, records created or not created, etc., as relevant. If a test is calling a void method that sends an email, you might assert that an email was indeed sent by querying the Messaging.SendEmailResult or checking a record that would only be created if the email logic ran, etc. Find something measurable to assert.

Testing for Governor Limits and Bulk

Purpose: Ensure that our code not only works for single records but also in bulk scenarios and respects governor limits. Tests should validate bulk behavior and limit compliance where possible.

Standard: Write tests that cover bulk scenarios and different governor limit conditions. At minimum, any trigger-related functionality must be tested with more than one record (e.g., 200 records) to ensure the trigger is bulkified. If you have a method that processes a list, test it with an empty list, a list of one, and a list of many.

Use Test.startTest() and Test.stopTest() to properly test asynchronous behavior and reset governor limit counting within a test method. For example, if you're testing a future or batch, you would call Test.startTest(), then invoke the async method, then call Test.stopTest(), and then do your asserts, because between start/stop Salesforce will run the async job.

Also, you can use the Limits class to assert that your code did not exceed a certain number of queries or DML operations if you want to enforce that (this is more advanced). For instance, if you expect a method should only do 1 query even when given 100 records, you could check Limits.getQueries() difference before and after. However, this is rarely done in tests; it's more common to rely on code review or static analysis for that. But it's possible to include in tests as a safeguard (again, making sure not to depend on absolute values that could change if code changes, unless they must not change beyond a threshold).

Example:

```
@isTest

static void testBulkOrderProcessing() {

  // Create 200 orders to simulate bulk trigger context

  List<Order__c> orders = new List<Order__c>();

  for (Integer i = 0; i < 200; i++) {

    orders.add(new Order__c(Name='Test Order ' + i, Status__c='New'));

  }

  insert orders; // if Order__c has an after insert trigger, this will fire with 200 records

  // Now verify that all orders were processed by the trigger logic:

    for (Order__c o : [SELECT Id, Status__c, Processed_Flag__c FROM Order__c WHERE Id IN :orders]) {

     System.assertEquals('New', o.Status__c, 'Status should remain New');

          System.assertEquals(true, o.Processed_Flag__c, 'Processed_Flag should be true after trigger processing');

  }
```

```
}
@isTest
static void testBatchJobProcessing() {
  // assume BatchJob processes large data volumes
  // Schedule or call a batch job and use start/stop test to execute it
  Test.startTest();
  MyBatchJob batch = new MyBatchJob();
  Database.executeBatch(batch, 200); // batch size 200
  Test.stopTest();
  // After Test.stopTest, the batch has completed
  // Assert outcomes of batch, e.g., that certain records were updated
  Integer count = [SELECT COUNT() FROM Huge_Object__c WHERE Processed__c = true];
  System.assertEquals(10000, count, 'All 10k records should be marked processed by the batch job');
}
```

In testBulkOrderProcessing, we inserted 200 orders at once to simulate a bulk insert scenario. The trigger on Order (not shown) presumably sets Processed_Flag__c to true for each new order. The test confirms that after insert, all 200 records have Processed_Flag__c = true. If the trigger was not bulkified and only processed one record, this test would catch it because many of them would remain false.

In testBatchJobProcessing, we show the usage of Test.startTest() and Test.stopTest() around executing a batch. This ensures the batch actually runs within the test context. We then verify an outcome – e.g., maybe the batch was supposed to mark 10k Huge_Object__c records as processed. We check the count. The numbers are just illustrative. The important part is using the start/stop pattern for async.

Also, consider testing negative scenarios: e.g., if code should throw an exception when called improperly, use System.assertThrows (with a specific exception type and maybe message) to ensure it does throw. This is part of verification – e.g.:

System.assertThrows(SomeCustomException.class, () -> MyService.doRestrictedAction(userWithoutPermission));

(This assumes having Apex supports this lambda syntax – if not, the alternative is try-catch around doRestrictedAction and then assert that catch got called.)

Prohibited Patterns in Tests

Purpose: Identify things not to do in tests that can undermine their reliability or the stability of the testing process.

Standard: Do not use SeeAllData=true in test classes or methods, except in the rare scenario where you must test something with actual org data (which usually indicates a design issue). We forbid @isTest(SeeAllData=true) in our coding standards because it breaks test isolation and can cause tests to pass or fail depending on org data. Instead, fabricate the data within the test.

Avoid hardcoding IDs of real records (like a specific record Id that exists in your dev org) in tests. If you need a specific configuration, create it in test (e.g., create a User or Pricebook as needed). Salesforce provides some global accessible data like Standard Pricebook or certain user info, but if you use those, use the provided methods to get them (e.g., Test.getStandardPricebookId() for pricebook).

Don't write tests that depend on the order of execution of triggers or batch chaining beyond what is guaranteed. For example, if two triggers fire on the same object, you cannot guarantee order, so your test should be written to accommodate either order or (preferably) ensure you only have one trigger per object as per our coding standards.

Do not perform callouts in tests without a mock. If your code does HTTP callouts, you must use Test.setMock with a HttpCalloutMock implementation, or mark the test method with @IsTest(isParallel=false) and Test.startTest()/stopTest() to allow the callout (with callout to an actual service, which is not recommended in tests). The better approach is always to use a mock. This will be discussed more in Integration standards, but from a test perspective: *every callout in code must be mocked in tests to avoid external dependency*. If not, your test will fail with an exception that a callout was attempted without a mock.

Example:

```
@isTest

static void testDisallowSeeAllData() {

  // This test intentionally does not use SeeAllData.

  // We create what we need:

  List<Account> accts = new List<Account>{

    new Account(Name='A'), new Account(Name='B')

  };

  insert accts;

  //... proceed with test...

}

// Bad Example (should not be done):

@isTest(SeeAllData=true) static void testBadPractice() {

  // This is not allowed by our standards.

  Account anyAcct = [SELECT Id FROM Account LIMIT 1];

  System.assertNotEquals(null, anyAcct);

}
```

The second testBadPractice would rely on an Account being in the org (outside test context) – this is clearly against our standard and would be caught in code review or by static analysis as a violation. The first testDisallowSeeAllData demonstrates the correct approach: we created accounts within the test instead of trying to use existing ones.

Summary (Apex Testing Standards): This chapter mandates how we write tests: test classes and methods must be clearly named and annotated, tests must create their own data and not rely on org data (SeeAllData=false always), and they must include assertions to validate outcomes. We emphasized testing bulk scenarios and using the test framework features (startTest/stopTest) for async processes. By following these standards, our test suite will be robust and trustworthy – catching regressions when code changes and ensuring our code handles all required scenarios (single, bulk, positive, negative). A strong test suite is the safety net for deployment, and these standards ensure that safety net is well-knit.

SECURITY AND SHARING STANDARDS

Introduction: Security is paramount in Salesforce applications. Apex runs in a privileged context, but we often need to enforce the platform's security (like sharing, CRUD, and field-level security) in our code to protect data. This chapter outlines standards for writing Apex code that respects Salesforce security models: record sharing rules, object/field permissions, and general secure coding practices (like avoiding SOQL injection). Following these standards helps prevent security vulnerabilities and ensures our code doesn't expose data inappropriately or violate user permissions.

Enforcing Sharing Rules

Purpose: Control record-level data access in Apex to honor the current user's sharing rules when appropriate.

Standard: Apex classes must explicitly declare the sharing model (with sharing, without sharing, or use inherited sharing as needed) and generally default to with sharing unless there's a specific reason to run in system mode. This makes it clear whether the class respects sharing. Not specifying means "without sharing" by default , which can unintentionally bypass security. Therefore, you must include one of these keywords in all class definitions (except triggers, which run with sharing rules based on the context of the calling class or if none, default to without sharing).

Use with sharing for classes that handle data viewing/editing in user context (e.g., classes called from Lightning UI or REST endpoints where user permissions should apply).

Use without sharing for classes that need to ensure full access (e.g., a utility that must query all records regardless of user, perhaps for an admin operation), but only after careful consideration.

inherited sharing is useful for classes, often services, that might be called from both with-sharing and without-sharing contexts; it allows the sharing to follow the caller's context (introduced in later API versions).

If a class is marked without sharing, consider security implications inside it (then CRUD/FLS checks become even more crucial, as it can see all data). Generally, default to with sharing if unsure.

Example:

```
public with sharing class AccountService {

  public static List<Account> getUserAccounts() {

    // This will only return accounts the running user has access to, due
to "with sharing"

    return [SELECT Id, Name, Industry FROM Account];

  }

}

public without sharing class AdminDataFetcher {

  public List<User> getAllUsers() {

    // Runs in system context, returns all users

    return [SELECT Id, Name, Profile.Name FROM User];

  }

}
```

AccountService is declared with sharing, so any queries inside respect the current user's record access (sharing rules, ownership, etc.). Admin-

DataFetcher is without sharing, meaning it will run with full access – presumably this might only be used by an admin context or has its own authentication around its usage.

By declaring these, we avoid the situation of accidentally running as system (without sharing) when we intended to enforce sharing (the default if not declared is without sharing for top-level classes). Conversely, if something truly needs system mode, being explicit with without sharing communicates that.

Note: Inner classes inherit sharing from their outer classes. If you use the inherited sharing keyword (supported in Apex, not shown above), it means if this class is called by another class, it will run with the sharing of that context; if executed as entry point (e.g., via REST), it runs with sharing. This is often a good default for utility classes called by both system and user contexts.

Object and Field-Level Security (CRUD/FLS)

Purpose: Ensure Apex code does not bypass user permissions on objects and fields inadvertently when running in user context (and even in system context for compliance, especially if exposing data via UI or API).

Standard: Check CRUD (Create, Read, Update, Delete) and FLS (Field Level Security) before performing DML or querying sensitive fields in Apex that runs in a user context. This is especially necessary in with sharing classes or generally classes used by end-user actions (like AuraEnabled, controllers, etc.). Use the Schema.sObjectType and DescribeFieldResult methods to verify permissions:

Before a SOQL or SOSL query, ensure the user has at least read access on the object and fields being queried.

Before a DML operation (insert/update/delete), ensure the user has appropriate access: e.g., isCreateable on object and fields for insert, isUpdateable for update, etc.

If user lacks permission, handle gracefully: e.g., throw a System.NoAccessException or a custom exception, or remove those fields from the operation (like using Security.stripInaccessible to strip out fields the user shouldn't set or see).

Salesforce has a Security library method Security.stripInaccessible(AccessType, sObjectOrList, schemaDescribe) which can remove fields the user isn't allowed to access from records (useful to avoid accidentally exposing them). We should use this when returning SObjects to UI or external, to sanitize them.

For custom services, you might centralize these checks in a utility. But as a standard, *every code path that queries or DMLs in response to user action should enforce CRUD/FLS*. Code running as part of internal automation (like in triggers, you might not need to check because that automation is system context and not directly user-initiated, but if that trigger then surfaces data to user, consider it).

Example:

```
public with sharing class CaseController {

  public static Case getCaseDetails(Id caseId) {

    // Check read access on Case and specific fields

    if (!Schema.sObjectType.Case.isAccessible()

      || !Schema.sObjectType.Case.fields.Subject.isAccessible()

      || !Schema.sObjectType.Case.fields.Description.isAccessible()) {

          throw new SecurityException('Insufficient access to Case or its fields.');

    }

     Case c = [SELECT Id, Subject, Description FROM Case WHERE Id = :caseId];

    return c;

  }

  public static void updateCaseDescription(Id caseId, String newDesc) {

    // Check field update access
```

```
    if (!Schema.sObjectType.Case.isUpdateable()
      || !Schema.sObjectType.Case.fields.Description.isUpdateable()) {
          throw new SecurityException('No permission to update case description.');
    }
    Case c = new Case(Id = caseId, Description = newDesc);
    update c;
  }
}
```

In getCaseDetails, we used Schema.sObjectType.Case.isAccessible() to check object read permission and fields.Subject.isAccessible() etc. for each field we plan to expose. Only then we query. If the user lacks access, we throw an exception (could also just return null or an error message – depends on design, but exception is clear).

In updateCaseDescription, we ensure the user can update Case and specifically the Description field. If not, we throw a security exception. Otherwise, we perform the update. This prevents a user without edit rights from modifying a case's description via our code.

Using Security.stripInaccessible alternative example for queries:

```
List<Account> results = [SELECT Id, Name, Secret_Field__c FROM Account];
```

```
results = Security.stripInaccessible(AccessType.READABLE, results, Schema.SObjectType.Account);
```

This would remove Secret_Field__c from each Account record in results if the running user doesn't have permission to see it. This way, you don't accidentally send that field's value to the client.

Protected Data and Secrets

Purpose: Outline how to handle sensitive information within Apex so that it's not exposed or logged insecurely.

Standard: Never hard-code sensitive information (like passwords, tokens, secrets) in Apex code, and never expose them in logs or debug statements. Use Protected Custom Settings/Custom Metadata or Named Credentials for storing such info. If your code handles sensitive data (like a user's PII or a secret key), avoid doing System.debug on it or building error messages that reveal it. Mark variables containing sensitive data as Transient in Visualforce controllers (to not send to view state) if applicable, though that's more VF-specific.

When logging errors, sanitize messages. For example, if an exception contains an SOQL query that had user input (possible injection attempt), don't log the raw query with input – or use binding to avoid that scenario altogether (which you should).

Example:

```
public without sharing class PaymentGateway {

  // Assume an encryption key is stored in Custom Metadata (or as an encrypted field in settings)

    private static final String ENCRYPTION_KEY = MySettings__c.getInstance().EncryptionKey__c;


  public static String encryptCardNumber(String plainCard) {

    if (String.isBlank(ENCRYPTION_KEY)) {

      throw new SecurityException('Encryption key not configured.');

    }

    // perform encryption (dummy example)

          String cipher = Crypto.encryptWithManagedIV('AES128', EncodingUtil.base64Decode(ENCRYPTION_KEY), Blob.valueOf(plainCard));

    return EncodingUtil.base64Encode(cipher);

  }
```

```
    public static void processPayment(Decimal amount, String cardNumber) {
        // DO NOT log the card number or sensitive info
        System.debug('Processing payment of ' + amount);
        //... process using encrypted or tokenized card...
    }
}
```

In this example, the encryption key is not hardcoded but fetched from a custom setting (which could be protected). The processPayment method deliberately does not log the card number (which is sensitive). Logging only the amount is fine.

Another example: if you catch an exception that includes sensitive info in the message, you might want to log a generic message or scrub it. But mainly, avoid printing out things like user credentials or personal data.

SOQL Injection Prevention

Purpose: Prevent malicious user input from altering queries (SOQL injection) when dynamic SOQL is used.

Standard: Avoid dynamic SOQL when possible (as per earlier chapter, use static with binding). If dynamic SOQL/SOSL must be used (like constructing flexible search queries), **never directly concatenate user input into the query string.** Use binding variables in static SOQL or use String.escapeSingleQuotes(userInput) for dynamic SOQL to sanitize apostrophes in user input. Better yet, use type-safe filtering: e.g., if user picks a field and value from UI for a query, verify the field name against a whitelist and then use bind for the value.

Example:

```
public without sharing class AccountSearch {
    public static List<Account> searchAccountsByName(String nameInput) {
```

```
    // Safe: using binding in static SOQL, no injection possible.
      String searchKey = '%' + String.escapeSingleQuotes(nameInput) +
'%';
      return [SELECT Id, Name FROM Account WHERE Name LIKE
:searchKey];
  }

  public static List<Account> searchDynamic(String field, String value) {
      // Suppose we allow dynamic field filtering on Account by specific
fields
    Set<String> allowedFields = new Set<String>{'Name','Industry'};
    if (!allowedFields.contains(field)) {
      throw new SecurityException('Field not allowed for search');
    }
      String soql = 'SELECT Id, ' + field + ' FROM Account WHERE ' +
field + ' = :val';
    // Bind the value safely using Database.query with a binding context
    // Actually, Database.query doesn't support bind this way, so instead:
    soql = soql.replace(':val', '\'' + String.escapeSingleQuotes(value) + '\'');
    return Database.query(soql);
  }
}
```

In searchAccountsByName, we use static SOQL with a bind variable :searchKey. We still escape single quotes in the user input just to be safe (though in a bind it's handled, but if nameInput contained %, that we allow, but a quote could break, so escapeSingleQuotes is a good measure).

In searchDynamic, which is trickier because field is dynamic, we maintain a whitelist of allowed field names. We then construct the SOQL string. We cannot directly bind value in a string (Database.query doesn't allow passing bind separately except through dynamic binding in QueryLocator maybe), so we replace a placeholder with the escaped value. We ensure to escape quotes in the value. If value itself came from user, that prevents breaking out of the quotes context. Additionally, we protected the field by whitelist to avoid things like field = "Name FROM Account WHERE IsActive__c = true OR Name" injections.

Essentially, the standard is to treat any user-supplied string that goes into a query carefully. Using static queries with binds is easiest and strongly recommended.

The Database.queryWithBinds method enables developers to safely execute dynamic SOQL queries using bind variables, which helps prevent injection vulnerabilities by separating user input from the query string. This approach eliminates the need for manual escaping of user-supplied values, as the binding mechanism automatically handles special characters and quotes. Using Database.queryWithBinds is recommended whenever possible for dynamic queries, as it provides both security and convenience compared to manual string manipulation.

General Secure Coding Practices

Purpose: Cover additional general standards for writing secure Apex.

Standard: Use Apex managed sharing for programmatic sharing of records when needed, instead of creating Share records without understanding. If business requires granting record access via Apex beyond OWD, use the proper mechanism (like creating Object__Share records or using Share object for standard objects) – but ensure that such code is only run by privileged users (system context or has proper checks).

Avoid storing sensitive business logic on the client side (that's more about client code, but in Apex context: calculations that should be secure should remain on server side, etc.). This is just a note that sometimes devs might be tempted to expose too much to client scripts – not directly an Apex standard, but related.

Use with sharing for @AuraEnabled and remote methods: It's worth reiterating that any Apex method exposed to UI (Aura, LWC via wire or imperative call, or global webservice) should either be in a with sharing class or carefully ensure security within. Salesforce actually updated that Lightning context controllers default to with sharing if not specified , but we still explicitly declare.

Monitor for security issues: e.g., if calling Crypto functions, ensure key handling is correct (don't expose encryption keys in logs or storage unencrypted), if doing file or attachment processing, ensure no malicious content is unintentionally served, etc.

While not an exhaustive security guide, the standards above (sharing, CRUD/FLS, no secrets in code, injection prevention) cover the main enforceable points.

Summary (Security and Sharing Standards): In this chapter, we set rules that Apex code must explicitly handle security: always declare sharing in classes (favor with sharing to enforce record access), always check user permissions for objects and fields when performing queries or DML in a user context (or use utility methods like stripInaccessible), never expose sensitive info in code or logs, and guard against injection by using bind variables or escaping input. These standards are critical to protect data. Failing to follow them could result in users seeing data they shouldn't, or hackers manipulating input to get unauthorized access. By adhering to these, we align with Salesforce's security model and ensure our custom code doesn't become a security weak point.

STANDARDS FOR GOVERNOR LIMITS AND RESOURCE MANAGEMENT

Introduction: Salesforce's multitenant architecture imposes governor limits on resource usage (SOQL queries, DML statements, CPU time, heap, etc.). This chapter defines standards to ensure our Apex code stays well within those limits and uses resources efficiently. The core theme is **bulkification** and efficient algorithms. We focus on rules that prevent hitting limits like "100 SOQL per transaction" or "50k records query limit", and guidelines for writing scalable Apex that can handle large data volumes by using best practices like bulk queries, collections, and queueable/batch where appropriate.

Bulkify All Code (No SOQL/DML in Loops)

Purpose: Ensure that Apex code can handle collections of records (bulk operations) without degradation or limit failures, which is a fundamental Salesforce requirement.

Standard: All Apex code (triggers, loop logic, etc.) must be bulkified: never perform SOQL queries or DML statements inside a loop. Instead, gather the data needed and perform set-based operations. This applies to loops over records as well as loops over collections in general. The pattern to avoid is:

```
for (...) {
... = [SELECT...]; // BAD (SOQL in loop)
}
```

or

```
for (...) {
  update...; // BAD (DML in loop)
}
```

The correct approach is to move those out of the loop:

If you need to query related data for a list of records, first collect the keys and do one query outside the loop (e.g., gather all needed parent Ids then query parents in one go).

If you need to perform DML on a list of records that you are assembling in a loop, add them to a list and do one DML on the list after the loop.

Additionally, when querying large datasets, use query limits or batchable processing rather than one huge query in memory if approaching limits (this is more design, but rule of thumb: queries returning thousands of records should be carefully handled).

Example (Trigger bulkification):

```
trigger ContactUpdate on Contact (before update) {
  // BAD practice: querying in loop (not allowed by standard)
  for (Contact c : Trigger.new) {
        Opportunity opp = [SELECT Id, Amount FROM Opportunity WHERE Id = :c.Last_Opportunity__c];
    //... do something with opp...
  }
  // Good practice: bulk query outside loop
  Set<Id> oppIds = new Set<Id>();
  for (Contact c : Trigger.new) {
        if (c.Last_Opportunity__c != null) oppIds.add(c.Last_Opportunity__c);
  }
  Map<Id, Opportunity> oppMap = new Map<Id, Opportunity>(
    [SELECT Id, Amount FROM Opportunity WHERE Id IN :oppIds]
  );
  for (Contact c : Trigger.new) {
    Opportunity opp = oppMap.get(c.Last_Opportunity__c);
    //... use opp (if not null)...
  }
}
```

In the "Good practice" part, we moved the SOQL out of the loop by first collecting all Last_Opportunity__c Ids from the contacts being updated, then performing one query. This prevents hitting the SOQL limit (which is 100 per transaction). The "BAD practice" loop would perform a query for each contact, easily exceeding limits if many contacts are updated at once, and is therefore disallowed.

For DML, a similar transformation:

```
// BAD:

for (Account a : accountsToUpdate) {

  update a;

}

// GOOD:

update accountsToUpdate;
```

Though trivial, it's crucial especially if that loop could run many times. Even if the loop size is small, standard is to not do it.

Efficient Querying and Selectivity

Purpose: Write queries that are efficient and selective to avoid performance issues and query limits (like 50k record limit).

Standard: Only query the fields that are needed (avoid SELECT * or unnecessary fields). This reduces CPU and heap usage. **Filter queries to retrieve only necessary records** – use WHERE clauses that make use of indexed fields (like Id, External Ids, or fields with indexes) to avoid full table scans that might hit limits or be slow. If expecting large result sets, ensure you have proper filters or consider batch processing.

Do not rely on queries without filters that could return huge numbers of records (e.g., querying all Accounts in a trigger context would be bad if there are thousands of accounts). If a use case requires processing many records, consider Batch Apex.

When iterating large results, use SOQL for loops which under the hood handle query batches of 200 records at a time to avoid hitting heap limits:

```
for (Account acct : [SELECT... FROM Account WHERE...]) {
```

```
  // this is a SOQL for loop that fetches in chunks
}
```

This pattern is recommended by Salesforce for memory efficiency.

Example:

```
// Query with only needed fields and proper filtering
List<Case> recentHighPriorityCases = [SELECT Id, Subject, Status
                    FROM Case
                    WHERE IsClosed = FALSE AND Priority = 'High'
                    LIMIT 100];
// Using a SOQL for loop for potentially large results
Integer openCount = 0;
for (Account acct : [SELECT Id FROM Account WHERE IsActive__c =
true]) {
  openCount++;
}
System.debug('Active Account count = ' + openCount);
```

In the first query, we filter by IsClosed = FALSE and Priority = 'High' which likely are selective criteria (assuming not all cases are high priority). We also limit to 100 in this scenario because maybe we only need the most recent or first 100 high priority open cases for some logic. Fetching only Id, Subject, Status avoids pulling large fields like Description or attachments if not needed.

In the loop example, by iterating directly over the query with for (Account acct : [SOQL]), Salesforce automatically batches the retrieval in 200-record chunks (if there were thousands of active accounts, it won't hit heap limit as easily as doing one large list = [SOQL] with 10k results). This is a recommended practice for large queries.

Governed Iterations and Recursion Control

Purpose: Avoid runaway code that might inadvertently cause too many operations or infinite recursion.

Standard: Implement guards to prevent excessive looping or recursive trigger calls. For instance, if your trigger might update records that cause the trigger to fire again, use a static variable (in a handler class) to ensure one-time execution per trigger context (so-called trigger recursion control). This is a standard practice: e.g., a static Set of processed record Ids or a static Boolean flag.

Also, if you have loops within loops (nested loops), be cautious – a nested loop over two collections of size N and M results in N*M iterations, which can blow up. Often nested loops can be transformed into a single loop with a map lookup (which relates to bulkification using maps). So the standard is to avoid nested iteration over large collections; use data structures instead.

If performing a potentially heavy loop, check governor methods like Limits.getIterations() or simply ensure the logic is optimized.

Example:

```
public class ContactTriggerHandler {

  private static Boolean isContactTriggerRunning = false;

    public static void onAfterUpdate(List<Contact> newList, Map<Id,
Contact> oldMap) {

    if (isContactTriggerRunning) {

      return; // Prevent recursion

    }

    isContactTriggerRunning = true;

    //... perform actions that might update contacts again...

    isContactTriggerRunning = false;

  }
```

}.

Here, isContactTriggerRunning prevents the trigger logic from re-entering if an update on Contact triggers the same logic again (maybe the logic updates contacts causing the trigger to fire again). By setting a flag, we avoid infinite recursion or double-processing. The standard is that any trigger or batch that could re-invoke itself should have such a guard.

For nested loops, consider:

```
// Instead of nested loops:

for (Account a : accounts) {

  for (Contact c : contacts) {

    if (c.AccountId == a.Id) {

      // link or process

    }

  }

}

// Use a map to avoid nested loop:

Map<Id, List<Contact>> contactsByAccount = new Map<Id, List<Contact>>();

for (Contact c : contacts) {

  contactsByAccount.computeIfAbsent(c.AccountId, k -> new List<Contact>()).add(c);

}

for (Account a : accounts) {

  List<Contact> relatedContacts = contactsByAccount.get(a.Id);

  // process contacts in one go

}
```

(This uses a hypothetical computeIfAbsent for brevity - in Apex, just do the check and initialize list if null.)

Managing Large Data Volumes

Purpose: Provide standards for operations that might deal with tens of thousands or millions of records, where normal Apex transaction limits might not suffice.

Standard: If an operation must handle more than a few thousand records, it should be implemented with a Batch Apex or Queueable chaining rather than a single transaction. This is partly an architectural decision, but as a standard: for example, a method that might potentially update 50k+ records must not try to do so in one go (which would hit DML rows limit), but instead should chunk the work.

So the enforceable part: if you find code that queries or processes extremely large lists without using batch, treat it as a red flag unless it's known the dataset is smaller. As a coding standard, document expected sizes and ensure code either defends against going beyond or explicitly uses batch.

Also, use Limits.getLimitDMLRows() and Limits.getDMLRows() if you want to proactively stop if nearing limits (rarely used, but possible). More practically, design to avoid approaching the limit in first place.

If using batch, ensure the batch size is optimal (standard 200 is usually fine; if records are very large in terms of fields, maybe smaller batch size to avoid heap issues).

Example:

```
public without sharing class DataArchiver {

  public static void archiveOldRecords() {

    // Instead of doing one massive query and DML, we invoke a batch

    Database.executeBatch(new ArchiveOldRecordsBatch(), 100);

  }

}
```

```
public without sharing class ArchiveOldRecordsBatch implements
Database.Batchable<SObject> {

  public Database.QueryLocator start(Database.BatchableContext bc) {

    // Query potentially large number of records

    return Database.getQueryLocator([

        SELECT Id, CreatedDate, Status__c FROM Custom_Object__c
WHERE CreatedDate < LAST_YEAR

    ]);

  }

  public void execute(Database.BatchableContext bc, List<Custom_Ob-
ject__c> scope) {

     List<Custom_Object__c> toArchive = new List<Custom_Object__c>
();

    for (Custom_Object__c rec : scope) {

      rec.Status__c = 'Archived';

      toArchive.add(rec);

    }

    update toArchive;

  }

  public void finish(Database.BatchableContext bc) {

    // maybe post-processing

  }

}
```

In DataArchiver.archiveOldRecords(), rather than trying to update all old records at once (which if old records are, say, 100k+, would break limits), it delegates to a batch class. The batch uses a QueryLocator which can handle up to 50 million records, processing 100 at a time here. This is the correct way to handle large volumes.

If for some reason batch is not used (maybe due to restrictions), then the code should at least implement its own chunking (e.g., processing in loops of 200 records with separate transactions via Queueable chaining). But using batch or scheduled batch is the standard approach.

Proper Use of Collections and Caching

Purpose: Emphasize reusing results within a transaction to avoid duplicate work and managing collections for efficiency.

Standard: Cache results of common queries or calculations in static variables or maps if they will be needed multiple times in a single transaction. This prevents doing the same SOQL twice in one execution. For example, if a trigger handling multiple objects needs a lookup list of configuration records, query it once and store it for reuse.

Also, if code calls an external service or expensive computation, consider caching the result in a transient store for that execution.

However, be careful to avoid holding onto large collections longer than needed (to conserve memory). Clear or let references go out of scope if no longer used, so they can be garbage-collected (Apex is auto GC, but deterministically setting to null can hint it).

Example:

```
public class PriceCalculator {

    // Static cache for product prices to avoid multiple queries in one transaction

  private static Map<Id, Decimal> pricebookCache;

  public static Decimal getProductPrice(Id pricebookEntryId) {

    if (pricebookCache == null) {

      pricebookCache = new Map<Id, Decimal>();

        for (PricebookEntry pbe : [SELECT Id, UnitPrice FROM PricebookEntry WHERE Pricebook2.IsStandard = true]) {

        pricebookCache.put(pbe.Id, pbe.UnitPrice);
```

```
            }
        }
        return pricebookCache.get(pricebookEntryId);
    }
}
```

In this PriceCalculator, the first time getProductPrice is called in a transaction, it fills pricebookCache with all standard pricebook prices. Subsequent calls simply retrieve from the map, avoiding additional queries. This is acceptable as long as the number of entries is not huge (standard pricebook usually limited products or you could filter further). This kind of caching is helpful to reduce redundant queries.

Another simple caching: storing UserInfo.getUserId() or UserInfo.getProfileId() if needed repeatedly, though those calls are cheap so not necessary. More useful is caching something like a custom setting value (so you query CustomSetting__c.getInstance() once and reuse it).

Summary (Governor Limits and Resource Management): This chapter's standards ensure our code runs efficiently within Salesforce's limits. We enforce bulkification: absolutely no SOQL/DML in loops. We stress selective queries and using proper structures like maps and sets to avoid inefficiencies. We also cover using batch apex for large volumes, caching repeated data, and preventing recursion or runaway loops. By following these, our code should scale to handle large datasets and high throughput without hitting governor limits or causing slowdowns. These standards are often automatically checked by static analysis tools (for example, PMD rules can catch SOQL in loop, etc.), and they are fundamental in any code review.

STANDARDS FOR ASYNCHRONOUS APEX

Introduction: Asynchronous Apex (Future methods, Queueable, Batch Apex, Scheduled Apex) is used for operations that need to run in the background, handle larger volumes, or be scheduled. This chapter defines standards for writing and using asynchronous Apex effectively. Key considerations include proper method signatures, resource limits for async (like not chaining indefinitely), and patterns for using each type of async job responsibly. These standards help ensure asynchronous processes are reliable and maintainable.

Future Methods

Purpose: Provide guidelines for using future methods (@future) appropriately, including method signature and when to use vs other async.

Standard: Future methods must be static and void, with only primitive or collection-of-primitive parameters (no sObjects or complex types). This is actually a requirement by Salesforce, but we restate as a standard: always mark them public static void <methodName>(<primitive types>) @future. The annotation goes before the method, and if the future needs to do a callout, include @future(callout=true).

Use future methods for scenarios: callouts from triggers (to avoid mixed DML issues or to offload callout), or quick one-off async tasks that don't need chaining. Do not use future for long-chains or complex processes; use Queueable instead for better control.

Also, limit the number of future calls. There's a limit of 50 futures per transaction, so code should not design a loop that calls a future each iteration – it should batch the work to within 50 (or ideally fewer). If needing more, consider an alternate pattern (like one future that invokes a batch, etc.).

Naming convention: method name should clearly indicate an async action, often including a verb like sendAsyncEmail or suffix Async to differentiate from sync methods.

Example:

```
public class OrderNotifier {
  @future(callout=true)
  public static void sendOrderConfirmationEmail(Set<Id> orderIds) {
    // retrieve necessary data
```

```
    List<Order__c> orders = [SELECT Id, Customer_Email__c, Status__c
FROM Order__c WHERE Id IN :orderIds];

    for (Order__c o : orders) {

            // call an external service or send email - allowed since call-
out=true

        HttpRequest req = new HttpRequest();

        //... prepare request...

        Http http = new Http();

        HTTPResponse res = http.send(req);

        // handle response...

    }

  }

}
```

This sendOrderConfirmationEmail future method is static void, takes a Set of Ids (collection of primitive). It's marked callout=true because it performs HTTP callouts. This adheres to the signature requirements and usage. We use a Set of Ids to allow batching multiple orders into one call (so if a trigger has many orders, we pass their Ids into one future call, instead of one call per order, staying under the 50 future limit).

We should ensure the future does not call another future (future from future is not allowed), nor should it do too heavy work repeatedly. If a future might handle a large set, consider splitting or using queueable.

Queueable Jobs

Purpose: Encourage use of Queueable for async tasks that require more complex coordination or chaining, and provide standards for implementing them.

Standard: Queueable classes must implement the Queueable interface and their execute method should handle a discrete unit of work. They can use instance state to pass data (Queueable allows non-primitive fields in the class, unlike future which only allows primitives as parameters). This is useful for passing sObjects or large sets to async context.

Queueable job classes should be named clearly (e.g., end with "Queueable" or "Job"). They should include only the minimal necessary state to perform the job. Leverage the fact that you can chain queueables: if chaining, ensure to avoid infinite or uncontrolled chaining loops. There's a limit: a Queueable can enqueue at most one other Queueable within it.

If using the new Queueable Finalizer (IFutureCallback interface) to handle post-queueable completion, ensure to implement that in a separate class and attach via System.attachQueueable() as needed – that's advanced, but a standard could be to use finalizers for cleanup if multiple queueables are in a chain and you need a final step (monitors all complete).

Example:

```
public class DataSyncQueueable implements Queueable, Database.AllowCallouts {

  private List<Id> recordIds;

  public DataSyncQueueable(List<Id> recordIds) {

    this.recordIds = recordIds;

  }

  public void execute(QueueableContext context) {

        // Process up to e.g. 100 records from recordIds (if more, chain another)

        List<Id> toProcessNow = (recordIds.size() > 100) ? recordIds.subList(0, 100) : recordIds;

      List<Id> remaining = (recordIds.size() > 100) ? recordIds.subList(100, recordIds.size()) : new List<Id>();

     //... perform callouts or processing on toProcessNow...
```

```
    if (!remaining.isEmpty()) {
      // Chain next queueable for remaining records
      System.enqueueJob(new DataSyncQueueable(remaining));
    }
  }
}
```

In this DataSyncQueueable, we implemented Queueable and also Database.AllowCallouts because it might call out. We chunk the work by taking at most 100 record Ids to process in one execute (just as an example; could be determined by callout limits or whatever). If there are more, we enqueue a new DataSyncQueueable with the remainder. This chaining must eventually terminate (here it will after processing all IDs). We ensure only one new job per execute to respect the platform rule.

Queueables are preferable to futures for many things now because they can be monitored (with job Id) and can be chained. So our standard might suggest: use Queueable over future if you need to chain or need non-primitive params.

Batch Apex

Purpose: Guide the use of Batch Apex for large-scale operations and set standards for implementing batch classes.

Standard: Batch classes must implement Database.Batchable<T> and optionally Database.Stateful if maintaining state between batches. The batch size should be set according to data size (default 200 is fine in most cases). If a batch is processing related records and needs to avoid mixed DML, design accordingly (maybe do only one object type per batch or use the stateful to collect and post-process later in finish).

Batch classes should be stateless unless necessary – using Database.Stateful can increase memory usage, so only use it if you need to keep aggregate info (like counting records processed).

Always implement error handling inside batch execute: catch exceptions around each chunk's logic so that one bad record doesn't fail the whole batch or, if you intentionally want to fail all on first error, then handle accordingly. Perhaps accumulate errors in a stateful list to report in finish.

Batch class naming: typically ends with Batch or similar (like ArchiveOldRecordsBatch as earlier).

Batch jobs can be scheduled. If scheduling, ensure the schedule is documented and the class is idempotent (multiple runs either okay or prevented if not needed concurrently).

Example:

```
global class LeadConvertBatch implements Database.Batchable<sObject>, Database.Stateful {

  global Integer successCount = 0;

  global Database.QueryLocator start(Database.BatchableContext bc) {

    return Database.getQueryLocator(

      [SELECT Id, Status FROM Lead WHERE Status = 'Qualified']

    );

  }

    global void execute(Database.BatchableContext bc, List<SObject> scope) {

    List<Lead> leads = (List<Lead>)scope;

    for (Lead l : leads) {

      try {

        // Suppose we convert leads - call some service or DML

        l.Status = 'Converted';

        // (In reality use Database.LeadConvert for actual conversion)

        update l;
```

```
            successCount++;
        } catch (Exception e) {
            // Log error for this record and continue
            System.debug('Failed to process Lead ' + l.Id + ': ' + e);
        }
      }
  }
  global void finish(Database.BatchableContext bc) {
      System.debug('LeadConvertBatch completed. Success count: ' +
successCount);
      // Could send an email or do post-cleanup
  }
}
```

In LeadConvertBatch, we mark it global (batchable needs to be global if called via global context like schedulers, or just public if invoked internally), and stateful because we keep a successCount. We properly catch exceptions inside execute loop to handle any failing lead conversion without stopping the batch. The finish method logs the successes, could also notify admin.

The query in start uses QueryLocator for potentially many leads. We filtered by Status = Qualified for some selectivity.

Scheduled Apex

Purpose: Lay out how to write schedulable classes for regular jobs, and what standards to follow.

Standard: Scheduled classes must implement Schedulable and the execute(SchedulableContext) method. They should be used to schedule batch or other jobs, not to do large logic directly in the execute (to avoid limits in the single schedule context if job is heavy). Essentially, the scheduled job is often just a trigger to call a batch or queueable.

When scheduling via System.schedule, use a meaningful job name and CRON expression. Standard: define CRON expressions carefully to avoid collisions (like if class should only run one at a time, ensure schedule frequency doesn't overlap with potential runtime).

Also, maintain scheduling in either code or metadata (like perhaps document schedule times in an admin guide). The coding standard is mostly: keep the scheduled execute lean.

Example:

```
public class DailyDataCleanupScheduler implements Schedulable {

  public void execute(SchedulableContext sc) {

    // Simply invoke the batch job for data cleanup

    Database.executeBatch(new DataCleanupBatch(), 200);

  }

}

// Elsewhere, schedule this in code or via UI: e.g.

// System.schedule('Daily Data Cleanup', '0 0 2 * * ? *', new DailyDataCleanupScheduler());
```

DailyDataCleanupScheduler doesn't do cleanup itself; it calls DataCleanupBatch. This keeps the schedule job quick. If DataCleanupBatch is large, it will chunk out as per batch logic.

The schedule string '0 0 2 * * ? *' means run at 2:00 AM daily. We used a descriptive job name 'Daily Data Cleanup'. Standards might include always prefixing scheduled job names with environment or project if needed to avoid confusion.

Governing Asynchronous Operations

Purpose: Provide rules to avoid misuse of async, like too many jobs or uncontrolled parallelism.

Standard: Do not queue excessive asynchronous jobs in a single context. For example, if processing 100 records, don't enqueue 100 queueables; instead, batch them into fewer jobs (as we did with chunking). There's a limit of 50 queued jobs per transaction (for future and queueable combined) and 250k jobs queued org-wide at a time. Code should avoid designs that approach these limits.

If using batch, know that up to 5 concurrent batches can run by default; if more, they wait. So scheduling 10 batches at the exact same time doesn't run them all at once, which is okay, but consider staggering if possible.

Always test asynchronous code with enough data to ensure it scales (maybe mentioned in testing, but as standard, ensure your asynchronous logic itself does not single-thread something that could be multi-threaded by platform causing conflicts. For example, two queueables running in parallel might conflict if both update same data – design to avoid or handle locking via FOR UPDATE or such if needed).

Avoid long chains: If queueable or schedule chaining is used, ensure there's an exit condition. For example, do not have a queueable that always enqueues a new one regardless (that would spin indefinitely until hitting limit of chain depth or number of jobs).

Summary (Asynchronous Apex Standards): We set standards for each asynchronous Apex type: future methods must follow signature rules and be used sparingly; queueables should be favored for complex async and properly handle chaining; batch classes should be used for heavy data jobs and be well-structured with error handling and reasonable batch sizes; scheduled Apex should only trigger processes at set times and not do heavy work inline. We also emphasized not exceeding limits like too many jobs. By following these standards, our async processes will be robust, not over-consume resources, and easier to manage (e.g., knowing where to find chain logic or schedule definitions). Async code often deals with large scale operations, so these standards tie closely with governor limit management as well.

STANDARDS FOR APEX INTEGRATIONS

Introduction: Integration code connects Salesforce with external systems or processes, often involving callouts (HTTP or SOAP) or inbound web services. This chapter outlines standards for writing Apex that integrates with external systems. Focus areas include callout structure, handling limits around callouts, using named credentials, and formatting Apex REST/SOAP classes. The goal is to make integration code secure, clear, and resilient.

Outbound Callouts (HTTP/REST)

Purpose: Define how to write Apex that calls external RESTful services or HTTP endpoints in a consistent and safe manner.

Standard: Use the Http classes (Http, HttpRequest, HttpResponse) for callouts and configure them properly: method, endpoint, headers, timeouts. Always set a reasonable timeout on HttpRequest (the default is 10 seconds, max is 120 sec; choose what makes sense). Use Named Credentials or Custom Settings to store endpoints and credentials, do not hardcode absolute URLs or authentication tokens in code.

Structure callout code to handle responses robustly:

Check the status code of the response (response.getStatusCode()).

Parse the response body only if expected (e.g., if status 200).

Handle error HTTP statuses (log or throw a custom exception with details).

Perform callouts in @future or queueable if calling from a trigger context or if multiple callouts needed to avoid mixed DML or to stay within callout limits (100 callouts per transaction).

For SOAP callouts, use generated Apex classes from WSDL2Apex if applicable and follow similar patterns (set input, call the generated stub, handle exceptions).

Example:

public class ShippingServiceClient {

```
private static final String ENDPOINT = 'callout:ShippingService'; // Using Named Credential 'ShippingService'

public static ShippingStatus getShipmentStatus(String trackingNumber) {

HttpRequest req = new HttpRequest();

req.setEndpoint(ENDPOINT + '/track?number=' + EncodingUtil.urlEncode(trackingNumber, 'UTF-8'));

req.setMethod('GET');

req.setTimeout(10000); // 10 seconds

Http http = new Http();

try {

  HttpResponse res = http.send(req);

  Integer code = res.getStatusCode();

  if (code == 200) {

    // parse JSON

    ShippingStatus status = (ShippingStatus) JSON.deserialize(res.getBody(), ShippingStatus.class);

    return status;

  } else {

    System.debug('ShippingService callout failed: ' + code + ' ' + res.getStatus());

    throw new CalloutException('Failed to get shipping status: HTTP '+code);

  }

} catch (CalloutException e) {

  throw e; // rethrow known exceptions

} catch (Exception e) {
```

```
        // Catch any unanticipated exceptions

        throw new CalloutException('Unexpected error during callout: ' +
e.getMessage());

    }

  }

}
```

In this example, ENDPOINT uses a named credential alias callout:ShippingService so the actual protocol/host and authentication are stored in Salesforce rather than code. We URL-encode the tracking number to be safe in the query param. We set a timeout. We handle the HTTP response by checking status code. For 200, we parse JSON via JSON.deserialize into a custom Apex class ShippingStatus (which would be defined to match JSON structure). If non-200, we throw a CalloutException (a custom exception class, presumably defined). We catch exceptions: specifically rethrow CalloutException or wrap generic exceptions into a CalloutException for consistent error handling by callers.

Standard also: ensure the method is annotated @future(callout=true) or called from one, if it's used in contexts that require async (the example is synchronous static; it would work if called from an async context or from a Visualforce controller etc., but triggers should not call this directly except in a future or queueable due to callout restrictions).

Inbound Apex REST Services

Purpose: Provide standards for writing Apex classes that expose REST endpoints (using @RestResource and @Http* annotations) for external systems to call Salesforce.

Standard: Apex REST classes should be global or public and annotated with @RestResource(urlMapping='/YourPath/*'). Within them, methods handling HTTP verbs should be static and annotated with @HttpGet, @HttpPost, etc. These methods should use descriptive parameter names and return types as appropriate. They should enforce security (sharing, user auth is via session or OAuth anyway, but within method check user perms if doing sensitive things).

Use JSON serialization/deserialization properly: accept parameters as method arguments if simple, or as request body (then you can use RestContext.request to get input stream and parse JSON manually or via framework). When returning, you can return a list or object and Apex will serialize it to JSON.

Document the REST API with comments (so it's easier for consumers to know what to send/receive).

Follow naming: class might end with "Service" or "Controller" or something indicating it's a REST endpoint provider.

Example:

```
@RestResource(urlMapping='/inventory/*')

global with sharing class InventoryService {

  @HttpGet

  global static List<Product__c> getProducts() {

    // Returns a list of products (for demo, limit to some number)

    RestRequest req = RestContext.request;

    String category = req.params.get('category');

          String q = 'SELECT Id, Name, Quantity__c, Price__c FROM
Product__c';

    if (category != null) {

      q += ' WHERE Category__c = :cat';

    }

      // Note: above uses bind, but since constructing string, might use
escapeSingleQuotes for category if dynamic

    List<Product__c> prods = Database.query(q);

    return prods;

  }
```

```
  @HttpPost
  global static Order__c createOrder(OrderWrapper input) {
    // Accepts an order via JSON, creates an Order__c record
    if (input == null || input.products.isEmpty()) {
      RestContext.response.statusCode = 400;
      return null;
    }
    Order__c ord = new Order__c(Name = input.name, Total_Amount__c
= input.total);
    insert ord;
    // Could also insert order line items...
    return ord;
  }

  global class OrderWrapper {
    public String name;
    public Decimal total;
    public List<Id> products;
  }
}
```

In InventoryService, we used @RestResource with a URL mapping. The class and methods are global because external systems will call them (global is required for OAuth access to them). We used with sharing to enforce sharing for whatever it does.

The @HttpGet getProducts takes an optional query param for category. We demonstrated building a query string (though better to use binding; our concatenation with bind is pseudo-codey, but one must actually use escape for dynamic field values if present – as a standard, we'd avoid dynamic SOQL if possible here, maybe have separate endpoints or a limited set of filters). It returns a list of Product__c; the framework serializes that to JSON automatically.

The @HttpPost createOrder accepts a custom wrapper class OrderWrapper which will be populated from the JSON body of the request (this is an Apex feature: method params for REST can be sObject or user-defined types to bind JSON). We check if input is valid, otherwise set response status to 400 (Bad Request) and return null. We then create an Order__c record. Note we didn't do extensive error handling here due to brevity, but normally we'd catch DML exceptions and set response code accordingly.

Standard points: always set appropriate HTTP status code in RestContext.response.statusCode when returning errors or even success (default 200 for GET, 201 for creation maybe). Don't leak internal exceptions directly to response without formatting.

Inbound Apex SOAP Services

Purpose: Provide standards for global Apex classes with webService methods (legacy SOAP integration).

Standard: If using SOAP (webService keyword and global classes), ensure the class is global and methods are global static, and use complex types (global inner classes) as needed for parameters or return types. Essentially follow WSDL2Apex style guidelines. Document the methods with comments.

Since SOAP is less common nowadays compared to REST, often WSDL2Apex is used to call external SOAP. But for exposing, you use webService annotation. The standard: try to use REST unless there's a specific reason to use SOAP (some older integration requirement).

If you do SOAP, treat it similar to REST in terms of security and input validation.

Example:

global class AccountWebService {

```
        webService static AccountInfo getAccountInfo(String accountNumber) {
    // Validate input
    if (accountNumber == null || accountNumber.length() == 0) {
        throw new webServiceCalloutException('Account number is required');
    }
    Account acct = [SELECT Id, Name, AnnualRevenue FROM Account WHERE AccountNumber = :accountNumber LIMIT 1];
    AccountInfo info = new AccountInfo();
    if (acct != null) {
      info.name = acct.Name;
      info.revenue = acct.AnnualRevenue;
    }
    return info;
  }

  global class AccountInfo {
    webService String name;
    webService Decimal revenue;
  }
}
```

This defines a SOAP web service method getAccountInfo that returns a custom structure AccountInfo. We marked the class global and the inner class and fields webService so they appear in the WSDL. We throw a webServiceCalloutException (which is how to throw SOAP faults). The pattern is straightforward. (Though note: because it's a webService method, sharing defaults to without sharing unless defined otherwise; could do with sharing global class AccountWebService if needed.)

Handling Integration Exceptions and Retries

Purpose: Ensure that integration failures are handled gracefully and possibly retried if needed.

Standard: When an outbound integration (callout) fails (network issue, timeout, non-200 response), the code should handle it. This could mean:

Logging the failure (maybe to a custom object or an email to admins).

Potentially implementing a retry mechanism: e.g., queue up the record ID to try again later (perhaps via a scheduled job or a smaller queueable chain).

At minimum, do not silently ignore the failure.

For inbound, if an external system calls our REST/SOAP and something fails (like DML exception inside), we should return an appropriate error message and status code or fault so the external system knows it didn't succeed.

Any integration that updates external systems should consider idempotency (e.g., if a call is retried, ensure it doesn't double-process something like double create an order). This often involves sending unique identifiers or checking if an external ID has already been processed.

Standard practice: For critical integrations, implement a small persistent log (custom object like Integration_Log__c) recording requests and responses or at least statuses (especially for asynchronous callouts or incoming messages), so there's an audit trail. This isn't a coding standard per se, but encourage capturing enough info for troubleshooting.

Example (retry logic):

Imagine a callout that fails, we could do:

HttpResponse res = http.send(req);

if (res.getStatusCode() >= 500) {

```
// server error, schedule a retry
System.enqueueJob(new IntegrationRetryQueueable(orderId));
}
```

This enqueues a queueable to retry later (you might add a delay by scheduling in X minutes via a Schedule if needed because enqueue will run immediately usually).

Our standards should caution not to create infinite retry loops. Maybe attempt a retry 3 times then give up and alert someone.

Summary (Apex Integrations Standards): We established that integration code should be written carefully: outbound callouts must use proper patterns (no hardcoded credentials, handle responses, call asynchronously if needed), inbound endpoints should be well-defined and secure. We stress using Named Credentials for external endpoints (so credentials are not in code), checking and parsing responses properly, and handling errors (like HTTP errors) explicitly. We also mentioned ensuring not to expose data via integrations without respecting sharing or perms. By following these standards, integration points will be robust and maintainable. Integrations are often points of failure due to network or external issues, so our code needs to gracefully handle those eventualities.

METADATA AND ANNOTATION STANDARDS

Introduction: Salesforce Apex provides various metadata-driven features and annotations that add metadata to code elements (such as @AuraEnabled, @Deprecated, custom metadata usage, etc.). This chapter covers standards for using these annotations and metadata features consistently. It also addresses how to externalize configuration (no hardcoding of environment-specific values, use custom metadata/settings and labels).

No Hard-Coded IDs or Org-Specific Data

Purpose: Ensure code is portable between orgs (e.g., sandbox to production) and configurable without edits.

Standard: Do not hard-code record IDs, profile IDs, user IDs, or any org-specific identifiers in Apex code. Instead, use Custom Settings/Custom Metadata to store references to particular records (like an ID of a special record) or query by some unique field value (like find by Name or External Id). For example, if you need the Id of a specific record (say a particular Account used as a reference), store an external Id on that record and query by it, or store the Id in a Custom Setting.

Similarly, for profile or user references, use profile names or user attributes in queries if needed (though even that can break if names change; better to avoid logic based on specific users/profiles in code, handle via config or permission sets).

This standard prevents the common error where code works in one org but not in another due to different IDs.

Example:

Instead of:

Id specialAccountId = '001ABC000123456'; // Hard-coded (BAD)

Account special = [SELECT Id, Name FROM Account WHERE Id = :specialAccountId];

Do:

String specialAccountName = 'Special Account';

Account special = [SELECT Id, Name FROM Account WHERE Name = :specialAccountName LIMIT 1];

// or use a Custom Metadata that stores 'SpecialAccountId' and retrieve from there

Or have a Custom Setting/Metadata CustomSetting__c.SpecialAccountId__c and use that.

Custom Labels for User-Facing Text

Purpose: Externalize strings that are shown to users (for translation or easy editing without code change).

Standard: Use Custom Labels for all end-user facing text, such as error messages, UI messages, etc. The Apex code should reference labels via System.Label.LabelName. This allows changing text or translating without modifying code. Exceptions: truly static technical strings that are never user-visible (like a Map key or an integration constant) need not be labels.

Name labels meaningfully, e.g., lbl_Error_InsufficientPermissions. In code:

throw new CustomException(Label.lbl_Error_InsufficientPermissions);

Example:

```
if (!hasAccess) {
   // Using a custom label for the error message
   throw new SecurityException(System.Label.Error_No_Access);
}
System.debug('Processing ' + System.Label.Info_StartProcess);
// Suppose Info_StartProcess label = "Starting processing..."
```

This ensures if we want to reword "Starting processing..." or translate it, we do so in labels, not code.

Apex Annotations Usage

Purpose: Provide guidelines for the various Apex annotations (beyond test and future which we covered, focusing on others: @AuraEnabled, @InvocableMethod, @TestVisible, @Deprecated, etc.).

Standard:

@AuraEnabled: Use this annotation on methods or properties that need to be accessed by Lightning Components (Aura or LWC via wire/imperative). Only annotate what is necessary to expose. If a method is @AuraEnabled, ensure the class is either with sharing or enforces security internally. Also consider using @AuraEnabled(cacheable=true) on get methods that only retrieve data and can be cached on client side for performance.

Example:

```
public with sharing class AccountController {

  @AuraEnabled(cacheable=true)

  public static Account getAccount(Id accountId) {... }

}
```

The cacheable implies this method does not change data and can be called from LWC wire for caching.

@InvocableMethod and @InvocableVariable: Use these for exposing methods to Flow. Standard: methods should be static and either global or public (global if packaged). Use the label and description attributes in the annotation to provide user-friendly info in Flow Builder. Only one invocable method per class (Flow requirement), so follow that.

Example:

```
public class LeadUtils {

   @InvocableMethod(label='Convert Leads' description='Converts a list
of lead IDs to contacts')

  public static List<ConvertResult> convertLeads(List<Id> leadIds) {... }

}
```

And any complex input can be defined as a class with @InvocableVariable fields.

@TestVisible: Use this annotation sparingly to expose private variables or methods to test classes. Only use it when there's no clean public interface to test necessary internal state. Standard: you may mark a private method as @TestVisible instead of making it public just for tests. This is acceptable but if you find you use it too often, consider refactoring code for better testability. Marking something @TestVisible does make it available to any test running in that org, but not in production runtime. It's fine as a tool.

Example:

```
private class OrderHelper {

  @TestVisible private static Integer DEFAULT_DISCOUNT = 5;

...

}
```

so that tests can read or even set DEFAULT_DISCOUNT if needed (TestVisible allows tests to set it too).

@Deprecated: Use this annotation to mark classes or methods that should no longer be used. If an API (global or public method) is being phased out, mark it @Deprecated and ideally add a comment what to use instead. This helps other devs to avoid using it (and some tools warn when calling deprecated). Standard: Provide alternative in documentation. Eventually, you might remove it in a future version. But note in managed packages, removing global things is tricky, so deprecate first.

Example:

```
@Deprecated

global static void oldMethod() { /* Use newMethod instead */ }
```

@SuppressWarnings: (rare, but exists to suppress PMD or compiler warnings). Standard: If you use @SuppressWarnings('PMD') or specific rule, document why in a comment. Use it extremely sparingly - it's better to fix the code triggering the warning, but if a warning is false positive or negligible, you can suppress with a clear comment. This ensures our code analysis remains clean.

Example:

```
@SuppressWarnings('PMD.ApexUnitTestsShouldHaveAsserts')

static void testExampleWithoutAssert() {

  // This test intentionally has no assert because...

}
```

But you'd rarely justify no asserts.

Usage of Custom Settings/Metadata for Config

Purpose: Encourage externalizing configuration (like thresholds, toggle flags, etc.) to custom settings/metadata.

Standard: Use Hierarchy Custom Settings or Custom Metadata Types for configuration that might change per org or needs to be tweaked without a code deploy. For example, threshold values, feature toggles, default values, integration endpoints (if not using Named Credential), etc.

If using List Custom Settings (non-hierarchy), treat them as data that can be manipulated. If using Custom Metadata, you get package deployable, not modifiable in subscriber (unless marked as such).

Access them in code via the proper methods: e.g., MySetting__c.getInstance().Field__c (for hierarchy, automatically picks current user or default) or MyCustomMetadata__c.getAll() then by key.

Standard: avoid frequent queries to settings in loops, because they can be cached by getInstance but ensure not in loop anyway.

Example:

```
Decimal maxDiscount = MySettings__c.getInstance().Max_Discount__c;

if (discount > maxDiscount) {

  discount = maxDiscount;

}
```

Here Max_Discount__c can be set by admin in the custom setting instead of hardcoding a number.

Another:

```
Boolean featureOn = FeatureToggles__mdt.getInstance('NewCalcFeature').Enabled__c;

if (featureOn) {

  // use new calculation

} else {
```

```
    // old calculation
}
```

API Versioning in Apex Classes

Purpose: Ensure awareness of Apex class API versions for consistency and that all classes use an appropriate version.

Standard: All Apex classes and triggers should be created at a uniform API version (preferably the latest stable version available when the project started). When updating a class to use new features, you might need to bump its version. While not usually manually edited, it's something to monitor. The standard is simply: do not leave classes on extremely old API versions unless there's a known compatibility reason. Align them to avoid unpredictable differences in behavior.

Also, if you intentionally use a slightly older version to maintain a behavior (rare cases), document it in the class header.

This is not a code style but a metadata standard: if using SFDX, ensure sfdx-project has sourceApiVersion matching your desired version.

Summary (Metadata and Annotation Standards): In this chapter, we reinforced that code should not be tied to a specific org's data through hardcoded IDs or secrets – those should be in custom settings/metadata and labels. We described how and when to use Apex annotations like @AuraEnabled, @InvocableMethod, @TestVisible, etc., to keep code clean and properly exposed. We stressed externalizing user-visible strings to labels for localization. By following these standards, our code-base becomes more flexible (e.g., can deploy to a new org without breaking, can have features toggled via config, and integrates well with Salesforce features like Flows and Lightning via proper annotations). This leads to a more admin-friendly and maintainable solution.

SOURCE CONTROL AND DEPLOYMENT STANDARDS

Introduction: In an enterprise setup, Apex code is managed via source control and deployed through automated processes. This chapter outlines standards around using version control (like Git) and deploying Apex and related metadata in a consistent manner. While not code style per se, it's important to standardize how code is moved between orgs and how we ensure integrity during deployments.

Version Control Usage

Purpose: Ensure that all Apex code (and metadata) is tracked in a source control system, enabling collaborative development and history tracking.

Standard: All Apex classes, triggers, and related metadata (like profiles, permission sets that grant Apex permissions, etc.) must be stored in the team's source control repository. Developers should not write Apex directly in production or outside of this process. Every code change should go through version control (via commits and code review if possible) before deployment to higher environments.

We enforce using a specific branching strategy (e.g., feature branches merged into dev branch, then into main for release). The code repository should mirror the Salesforce metadata structure (if using SFDX, a package directory; if using Metadata API format, then folders by type).

The standard is also to include the -meta.xml files for each Apex class/trigger in source control. Those define API version and other attributes.

No code should be considered "final" unless it's in version control. If emergency prod hotfixes are made (avoid if possible), they must be back-committed to the repo to maintain consistency.

Deployment Process

Purpose: Standardize how code is deployed to different orgs to minimize issues.

Standard: Use automated deployment tools (like Salesforce DX CI/CD pipelines or ANT scripts) to deploy Apex code from source control to orgs. This ensures that what's in source is what goes to the org. No manual changes in the target org after deployment.

All deployments must run all tests (or at least all relevant tests) and achieve no test failures and the required coverage. Our standard is to aim for > 85% coverage overall even though 75% is the minimum, to have a safety margin.

Deployment scripts or pipelines should validate code style (maybe run static analysis), run tests in a staging environment, and then deploy to production in a controlled window.

Also, maintain an explicit manifest (package.xml or sfdx source tracking) of what to deploy, rather than deploying random unmanaged changes.

If any Apex class is dependent on certain metadata (like a Custom Setting), ensure that metadata is deployed together, or documented that it must exist.

Continuous Integration and Quality Gates

Purpose: Integrate quality checks into the development process.

Standard: Set up Continuous Integration (CI) to run on each commit or pull request, which runs static code analysis (PMD, CodeScan, SonarQube) and Apex tests. The standards for code (like no SOQL in loop, naming etc.) can be enforced by PMD rules, which should pass before merging code. Similarly, code coverage gates (e.g., fail build if coverage < 80%) can be in place.

While this is more process, it's part of standards: no code goes to main branch unless it passes CI checks.

Environment Consistency

Purpose: Ensure that developers and CI are using the same configurations.

Standard: Use the same Apex compiler version across environments (by aligning API versions and using SFDX with a specified sourceApiVersion). Also, devs should all use the same code formatter config if auto-formatting (like Prettier Apex with same settings) to avoid churn in diffs. The repository can include editor config or prettier config.

Also, maintain a consistent project structure (no one should introduce a different directory for classes; they go in the standard folder).

Git Conventions (if using Git)

Purpose: Standardize commit messages and branch names for clarity.

Standard: Use descriptive commit messages that include reference to work items (like "PROJ-123: Implemented validation in OrderTrigger"). This links code changes to requirements or bug fixes. Avoid committing large unrelated changes in one go; keep commits focused.

Branch naming: for feature development, use a prefix like feature/ or bugfix/ followed by issue number or description, e.g., feature/PROJ-123-order-validation. This makes it clear in the repository what's on that branch.

Pull requests should be required and at least one peer reviewer should approve changes, including checking adherence to standards.

Summary (Source Control and Deployment Standards): We outlined that all Apex code must live in source control and be deployed through an automated pipeline. No direct org edits for Apex. Branch and commit conventions should be followed to tie code to context. We emphasize running tests and static analysis as part of CI to enforce the standards in earlier chapters. By following these processes, we reduce the risk of code discrepancies between environments and ensure higher code quality and traceability of changes.

STANDARDS ENFORCEMENT TOOLS AND CONFIGURATION

Introduction: Consistency with coding standards can be enforced using various tools and configurations. This chapter covers how to use static code analysis, linters, formatters, and other tools to automatically check and enforce the Apex standards described. It also includes how to configure these tools (like PMD rulesets) to match our standards.

Static Code Analysis (PMD/SonarQube)

Purpose: Use automated analysis to catch standard violations early (like style issues or potential bugs).

Standard: Our codebase should be analyzed with a static analysis tool (e.g., PMD with an Apex ruleset) as part of CI and optionally in developer IDEs. We will maintain a PMD ruleset XML that reflects our standards:

Include rules like Avoid SOQL In Loops, BracesRequired, NamingConventions (for class, method, variables) , etc.

If certain rules are not relevant or too strict, configure or suppress them in the ruleset rather than ignoring them.

For SonarQube, have a quality profile for Apex that includes rules for the above. Set a quality gate that code must meet (e.g., no new critical issues).

Developers should run these checks locally (via VSCode extensions or pre-commit hooks) to catch issues before pushing.

Example PMD Rule configuration snippet:

```
<rule ref="category/apex/bestpractices.xml/ApexUnitTestClassShouldHaveAsserts"/>

<rule ref="category/apex/performance.xml/AvoidDmlStatementsInLoops"/>

<rule ref="category/apex/codestyle.xml/ClassNamingConventions">

  <properties>

      <property name="regex" value="^[A-Z][A-Za-z0-9]+$"/> <!-- PascalCase -->

  </properties>

</rule>
```

This would enforce that test classes have asserts, no DML in loops, and class names match the regex (PascalCase starting capital). We would include many more rules similarly.

Code Formatting Tools

Purpose: Automatically format code to meet layout/spacing standards, so devs don't have to do it manually or argue over style.

Standard: Use the Prettier Apex plugin or similar code formatter with our agreed settings (4-space indent, etc.) in all development environments. Configure it in the repository (with a.prettierrc config) so that running the formatter yields consistent results. Optionally enforce formatting in CI (some projects fail the build if formatting is off, or simply auto-fix and commit).

If Prettier is used, ensure the team knows to run it (most IDE can auto-format on save). This handles a lot of our formatting standards (indentation, spaces, line breaks). Prettier uses its own style which is mostly aligned to common standards, and we can adjust printWidth (say 100).

Where the formatter cannot enforce (like naming), rely on static analysis or manual review.

IDE Setup and Plugins

Purpose: Provide guidance on dev setup to catch issues early.

Standard: Developers should install recommended plugins in VSCode (Salesforce Extensions, PMD, Prettier, etc.) configured to use our rules. Possibly include a project configuration file that VSCode can pick up (like.vscode/settings.json) to enable on-save formatting and linting.

Encourage using Illuminated Cloud or other IDE? Regardless, the environment should highlight violations (like a PMD linter will underline an SOQL-in-loop in red in real-time).

Code Review Checklist

Purpose: Use a checklist (maybe derived from this standards doc) to manually verify code during reviews beyond what tools catch.

Standard: Establish a code review checklist that reviewers must go through for each change. This could be a simple list in the PR template, e.g.:

Naming conventions followed?

Comments updated and present where needed?

No logic in triggers (only delegation)?

Bulk limits respected (no query in loop, etc.)?

Tests cover new code with asserts?

etc.

This ensures human oversight where tools might not fully catch intent issues (like whether the test actually covers logic or just exists).

Configuration Management

Purpose: Manage the configurations for these tools in a centralized way.

Standard: Keep tool configuration files (PMD ruleset, Prettier config, CI pipeline config) in source control. This way they version along with code. If we update a standard (say decide to allow something or tighten something), we update the config and all developers pull that change.

Also, if we find a certain static rule is too noisy or not applicable, adjust and document it rather than everyone suppressing individually.

Example repository structure:

/config/pmd/ruleset.xml

/.prettierrc

/.eslintrc (for LWC, if applicable)

/.vscode/settings.json (with editor.formatOnSave: true etc.)

By following these, enforcement becomes part of the development workflow.

Summary (Standards Enforcement Tools and Configuration): We will leverage static analysis tools like PMD or SonarQube configured with rules reflecting our standards, so many violations are caught automatically. Code formatting will be automated with a tool like Prettier to ensure consistency in layout. These tools integrated into CI mean that if code doesn't meet standards, the build fails or warnings are given, making sure issues are fixed early. With proper configuration in source control, enforcement is uniform across the team. The result is that adherence to this coding standard becomes part of our pipeline, not just a one-time read.

STANDARDS GOVERNANCE AND VERSIONING

Introduction: Coding standards themselves need maintenance. This chapter describes how we govern these standards: how changes are proposed, reviewed, and versioned, and how compliance is monitored in the long run. It ensures the standards stay relevant and up-to-date with Salesforce platform changes.

Ownership and Updates of Standards

Purpose: Define who is responsible for the coding standards and how to change them.

Standard: Appoint a "Standards Committee" or lead architect responsible for the Apex coding standards. This could be a group of senior developers/architects. Changes to standards (additions, modifications) should be discussed with this group and then communicated to the whole team.

Maintain this standards document in a central location (like a Confluence page or repository README) and update its version number whenever significant changes are made.

Encourage team members to suggest improvements or new rules especially when new Salesforce features come out (e.g., if a new annotation or pattern emerges, incorporate it).

Set a review cycle, e.g., revisit the standards every 6-12 months or when a major Salesforce release adds relevant features, to incorporate best practices.

Versioning the Standards Document

Purpose: Keep track of revisions to the standard so that it's clear which version a codebase is following.

Standard: Label this document with a version and date. For example, "Apex Coding Standards v1.0 (Jan 2025)". When updates are made, increment to v1.1, v2.0, etc., and highlight changes (maybe in an appendix or changelog section).

Optionally, include in code header comments a reference to the standards version (some organizations do like: @codingStandards Version 1.0 in class headers to indicate which version of the standard the code was written under).

At minimum, communicate to all devs when a new version is released, and possibly hold a short training or summary of what's new.

Training and Onboarding

Purpose: Ensure new developers quickly learn and adopt the standards.

Standard: All new developers must read this coding standards handbook and ideally have a kickoff session where a senior dev walks through key points. Provide examples of compliant vs non-compliant code as part of onboarding.

Possibly integrate some exercises or use real code review scenarios to reinforce the rules.

Ongoing, if code reviews find recurring misunderstandings, do a refresher training for the team on those specific standards.

Compliance Monitoring

Purpose: Continuously monitor that the code in repository remains compliant and address deviations.

Standard: Use metrics from static analysis (like number of rule violations) to gauge compliance trends. For example, SonarQube's static analysis can show if code quality is improving or regressing. We should aim for zero critical violations and gradually reduce minor ones.

Perform periodic manual audits of a random sample of classes to see if standards are applied (for things not easily tool-checked, like consistency of comments or effective test coverage meaning not just % but quality of tests).

If a pattern of non-compliance is found (e.g., many instances of missing ApexDoc comments), address it: perhaps create a work item for developers to add those comments, and update the CI to flag those explicitly.

Exception Process

Purpose: Allow flexibility in rare cases where deviating from a standard is necessary.

Standard: If a developer believes an exception to a rule is warranted, they must get approval from the standards committee or lead and document the rationale in the code (comment) and perhaps in a central log. For example, maybe a certain integration requires multiple callouts in a loop and cannot easily be batched; an exception might be allowed but documented as technical debt or special case.

Such exceptions should be revisited to see if they can be eliminated later.

Keep exceptions minimal to preserve the integrity of the standards.

Summary (Standards Governance and Versioning): We clarified that this coding standard document is a living artifact. There's a process to update it as Apex and our development practices evolve. A versioned approach ensures we know what rules applied at what time (useful if, for instance, older code doesn't meet a newer rule – we then decide to refactor or not). By assigning ownership of the standards and integrating them into onboarding and continuous learning, we ensure team buy-in. Finally, tracking compliance and having a way to handle necessary exceptions makes the standards regime practical and adaptable over time. The goal is a balance: strong consistency but with room to adapt when needed, all under version control for the standards themselves.

QUICK REFERENCE AND CHECKLIST

Introduction: This chapter provides a condensed checklist of the key coding standards from previous chapters. Developers and reviewers can use this as a quick reference to verify code compliance. It's organized by category (naming, formatting, etc.) for ease of scanning.

Naming Standards Checklist

Classes/Triggers: PascalCase, start with uppercase. Example: CustomerService, AccountTrigger. Test classes end with Test.

Methods: camelCase, start with lowercase verb, no underscores. Example: calculateTotal.

Variables: camelCase, meaningful names, no single letters (except loop indices), constants in ALL_CAPS with underscores.

Booleans: named to imply true meaning (e.g., isActive, hasPermission).

No name collisions: Avoid methods having same name as class, etc., to prevent confusion.

One Trigger per Object: Named <Object>Trigger ; logic encapsulated in handler class (naming like <Object>TriggerHandler).

Formatting Standards Checklist

Indentation: 4 spaces per level, no tabs.

Braces: K&R style – opening brace on same line, closing on its own line. Always use braces for if/for/while, even one-liners.

Line Length: ~100 characters max. Wrap lines that are longer, indent continued lines +4 spaces.

Blank Lines: Use to separate methods and logical code blocks, but no multiple blank lines.

Spaces: One space after commas, around operators (=, +, etc.), after control keywords if (...) (space after if). No trailing whitespace.

One Statement per Line: Do not put multiple statements on the same line.

Section Order in Class: static constants, static vars, instance vars, constructors, methods (public before private), inner classes last.

Commenting Standards Checklist

Class Header: Every class/trigger has a header comment with @description and optionally @author, @date.

Method Comments: Public/protected methods have ApexDoc comment with @param and @return explaining usage.

Inline Comments: Explain non-obvious logic (prefer "why" over "what"). No commented-out blocks of old code in repository.

No Redundant Comments: Don't state the obvious. ("// increment i" on i++ is not needed).

Use TODO/FIXME tags if needed to mark future work, with developer name/date if possible.

No sensitive info in comments: Don't include passwords, secret URLs, etc., in comments (or code).

Apex Code Style Standards Checklist

Access Modifiers: Every class and method declares access (public/private/protected). Avoid global unless necessary.

Bulkification: No SOQL or DML inside loops. Ensure all triggers/processes handle collections.

Collections: Use maps/sets to avoid nested loops. Use SOQL for loops for large queries.

If/Else: Always use braces. Chain else-if instead of separate ifs for related conditions.

Switch: Include a when else default branch.

Exception Handling: Catch specific exceptions if needed, no empty catch blocks. Log or handle errors appropriately. Do not swallow exceptions silently.

Use Properties: Use {get; set;} for trivial getters/setters rather than manual methods.

Avoid Hardcoding: No hard-coded record IDs or magic numbers (use constants or config).

CRUD/FLS: Check user permissions for any SOQL/DML on objects/fields accessible to end-users.

Sharing: Classes default to "with sharing" (explicitly declare) unless system context needed.

Utility Classes: Mark as inherited sharing or as appropriate, with private constructors to prevent instantiation if only static methods.

Tests: (covered in testing checklist next)

Apex Testing Standards Checklist

Test Class Naming: <TargetClass>Test. Annotate with @isTest.

Test Methods: Annotate with @isTest (or testMethod keyword), name clearly (testScenario_expectedOutcome).

No SeeAllData: Do not use SeeAllData=true. Always create the needed data within tests.

Assertions: Every test method contains at least one System.assert* to validate results.

Test Data: Use @testSetup for common data. Otherwise, create in each test. No dependency on org data.

Cover Bulk: Write tests for bulk scenarios (e.g., inserting 200 records to trigger bulk behavior).

Async Tests: Use Test.startTest()/Test.stopTest() when testing future, queueable, batch.

Isolation: Each test method shouldn't rely on others. Each can be run independently.

Governor Testing: (Optional) Test that large inputs don't break limits if feasible (like ensure a trigger properly handles 200 records without hitting SOQL limit – usually implicitly tested by not throwing an exception).

Don't Skip Tests: No Test.isRunningTest() conditions in code to skip important logic (unless absolutely needed for something like avoiding callouts in tests, and even then supply a mock alternative).

Security Standards Checklist

Sharing: All classes specify "with sharing" or "without sharing". Default to "with sharing" for user-facing logic.

CRUD/FLS: Check Schema.sObjectType and DescribeField for permissions before queries/DML on objects accessible to user. Or use Security.stripInaccessible on SObjects for output.

SOQL Injection: Use bind variables or escapeSingleQuotes for any dynamic SOQL that includes user input.

Secrets: No hard-coded sensitive info (passwords, tokens). Use protected custom settings/Named Credentials.

Logs: Don't log sensitive data (e.g., personal info or credentials).

TestVisible: Only used to expose private members to test when absolutely necessary.

@AuraEnabled/@InvocableMethod: Ensure such methods enforce security (either by sharing or explicit checks) since they can be entry points.

Governor Limits Checklist

SOQL/DML Limits: Max 100 SOQL, 150 DML per transaction – ensure code stays well below these by bulk operations.

No more than 10k records in one DML: If processing more, break into batches.

CPU/Heap: Avoid extremely heavy loops or large data in memory. If using large collections, consider processing in chunks or using Batch Apex.

Future/Queueable Limits: Max 50 futures per txn, 50 enqueued jobs per txn – ensure not to exceed (e.g., don't call future in a loop 100 times).

Callouts: Max 100 callouts per txn – if near this, consider batching callouts via queueables.

Batch Apex: Use for large data jobs to stay within limits (and see asynchronous section).

Query Selectivity: Use indexed fields in WHERE for large data sets to avoid governor issues with non-selective queries.

Asynchronous Apex Checklist

Future Method Signature: static void, primitive params only.

Queueable: Implement Queueable, if chain next job, only enqueue one job at end of execute.

Batch: Implement Batchable; use manageable batch size (e.g., 200). If stateful, declare Database.Stateful.

Scheduled: Schedulable classes should delegate to batch/queueable in execute, not do huge work inline.

Callout annotation: @future(callout=true) or implements AllowCallouts on queueable/batch if doing callouts.

Avoid async in async: Don't call future from batch, etc. (not allowed by platform). Only chain where allowed (queueable to queueable or batch to queueable in finish).

Idempotence: Ensure if a queued job is retried or runs twice, it doesn't double-process (e.g., use external IDs or checks to prevent duplicates).

Integration Checklist

Named Credentials: Use for external endpoint and authentication where possible. Reference with callout: prefix in endpoint URL.

HTTP Callouts: Set method, endpoint, headers, body, timeout. Check response status and handle errors.

WebService (SOAP) Apex: global classes and methods, use proper types and handle exceptions.

REST Apex: Use @RestResource and @HttpGet/Post appropriately. Ensure sharing is set to enforce security.

Integration Security: Never expose more data than needed. Validate inputs from external calls.

Retries/Failures: Implement retry logic or at least logging for callout failures. Don't silently drop.

Test Integration: Use HttpCalloutMock in tests for callouts, and test classes for inbound REST/SOAP (can simulate requests via RestRequest in tests).

Metadata & Config Checklist

Custom Settings/Metadata: All configurable values (thresholds, toggles) stored outside code. Code fetches them.

Custom Labels: All user-facing messages or errors use System.Label.

Annotations:

@AuraEnabled only on necessary methods/properties (prefer on static methods for LWC).

@InvocableMethod: one per class, properly labeled.

@Deprecated tag on obsolete methods/classes.

@TestVisible only to expose what's needed for testing.

API Version: All classes on consistent API version (document if using specific versions for compatibility).

Meta Files: Ensure -meta.xml for classes/triggers have correct attributes (like apiVersion, status active for triggers).

No Hardcoded Org Info: (repeat) No ids, no references to specific org names or URLs.

This quick reference should help ensure that during development and code reviews, nothing important is overlooked. By checking code against these points, we maintain high quality and consistency.

Conclusion: Adhering to the Salesforce Apex Coding Standards detailed in this handbook will result in a codebase that is consistent, reliable, secure, and maintainable. Use this checklist as a final pass when writing or reviewing code to ensure all standards are met. Compliance with these standards ultimately streamlines development and improves the robustness of our Salesforce implementations.

ABOUT THE AUTHORS

Matthew Morris is a Salesforce CTO, Salesforce MVP Hall of Fame member, and one of the most respected architectural voices in the global Salesforce community. With a career spanning architecture, engineering leadership, and enterprise transformation, he has become widely known for his ability to make complex Salesforce concepts clear, visual, and actionable for teams of all skill levels.

Over the past decade, Matthew has helped organisations modernise their Salesforce estates, embrace AI-enabled delivery models, and simplify sprawling architectures. His work focuses on bringing strategic clarity to large enterprise programmes—bridging the gap between technical depth, business priorities, and the realities of operating at scale. As a speaker at Dreamforce, London's Calling, and other major conferences, he has trained thousands of architects, developers, and admins in designing cleaner, more resilient, and more scalable Salesforce solutions.

Matthew is also a published author and course creator. His Pluralsight course Play by Play: Diagramming Salesforce Solutions is a widely used resource for architects who want to communicate design decisions with greater impact. His ongoing research into Data 360, AI-first delivery, and modernisation patterns continues to shape how architects think about the future of multi-org environments and enterprise data strategy.

Known for his authenticity, generosity, and commitment to demystifying complexity, Matthew brings a practitioner's mindset to everything he teaches. His contributions to this book reflect the same strengths he is famous for across the ecosystem: clarity, practicality, and a relentless focus on helping teams deliver better outcomes through smarter architecture and thoughtful use of AI.

Andy Forbes is an IT architect, program leader, and author with more than four decades of experience designing and delivering complex technology solutions across the public and private sectors. His career has been defined by a practical ability to connect business goals with architecture, data, and delivery models that actually work at scale.

Andy's work over the last two decades has focused on modernizing fragmented IT landscapes—consolidating systems, standardizing processes, improving data quality, and guiding users through enterprise transformation. He has developed a reputation for making complicated technology change feel structured, manageable, and grounded in real-world constraints rather than idealized theory. In the past several years, Andy has become a voice on how generative and predictive AI are reshaping the craft of software delivery. As coauthor of ChatGPT for Accelerating Salesforce Development, he has helped teams learn how to use AI to analyze designs, write and review code, improve documentation, strengthen testing, and speed up the entire delivery lifecycle. His current work focuses on building AI-first delivery methods—integrating tools like ChatGPT into analysis, architecture, Apex development, project governance, and team enablement.

Across his roles as architect, strategist, program manager, and author, Andy is known for a clear, plainspoken approach that cuts through complexity and emphasizes practical results. He is passionate about mentoring delivery teams, helping organizations adapt to the fast pace of AI-driven change, and giving practitioners the confidence to apply emerging technologies responsibly and effectively.

His contributions to this book reflect what he's known for throughout his career: pragmatic guidance, forward-looking insight, and a commitment to helping people do their best work.